Risk, Systems and Decisions

Series Editors

Igor Linkov
U.S. Army ERDC, Vicksburg, MS, USA

Jeffrey Keisler
College of Management, University of Massachusetts
Boston, MA, USA

James H. Lambert
University of Virginia, Charlottesville, VA, USA

Jose Figueira
University of Lisbon, Lisbon, Portugal

Health, environment, security, energy, technology are problem areas where man-made and natural systems face increasing demands, giving rise to concerns which touch on a range of firms, industries and government agencies. Although a body of powerful background theory about risk, decision, and systems has been generated over the last several decades, the exploitation of this theory in the service of tackling these systemic problems presents a substantial intellectual challenge. This book series includes works dealing with integrated design and solutions for social, technological, environmental, and economic goals. It features research and innovation in cross-disciplinary and transdisciplinary methods of decision analysis, systems analysis, risk assessment, risk management, risk communication, policy analysis, economic analysis, engineering, and the social sciences. The series explores topics at the intersection of the professional and scholarly communities of risk analysis, systems engineering, and decision analysis. Contributions include methodological developments that are well-suited to application for decision makers and managers.

More information about this series at http://www.springer.com/series/13439

Benjamin D. Trump • Christopher L. Cummings
Jennifer Kuzma • Igor Linkov
Editors

Synthetic Biology 2020: Frontiers in Risk Analysis and Governance

Springer

Editors
Benjamin D. Trump
US Army Engineer Research and
Development Center
Vicksburg, MS, USA

Jennifer Kuzma
North Carolina State University
Raleigh, NC, USA

Christopher L. Cummings
Nanyang Technological University
Singapore, Singapore

Igor Linkov
US Army Engineer Research
and Development Center
Vicksburg, MS, USA

ISSN 2626-6717   ISSN 2626-6725   (electronic)
Risk, Systems and Decisions
ISBN 978-3-030-27263-0   ISBN 978-3-030-27264-7   (eBook)
https://doi.org/10.1007/978-3-030-27264-7

This Springer imprint is published by the registered company Springer Nature Switzerland AG
The registered company address is: Gewerbestrasse 11, 6330 Cham, Switzerland

For Anna, who made this work possible.
~From Benjamin

Foreword

Engineered biological systems save and improve human lives every day. These benefits currently come largely via the healthcare field in the form of human insulin and other therapeutics produced in culture vessels or through agriculture in the form of improved crops. However, products generated by engineering biology are being explored and adopted by a wide array of industries, ranging from data storage to fast food. You can now buy an Impossible Burger™ containing yeast-based heme, the red, iron containing, component of blood, at Burger King™.

As these applications of technology grow, and the tools and techniques for engineering biology become more sophisticated and distributed. The International Genetically Engineered Machine (iGEM) competition is a bellwether of the field. In its first 13 years of existence, it grew from 31 participants in five teams to 5400 participants in 310 teams coming from six continents. While traditional players such as the United States and China still dominate this event, a map illustrating the geographic diversity of this group illustrates how truly global this technology has become (https://www.google.com/maps/d/viewer?mid=1RwdyeHgNKpViw10ITP Lses203JU).

The power and distribution of this technology increases the possibility that it will be subverted to cause harm. To ensure that this does not happen, it is essential that the practitioner community becomes and remains engaged with the security world, policy makers, and the general public. It is also critical that they learn to view their own research through the lens of those charged with protecting us all. There are a number of mechanisms and forums that can be used to facilitate this communication and learning. You hold in your hand one such tool.

The authors have assembled herein a valuable array of references and perspectives to highlight issues stemming from developing and bringing engineered biological systems to market. The chapters range from individual case studies to general analyses of the field. Each contains ideas and/or examples of how risk analyses and governance structures can or have been used to ensure that an engineered biological product or process has been vetted for consumer safety.

With issues ranging from the food-versus-fuel debate that surrounds the use of corn for the production of ethanol to who should be able to purchase synthetic DNA encoding deadly viruses, it is obvious that there is no one answer or approach to governing biotechnology. However, the case studies and background provided in this book provide touchpoints for many of the most important topics studied to date. I hope that researchers use this reference to assess and contextualize their work and that other stakeholders, including the general public, utilize it as a starting point to understanding the breadth of approaches that engineering biology community uses to assess the implications of their work.

Jeffrey "Clem" Fortman
Engineering Biology Research Consortium
Emeryville, CA, USA

Foreword

Synthetic biology today is at a critical inflection point. The stakes have never been higher. *Synthetic Biology 2020: Frontiers in Risk Analysis and Governance* is a particularly timely and insightful book with significant implications for the future trajectory of synthetic biology.

This ambitious volume comes at a significant juncture not only for how we think about synthetic biology but also for how we act to shape its future. The promise of synthetic biology for making the world a better place is dazzling, but so too are the risks, governance issues, and other difficult societal choices that will have to be made under conditions of uncertainty in order to realize its promise responsibly.

These issues only have become even harder, more complex, and important for all stakeholders over the last two decades. They are the subjects of this impressive volume.

A few of us are old enough to remember the "two cultures" debate advanced by noted British scientist and novelist C.P. Snow in *Two Cultures and Scientific Revolution* based on his Rede Lectures in 1959. For several decades after, it was a required reading in university classes and was broadly discussed in many settings. In essence, Snow posited that a societal divide existed between the sciences and the humanities (and, by implication, the social sciences, too) and that this created a rate-limiting obstacle to solving many of the world's societal challenges of the time.

Synthetic biology, in its initial two decades, has been unusual among transformative emerging technologies in addressing updated versions of the "two cultures" debate. First, from the outset, it has forged a diverse "community culture" that includes ethicists and religious scholars, social scientists, varied publics, and some thoughtful skeptics as active participants.

Second, as this volume makes clear, "unlike other innovations in the past, synthetic biology has not been insulated from social science inquiry during the innovation process." Nontechnical considerations, such as ELSI issues and public engagement, have played a significant role from the outset.

Third, most leading scientists, engineers, designers, and business innovators in synthetic biology accept both the legitimacy and urgency of addressing difficult questions about uncertainty, risk, governance, security, transparency, and public acceptance as essential building blocks for synthetic biology. Many of them were influenced by how addressing ELSI issues early on played a constructive role in enabling the initial human genome revolution. Older members of the community also witnessed the implications of failing to address them in a timely way during the GMO controversies. Treating ethics, governance, or other societal concerns merely as afterthoughts or second-order problems undermines trust and public acceptance and can introduce rate-limiting obstacles to new research and the pace of innovation.

One reason that *Synthetic Biology 2020 is* so significant and timely is its forward-looking focus on how to integrate new social science insights and more nuanced and context-specific considerations of risk, governance, and societal implications into the research and innovation process as new applications of synthetic biology proliferate in diverse domains. It also builds on important earlier social science work by Oye, Frow, Maynard, Florin, and others about the need for adaptive and anticipatory approaches. The editors have assembled a diverse set of perspectives that address decision making and societal choices for the expanding range of applications in synthetic biology.

Synthetic biology offers the promise to deliver solutions to a broad range of twenty-first-century societal grand challenges, to make biology easier to engineer for beneficial purposes, to enable more predictive and reliable applications of biology, and to support sustainable economic growth, including the expanding bioeconomy. But none of these benefits will be realized without a broadly accepted decision making that includes considerations of risk and uncertainty, governance and transparency, public trust, and trustworthiness.

The chapters that follow make a compelling case for shared collective responsibilities, adaptive and participatory governance and transparency, and more innovative and inclusive approaches to risk analysis and assessments, decision-making, and stakeholder involvement.

I first encountered synthetic biology in the spring of 2003 while serving on an MIT Corporation Committee. We received a short briefing about interesting courses and projects undertaken during MIT's Independent Activities Period (IAP) in January 2003. One immediately commanded my attention and ended up redirecting my own interests and activities in the intervening 16 years.

The course was called *Synthetic Biology Lab: Engineered Genetic Blinkers.* Offered to only 16 MIT students, it promised a "hands-on introduction to the design and fabrication of synthetic biology systems" based on a "standards parts list" of preexisting biological parts and de novo DNA synthesis.

The IAP was developed by four inquisitive MIT professors and senior researchers with diverse backgrounds in AI, environmental bioengineering, computer architecture, and electrical engineering—Tom Knight, Drew Endy, Randy Rettberg, and Gerald Sussman—who wanted to make biology easier to engineer. To some extent, it was modeled on the pathbreaking course 25 years earlier offered by Lynn Conway at MIT with Carver Mead from Caltech, which pioneered the development of VLSI in semiconductors, and showed the power of decoupling design from fabrication and collaborative infrastructure such as MOSIS.

The MIT IAP course struck me not only as super cool in the best MIT geek tradition but also as incredibly important for a much larger societal canvas. I immediately grasped the power of the convergence of biology and engineering (as well as the physical sciences). I could see how introducing an engineering mindset and methods to the living world could lay the groundwork for a transformational technology with broad applications for research, meeting societal needs, and economic competitiveness. And, finally, I foresaw that synthetic biology in many respects was a tool revolution—including in design, synthesis, and automation—to help us better understand the complexity of biology and, someday, to program living organisms to enable a broad array of societal benefits.

But I also had worked extensively with dual-use technologies and emerging technologies with significant risks, uncertainty, and complex security concerns. My varied experiences at the intersection of government policy, law, universities, international affairs, national security, and scientific research labs had convinced me about the need to integrate and address a broad range of safety, security, trust and trustworthiness, governance, and stakeholder perspectives into the mix from the outset—the sooner the better; the broader the better. Past misguided approaches, often rooted in hubris, elitism, and the convictions of some scientists that "if you only understood my research as well as I do then you would stop asking difficult questions," risked failure or suboptimal outcomes.

My views were reinforced soon after the initial MIT IAP synthetic biology course when I was talking with a senior US politician at that time. I was telling him about all the reasons why I was so enthused about this new field of synthetic biology and how I thought it could become so beneficial for advancing US national and global interests. He interrupted my narrative to say, "Rick, this all sounds great. But if something goes terribly wrong and people lose trust in it, or if it scares my colleagues in Congress, you won't have much. No investment, no buy in, no public acceptance." And, of course, I knew he was right.

Soon after, Drew Endy, then at MIT and now at Stanford, and I were invited to brief the National Academy of Sciences about this emerging field for the first time. Key leaders asked Drew and me not only to inform them about what synthetic biology was but also to look into our crystal balls. They were eager to get our views about its future and its implications not only for research but also for society.

Drew, of course, brilliantly addressed the key scientific and engineering aspects, including the "vision thing" about how synthetic biology could make biology easier to engineer and help change the world for the better. I focused on a litany of first-impression nontechnical issues that I foresaw emerging—decision making under conditions of uncertainty; legal issues related to freedom to operate and tort liability; novel regulatory issues and potential problems with lagging regulatory science or outdated frameworks; security concerns, including those of dual-use research; and a long list of international issues.

Looking back at my scribbled and often indecipherable notes from that initial National Academy briefing recently, I was struck by how much Drew and I got right—but also how much we missed largely because the pace of change in the last 15 years has been so dramatic. We have witnessed not only exponential scientific and engineering advances in synthetic biology but also equally important new research insights offered by social scientists, innovative new ways to think about risk assessments and governance, and many lessons about multistakeholder engagement.

Many of those social science insights and nontechnological innovations are reflected in *Synthetic Biology 2020*. This volume expands the breadth and depth of our understanding about how "collaboration between physical scientists and social scientists during the innovation process should provide valuable opportunities to question potential broader impacts and ensure that products are applied beneficially."

I often have commented about how, even in those early formative years, the relatively small synthetic biology community was deeply engaged with, and committed to, the subjects of this book. Fortunately, this ethos continues today, even as the synthetic biology community expands, renews, and rethinks what it should become.

It also is interesting to reflect about how the central issues of this book have been at the core of key synthetic biology thought leaders and influencers. The iGEM Foundation and the BioBricks Foundation are illustrative. iGEM long has celebrated ethics, societal implications, safety and security, and human practices as core values in synthetic biology. It integrates them as required components of the annual iGEM global synthetic biology competition jamboree that now attracts about 340 teams (university teams, as well as some high schools) from more than 40 countries. More than 40,000 iGEMmers, who have begun to think about the societal implications of their synthetic biology research, now constitute a robust "After iGEM" cohort of emerging leaders dispersed around the world.

The BioBricks Foundation (BBF) views its mission to "advance biotechnology in an open and ethical manner to benefit all people and the planet." It has been at the forefront of developing innovative tools that promote sharing, openness, capacity building, dialogue, and inclusivity. As the convener of the influential SB x.0 global series every few years, which brings together most of the key global players in the field, BBF has established a strong track record in celebrating diversity, enabling inclusion, and putting into practice the thematic priorities of this book. For example, the agenda of SB7.0 in Singapore in 2017, the most recent BioBricks Foundation global event, emphasized inclusion, diversity, and a broad range of ELSI issues as critical to the future of synthetic biology.

Newer, key umbrella organizations in synthetic biology, such as the Engineering Biology Research Consortium (EBRC), also have embraced a forward-looking and inclusive approach consistent with many of the recommendations in the chapters that follow. A number of EBRC's individual academic members are leaders in different social science fields, ethics and religion, safety and security, and governance. EBRC also has integrated the focal points of this book into all four of its core working groups—policy and international, education, security, and a detailed technical roadmap for synthetic biology.

Though many of the examples and discussions in this book are American-centric, this excellent volume compiled by Trump, Cummings, Galaitsi, Kuzma, and Linkov is not confined to an American audience. It addresses critical issues of consequence for an international one that reflects the truly global scope of synthetic biology.

Though national and international decisions about risk, governance, and ethics no doubt will continue to vary widely among countries, *Synthetic Biology 2020* provides a diverse toolkit and broadly applicable social science research to assist in making these often-difficult choices for society.

As Chairman of the OECD/BIAC Science, Technology, and Innovation Committee for an extended period, I have enjoyed a front row seat from which to observe and help shape the changing international landscape for synthetic biology. Europe, for example, integrated responsible innovation as a core element of synthetic biology research and innovation as part of its Horizon 2020 framework and plans for Horizon Europe. Japan's Smart Society 5.0 initiative seeks to incorporate the innovations of the fourth industrial revolution and other emerging technological innovations, such as synthetic biology and bio-digital convergence, into all aspects of Japanese life in responsible ways that improve the societal well-being of its citizens.

And China represents a particularly interesting case study to follow, given the rapid and widespread growth of all aspects of synthetic biology there. During the multiyear Six Academies initiative for synthetic biology among the US, UK, and Chinese national academies, I recall Chinese students and future young leaders in synthetic biology packing a large lecture hall in Shanghai in October 2011 to listen eagerly to ELSI talks by Sheila Jasanoff about "From ELSI to Responsible Innovation" and changing governance paradigms for synthetic biology by Anita Allen about the study of bioethical issues, and by Paul Gemmill about public engagement and the societal implications of synthetic biology research in the UK.

Chinese iGEM teams—there were more than 100 in the 2018 competition—routinely consider and address the societal implications of their synthetic biology research and projects for their local area, for China, and for the world as part of their iGEM projects. I attended a large meeting of Chinese iGEM teams from across China at Shanghai Tech in 2018. In a keynote presentation, Professor Guo-ping Zhao from the Shanghai Institute of Biological Sciences and the Chinese Academy of Sciences, who has played a central leadership role in advancing synthetic biology in China and globally, urged China's next generation of talented synthetic biologists to pay particular attention to risk and uncertainty, to safety and security, and to the societal implications of their work.

Both on the global stage and in each of the more than 40 countries with active synthetic biology communities, we will have to see how well the difficult and often complex subjects of *Synthetic Biology 2020* are integrated into synthetic biology research, innovation, policy making, and social discourse over the next two decades. This volume offers a diverse set of analytical tools, social science research, and policy lessons to guide them in making prudent and responsible choices, many of which will involve considerable uncertainty. The future of synthetic biology will depend, in large part, on the decision-making processes they follow and the choices they make.

Richard A. Johnson
Global Helix LLC
Washington, DC, USA

Preface

Synthetic biology offers powerful remedies for some of the world's most intractable problems, but these solutions may not be applied if the public perceives them to accompany unacceptable risks. The public forms opinions about tradeoffs for synthetic biology's risks and benefits, and already a small but notable population exists that favors banning the field outright until the risks are better understood. This book includes various perspectives of synthetic biology from the social sciences, such as with risk assessment, governance, ethics, and communication. Ultimately, we argue that synthetic biology is poised to provide valuable benefits to humanity that likely could not be achieved by alternate means, as well as to enrich the teams that create them. The incentives are prodigious and obvious, and the public deserves assurances that all potential downsides are duly considered and minimized accordingly. Incorporating social science analysis within the innovation process may impose constraints, but its simultaneous support in making the end products more acceptable to society at large should be considered a worthy trade-off.

Contributing authors in this volume represent diverse disciplines related to the development of synthetic biology applications and reflect on differing areas of risk analysis and governance that have developed for the field. In sum, the chapters of this volume note that while the first 20 years of synthetic biology development have focused strongly on technological innovation and product development, the next 20 should emphasize the synergy between developers, policy makers, and the public to generate the most beneficial, well-governed, and transparent technologies and products possible.

Many chapters in this volume provide new data and approaches that demonstrate the feasibility for multistakeholder efforts involving policy makers, regulators, industrial developers, workers, experts, and societal representatives to share responsibilities in the production of effective and acceptable governance in the face of uncertain risk probabilities. Such participation bestows responsibility and is a partial remedy for ignorance. More contributors not only ensure that the problem is examined from myriad perspectives representing distinct motives but also addresses public wariness to adopt new technologies. Industries engaging with the public can also foster transparency and address concerns as they arise. These steps may prevent a world of draconian regulations based on insufficient understanding and widespread fear. Simultaneous collaboration between physical scientists and social scientists during the innovation process should provide valuable opportunities to question potential broader impacts and ensure that products are applied beneficially. Thus, the unique situation of synthetic biology and its attention from the social science is an opportunity to demonstrate the value of collaboration and the security benefits it helps to provide.

Vicksburg, MS, USA   Benjamin D. Trump
Singapore, Singapore   Christopher L. Cummings
Raleigh, NC, USA   Jennifer Kuzma
Vicksburg, MS, USA   Igor Linkov

Acknowledgments

Edited volumes require substantial cooperation and support to execute, and we are thankful to everyone who had a hand in making this book happen.

We are grateful for the book's chief reviewer, Stephanie Galaitsi, who provided a critical and impartial eye throughout all the stages of the review process.

We also acknowledge the ideas and support of Dr. Elizabeth Ferguson, Dr. Ilker Adiguzel, Dr. Edward Perkins, Dr. Scott Greer, Dr. Christy Foran, and LTG Thomas Bostick who have each furthered our desire to pursue better research in synthetic biology in a unique way.

We are thankful for the support of George Siharulidze, Joshua Trump, and Miriam Pollock, who provided additional support via reviews and technical editing.

Dr. Trump would also like to acknowledge the considerable support he received during his field work in Singapore via the Institute of Occupational Medicine (Dr. Michael Riediker, Dr. Robert Aitken) and Nanyang Technological University (Dr. Ng Kee Woei).

We are grateful for the support of our respective institutions, including the US Army Engineer Research and Development Center, Nanyang Technological University, and North Carolina State University. We are also very thankful for the support of the host institutions of each of the authors included in this text—the contributions by this international audience make for a richer and more scientifically complete understanding of synthetic biology's risk analysis, governance, and communication scholarship.

Disclaimer

The ideas, research, and arguments represented within this book are the views of the chapter authors alone and may not represent the views of their affiliated organizations. No text herein should be taken as an official statement or position of any government, university, company, or other organizations.

Contents

About the Editors

Benjamin D. Trump is a Research Social Scientist for the US Army Engineer Research and Development Center. Dr. Trump's work focuses on decision making and governance of activities under significant uncertainty, such as emerging and enabling technologies (synthetic biology, nanotechnology) and developing organizational, infrastructural, social, and informational resilience against systemic threats to complex interconnected systems. Dr. Trump served as a delegate to assist US presence in OECD's Global Science Forum in 2017 and is the President of the Society for Risk Analysis' Decision Analysis and Risk Specialty Group in 2018–2019. He was selected as a Fellow of the Emerging Leaders in Biosecurity Initiative (ELBI), Class of 2019. Dr. Trump was also a contributing author of the International Risk Governance Council's Guidelines for the Governance of Systemic Risks, as well as their second volume of the Resource Guide on Resilience. Dr. Trump is also frequently active with several Advanced Research Workshops for the North Atlantic Treaty Organization's Science for Peace Programme, including his role as a Program Committee Lead for a workshop on Biosecurity for Synthetic Biology and Emerging Biotechnologies. Dr. Trump received his Ph.D. from the University of Michigan's School of Public Health, Department of Health Management and Policy in 2016. He received an M.S. (2012) in Public Policy and Management and a B.S. in Political Science (2011) from Carnegie Mellon University and completed his postdoctoral training at the University of Lisbon, Portugal.

Christopher L. Cummings serves as an Assistant Professor of Strategic Communication at Nanyang Technological University, Singapore—a top-ranked university in Asia and among the top 15 in the world. He is the Director of the International Strategic Communication Management program and has served in multiple leadership positions within the Society for Risk Analysis. Dr. Cummings' work focuses on advancing public engagement with science, developing risk communication theory, and improving public health decision making across the life span. An experienced social scientist and communication campaign scholar, Dr. Cummings uses a variety of quantitative, qualitative, and mixed methods, and his work has been featured in applied science and health venues, including the *Journal* *of Risk Research*; *Regulation and Governance*; *Nanotoxicology*; *PLOS One*; *Science, Technology, & Human Values*; and *Climate Research*, among others. Outside of academia, Dr. Cummings also consults with multiple government agencies across the Asia-Pacific on health risk communication issues, including dengue fever surveillance and response (Sri Lanka and Singapore), obesity and nutrition access (New Zealand), vaccine communication (Singapore, Australia, and United States), and public engagement with nanotechnology, synthetic biology, and geoengineering (Australia, United States, and Singapore). He also consults privately with various Fortune 500 companies on leadership training, strategic planning, and risk communication initiatives. He completed his B.A. in Communication Studies (California State University, Chico), his M.S. in Communication and Ph.D. in Communication, Rhetoric, and Digital Media (North Carolina State University), and his postdoctoral work under the supervision of Dr. Jennifer Kuzma at the Genetic Engineering and Society Center (North Carolina State University).

Jennifer Kuzma is the Goodnight-NC GSK Foundation Distinguished Professor in the School of Public and International Affairs and Cofounder and Codirector of the Genetic Engineering and Society (GES) Center (research.ncsu.edu/ges) at NC State University. Prior to her current position, she was Associate Professor at the Humphrey School of Public Affairs, University of Minnesota (2003–2013); Study Director for several NASEM reports related to biotechnology governance and bioterrorism (1999–2003); and AAAS Risk Policy Fellow at the USDA (1997–1998). She has over 120 scholarly publications on emerging technologies, risk analysis, and governance, including four articles in *Science* and *Nature* in the past three years. Kuzma served as a member of the WEF Global Futures Council on Technology, Values and Policy, the NASEM Committee on Preparing for Future Biotechnology, the FDA Blood Products Advisory Committee, and the UN WHO-FAO Expert Group for Agrifood Nanotechnologies and as SRA Secretary and Council member. She currently serves on the AAAS-ABA National Council of Scientists and Lawyers. She received the SRA Sigma Xi Distinguished Lecturer Award for her contributions to the field of risk analysis, was elected as a Fellow of AAAS in 2018 for her distinguished work in anticipatory governance of new technologies, and was awarded the Fulbright Canada Visiting Research Chair in Science Policy in 2017. She appears frequently in the media for her expertise in biotechnology policy, including *The New York Times*, *Science*, *Nature*, *NPR*, *The Washington Post*, *Scientific American*, PBS Nova, Wired, and ABC & NBC News. She obtained her Ph.D. in biochemistry from the U of CO Boulder in 1995, where she discovered bacterial isoprene production, and was awarded the first patent for methods of bioisoprene production. Her postdoctoral work at the Rockefeller University on plant drought and salinity tolerance led to an article in the journal *Science* in 1997.

Igor Linkov is the Risk and Decision Science Focus Area Lead with the US Army Engineer Research and Development Center and an Adjunct Professor with Carnegie Mellon University. Dr. Linkov has managed multiple risk and resilience assessments and management projects in many application domains, including cybersecurity, transportation, supply chain, homeland security and defense, and critical infrastructure. He was part of several Interagency Committees and Working Groups tasked with developing resilience metrics and resilience management approaches, including the US Army Corps of Engineers Resilience Roadmap. Dr. Linkov has organized more than 30 national and international conferences and continuing education workshops, including NATO workshops on Cyber Resilience in Estonia (2018) and Finland (2019), as well as Chaired Program Committee for 2015 and 2019 World Congresses on Risk in Singapore and Cape Town. He has published widely on environmental policy, environmental modeling, and risk analysis, including 20 books and over 350 peer-reviewed papers and book chapters in top journals, like *Nature*, *Nature Nanotechnology*, *Nature Climate Change*, among others. He has served on many review and advisory panels for DOD, DHS, FDA, EPA, NSF, the EU, and other US and international agencies. Dr. Linkov is Society for Risk Analysis Fellow and recipient of 2005 Chauncey Starr Award for exceptional contribution to risk analysis, as well as 2014 Outstanding Practitioner Award. He is an Elected Fellow with the American Association for the Advancement of Science (AAAS). Dr. Linkov has a B.S. and M.Sc. in Physics and Mathematics (Polytechnic Institute) and a Ph.D. in Environmental, Occupational and Radiation Health (University of Pittsburgh). He completed his postdoctoral training in risk assessment at the Harvard University.

Synthetic Biology: Perspectives on Risk Analysis, Governance, Communication, and ELSI

Benjamin D. Trump, Christopher L. Cummings, S. E. Galaitsi, Jennifer Kuzma, and Igor Linkov

Synthetic biology is a technology with incredible promise yet equally galling uncertainty. The United Nations Convention on Biological Diversity defines synthetic biology as "biotechnology that combines science, technology, and engineering to facilitate and accelerate the understanding, design, redesign, manufacture, and/or modification of genetic materials, living organisms, and biological systems" (Convention of Biological Diversity). Synthetic biology can produce entirely new organisms, some of which may pose risks to naturally existing ecosystems. While humans have been selectively breeding plants and animals for millennia, synthetic biology and its enabling technologies allow combining genetic material from organisms that cannot procreate in nature and grant more deliberate and precise control over the selection of genetic processes.

Synthetic biology innovations might support disease prevention, large-scale food production, and sustainable energy, as well as more dubious applications like eugenics and invasive manufactured organisms. The difference between highly beneficial and highly hazardous outcomes depends upon the decisions of the people funding, producing, and regulating synthetic biology projects. The new and unique qualities of synthetic materials and their complex intersections with existing biological, ecological, and sociotechnical systems raise the specter of unpredictable outcomes (Linkov et al. 2018) and can complicate these decisions. For established

B. D. Trump (✉) · I. Linkov
US Army Engineer Research and Development Center, Vicksburg, MS, USA
e-mail: bdtrump@umich.edu

C. L. Cummings
Nanyang Technological University, Singapore, Singapore

S. E. Galaitsi
US Army Corps of Engineers, Washington, DC, USA

J. Kuzma
North Carolina State University, Raleigh, NC, USA

© Springer Nature Switzerland AG 2020
B. D. Trump et al. (eds.), *Synthetic Biology 2020: Frontiers in Risk Analysis and Governance*, Risk, Systems and Decisions, https://doi.org/10.1007/978-3-030-27264-7_1

Fig. 1 Breakdown of
disciplines within social
science and implication-
related research pertinent
to synthetic biology

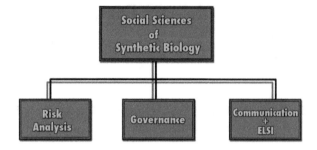

Fig. 1 Breakdown of disciplines within social science and implication-related research pertinent to synthetic biology

technologies, the current risk assessment and management paradigms are well-developed (Linkov et al. 2018), but there is uncertainty surrounding decisions in synthetic biology, including the scope of risks and the methods for monitoring them. This uncertainty should decrease as the field produces more data and stabilizes, which will require time, scholarship, and investment.

This book, *Synthetic Biology 2020: Frontiers in Risk Analysis and Governance,* examines the synthetic biology field after two decades of innovation. Within such a topic, the book includes perspectives of synthetic biology from the social sciences, such as risk assessment, governance, ethics, and communication (Fig. 1). Contributing authors in this volume represent diverse disciplines related to the development of synthetic biology's social sciences and consider different areas of risk analysis and governance that have developed over this time and the societal implications. The chapters of this volume note that while the first 20 years of synthetic biology development have focused strongly on technological innovation and product development, the next 20 must emphasize the synergy between developers, policymakers, and the public to generate the maximally beneficial, well-governed, and transparent technologies and products.

Making Sense of Synthetic Biology: Raw Opportunity and Uncertain Implications

The field is growing rapidly; estimates for 2020 equity funding forecast nearly $40 billion dollars to be directed to private synthetic biology companies (Polizzi et al. 2018), a 40-fold increase from funding in 2016. But the products of synthetic biology will not be demanded nor subsequently deployed if potential customers distrust their utility or safety. Fears of tragedies from synthetic biology applications are readily imaginable: privileged designer babies, bioterrorism, and disrupted ecosystems are all moral or physical calamities that could arise should synthetic biology development be inadequately regulated.

While there is usually risk in implementing new technologies, there is also risk in choosing to let existing hazards continue to control aspects of our environment. In that sense, unwarranted negative public perception of synthetic biology innovations could hinder societal advancement (Palma-Oliveira et al. 2018). While industry, government, and private actors have different priorities and motivations in producing and using synthetic biology, concerns over the safety and protection of their communities and the natural world that supports them are universally shared. However, some end users may disproportionally bear potential risks of synthetic biology applications and thus rationally perceive safety differently. These perceptions can be captured and addressed through social science assessments to guide safe and socially acceptable development of synthetic biology. Anticipating both physical and social outcomes enables insights to be integrated into revisions of previous decisions and improves the value of iterative processes of experimentation and innovation.

Technological development and assessment have historically occurred as two distinct steps, often separated by a time period measurable in years. First, technological breakthroughs rise within the physical and natural sciences, which are subsequently discussed by social science regarding the technology's societal implications, risks, and regulatory needs. For example, the growth of mass transportation technologies in the early nineteenth century brought risks of mechanical accidents and toxic emissions (Cummings et al. 2013). When dichloro-diphenyltrichloroethane (DDT) use spread globally in the 1940s, its deleterious environmental and human health effects were unknown. It took until 1972 for the United States to ban it, followed by a worldwide ban under the Stockholm Convention in 2001 (Cummings et al. 2013).

Inferring policy needs and recommendations for developing technology is an uncertain business. A product's impacts on society are difficult to assess when the product is underdeveloped and does not yet resemble the version that will ultimately be adopted. Many products may prove infeasible and therefore inconsequential (see Fig. 2), and their assessments can squander precious resources. Yet, waiting for a relatively finalized product disadvantages social scientists because their inquiry will be in its earliest stages while the physical scientists are finalizing their own and potentially beginning to market a product. Jasanoff (2009) writes that the responsive and reflexive nature of social science inquiry causes it to lag behind physical science research. But if the reflexive nature of social science were incorporated within the innovation process, the societal infeasibility of some products could be identified earlier and used to guide physical scientists to create more universally beneficial products.

Ideally, including social science in the innovation process can provide transparency that may reassure members of the public that the benefits of a new technology outweigh the risks. Such social science scrutiny promotes developing governance initiatives that can be operated in tandem with broader technological dissemination. However, the same critical inquiry of developing technologies may stoke fears of the new, uncertain, and unknown. Ideas can be presented out of context in ways that emphasize their risks without communicating their safeguards and may provoke public criticism and opposition. Too much opposition can hamper or even halt innovations before scientists can incorporate or address criticism in their products. A two-step process that separates innovation from evaluation precludes this type of

Fig. 2 Illustrating path dependency
or the winding road and various
choices that transform a scientific
idea into a commercial product

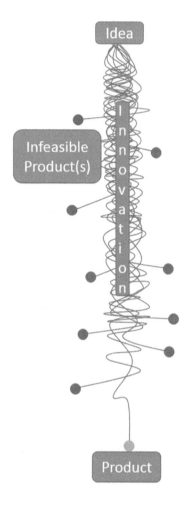

project derailment – but also can bring products to market without adequate safeties in place.

The social convulsions associated with emerging technologies could be less dramatic or harmful when better anticipated. Consider the automobile: delayed full-privileged licensure for teenage drivers, in combination with other factors, reduces crash rates for new drivers (Ferguson et al. 1996; Williams 2009). This information would have been useful in the 1940s when most US states picked the age 16 as the minimum driving age. States partially addressed the issue later by implementing graduated licensing laws, but the minimum driving age is now ingrained in the United States' car-dependent culture and is unlikely to change despite its recognized risks.

In recent decades, the time lag between physical science innovation and social science assessment and governmental mobilization has narrowed. Lessons learned from previous mistakes have prompted greater scrutiny and evaluation of technological impacts prior to their immersion in society. Synthetic biology in particular has not been insulated from social science inquiry during the innovation process. The physical and social science publications examining synthetic biology show nearly parallel trends in growth, indicating that social scientists have the ability to directly comment on emerging research (Trump et al. 2019). Torgersen and Schmidt (2013) and Shapira et al. (2015) attribute the contemporaneous, rather than lagged, growth of social science discourse of synthetic biology to the foundations laid by social science research on genetically modified organisms (GMOs), which had a controversial reception in the public sphere.

The simultaneous inquiry by both physical and social scientists augurs a process that will be fundamentally different than for previous innovations that developed outside of the public eye. Synthetic biology offers powerful remedies for some of the world's most intractable problems, but these solutions may not be developed or applied if the public perceives them to accompany unacceptable risks. Already a small but notable population exists that favors banning the field outright until the risks are better understood (Pauwels 2009; Pauwels 2013; Marris 2015). Such public mistrust and suspicion can be fueled by interest groups or misguided individuals (Linkov et al. 2018) who enjoy the public's attention. Calvert and Martin (2009) argue that the social concerns surrounding synthetic biology research through 2009 might have been addressed by "institutionaliz[ing]" social scientists' involvement in the field. A proactive and adaptive approach to risk management and governance can aid risk assessment in circumstances of limited experimental data (Oye 2012; Trump 2017), and social science inquiry can play a key role (Trump et al. 2018).

Since social science research of synthetic biology is already underway, physical and natural scientists have the opportunity to actively engage social scientists to evaluate innovations and help develop feasible products. In our modern era, physical scientists must understand that public perception matters and is a determinant in how applications of synthetic biology are ultimately funded, used, and governed. Because synthetic biology has the attention of social sciences so early in its innovation process, it has an opportunity to demonstrate the value of transdisciplinary collaboration in technological innovation as a way of providing secure benefits and a safe and socially acceptable forum for further exploration and development. Myriad perspectives around synthetic biology represent distinct motives and can directly address public wariness to adopt new technologies (Linkov et al. 2018). These steps may prevent a world of draconian policies based on insufficient understanding and widespread fear. Collaboration between physical scientists and social scientists during the innovation process should provide valuable opportunities to question potential broader impacts and ensure that products are acceptably safe.

Twenty Years of Synthetic Biology Development

Here, we present a short history of synthetic biology's development followed by brief descriptions of the chapters in this volume. As editors, we hope to provide a valuable and compelling resource that motivates the next generation of stakeholder collaborations to be resolute in envisioning a future that maximizes the potentials of synthetic biology while anticipating and respecting the needs and values of a diverse global citizenry.

Starting in the late 1970s, genetic engineers could blindly launch a novel gene into a host cell, hoping it landed in a good spot and worked in the new environment. After decades of incremental improvements in biochemical and genomic science, modern synthetic biology began to take root in the 1990s and early 2000s through engagement in more complex system engineering of viruses and bacteria. During the 1990s, "automated DNA sequencing and improved computational tools enabled complete microbial genomes to be sequenced, and high-throughput techniques for measuring RNA, protein, lipids and metabolites enabled scientists to generate a vast catalogue of cellular components and their interactions" (Cameron et al. 2014). This, coupled with a system engineering approach to biology, served as the core principles that made modern synthetic biology possible (Porcar and Pereto 2014; Cameron et al. 2014). In other words, genetic engineering around this time began to consider whether complex cellular networks could be viewed as engineered systems where biological engineering of a cell's DNA could yield complex changes to how those systems operate.

In 2000, *Nature* published two articles that discussed the deliberate creation of biological circuit devices (where biological parts inside a cell are designed to perform logical functions mimicking those observed in electronic circuits) by combining genes within *E. coli* cells. Gardner et al. (2000) constructed a genetic toggle switch to influence the expression of mutually inhibitory transcriptional repressors. Elowitz and Leibler (2000) engineered an oscillatory circuit that, when activated, produced the ordered and periodic oscillation of repressor protein expression. These publications encouraged the further development of research centered on circuit engineering and synthetic circuit construction to influence a cell's network design, including cell-to-cell communication and interactions (Weiss and Knight 2001).

During this time, the field of systems biology also emerged as a mature and independent field of inquiry pertaining to the computational and mathematical modeling of complex biological systems (Kitano 2002; Ideker et al. 2001). The field seeks to better understand the various properties of cells, tissues, and the systemic infrastructure that comprises living organisms (Hucka et al. 2004; Hood et al. 2004). This generally entails researching cell signaling networks or the signals and stimuli that govern and control cellular actions (Ingber 2003; Kitano 2002). For example, Park et al. (2003) published their work on posttranslational regulation using protein–protein interaction domains and scaffold proteins using *S. cerevisiae*. Coupled with earlier principles of genetic engineering, the technological and scientific advancements derived within systems biology serve as some of the driving forces behind the development of synthetic biology research (Andrianantoandro et al. 2006; Khalil and Collins 2010).

By 2004, synthetic biology had clearly evolved from a small number of biologists and engineers into a growing and unique field of emerging technology research in its own right. The Massachusetts Institute of Technology hosted "Synthetic Biology 1.0" in June 2004 as the first international conference explicitly dedicated to synthetic biology research (Ball 2004). At this meeting, an interdisciplinary collection of professionals encompassing the field of biology, chemistry, computer science, and others discussed the desire to design, build, and characterize biological systems and interactions (Ferber 2004). This conference spurred further international meetings known colloquially as the SBx.0, with the latest iteration as of this writing held in Imperial College London in 2013 (SB 6.0). This conference series advanced discussion around blending elements of engineering with molecular biology to determine whether synthetic biology could develop as an engineering field like electrical engineering or materials science (Cameron et al. 2014). Specifically, Endy et al. (2005) and Cameron et al. (2014) describe these early efforts as an attempt to produce a collection of modular parts and improve design pathways for engineered cells with the idea that modifying specific cell circuit designs could deliberately change the behavior or interactions of that cell with its local environment.

Between 2004 and 2010, "the second wave of synthetic biology" produced circuit design and metabolic engineering (Purnick and Weiss 2009; Isaacs et al. 2004). The former included attempts to expand RNA-derived cellular systems of biological circuit engineering from "transcriptional control" into posttranscriptional control vehicles and capabilities (Bayer and Smolke 2005). Generally accomplished using *E. coli*, various scientists sought to expand circuit and part designs, with one such circuit dedicated to the conversion of light into gene expression for a collection of *E. coli* cells (Levskaya et al. 2005).

For the developments in metabolic engineering, a group of scientists at the University of California, Berkeley, studied isoprenoid biosynthesis which produces artemisinic acid, or the component precursor to the wormwood *Artemisia annua* (Ro et al. 2006). Using a collection of organisms including *Saccharomyces cerevisiae* and *E. coli,* scientists under the leadership of Dr. Jay Keasling produced artemisinic acid using fermented yeast cells in controlled and pre-planned settings (Ro et al. 2006; Hale et al. 2007). The World Health Organization uses artemisinin combination therapies as the primary initial treatment for *P. falciparum* malaria. The drug destroys the majority of parasites in a patient's blood upon the drug's ingestion (Nosten and White 2007; Van Agtmael et al. 1999). However, the plant's erratic price points (ranging from $120 to $1200 USD per kilogram between 2005 and 2008) and fluctuating production levels have hindered naturally produced artemisinin antimalarial drug distribution in Africa and Southeast Asia (Mutabingwa 2005; White 2008; Kindermans et al. 2007). Natural artemisinin-based treatments may require subsidies and controlled crop development to ensure accessibility (Mutabingwa 2005; White 2008). However, synthetic production of artemisinic components provides a faster timeline and more efficient resource use and can obviate the reliance upon natural crop cycles for artemisinin plants. By 2013, the World Health Organization prequalified the use of semisynthetic artemisinin, allowing the pharmaceutical company Sanofi to begin its distribution with an initial shipment of 1.7 million artemisinin treatments in August 2014 (Singh and Vaidya 2015). This advancement in synthetic biology research demonstrated the technology's ability to yield therapeutic benefits for human health and commercial products (Hale et al. 2007; Westfall et al. 2012; Kong and Tan 2015).

Contemporary to these developments, since 2003, the nonprofit foundation International Genetically Engineered Machine (iGEM) has hosted annual competitions for teams of high school students, undergraduate students, graduate students, and entrepreneurs to build synthetic biological systems using pre-defined parts (Kelwick et al. 2015). Teams receive a kit of biological components from which to build biological systems and operate them in living cells (Kelwick et al. 2015; Mercer 2015; Stemerding 2015). Each fall, teams gather to demonstrate their creations and operate the pre-defined parts alongside biological parts they fostered for the competition (Kelwick et al. 2015; Stemerding 2015). The competition's membership grew to 130 teams worldwide by 2010 and 341 teams by 2019, with at least one team from every habitable continent on Earth (iGEM 2019). However, Tocchetti and Aguiton (2015) and Gronvall (2018) note that such "do-it-yourself" research raises concerns about biosafety and biosecurity risk. Though iGEM participants are screened and reviewed by multiple judges for safety concerns, some stakeholders in government and the lay public remain uneasy about the potential for risks, making the competition's biosafety and biosecurity practices a point of discussion for the synthetic biology community (Guan et al. 2013).

Following the rise in circuit design and eventual characterization alongside the growth and development of the synthetic biology research community, by 2008, synthetic biology's development had accelerated to creating more complex biological circuits and better controlling systemic biological behavior within cells. In this timeframe, the declining cost of gene synthesis alongside the development of high-throughput DNA assembly approaches advanced circuit engineering capabilities (Engler et al. 2008; Gibson et al. 2009; Cameron et al. 2014). This enabled greater control of genetic systems and novel gene expression such as light-sensing circuits within bacteria (Tabor et al. 2009) and faster and more complex pattern formation in *E. coli* swarms (Liu et al. 2011). Overall, this period drove greater connections between synthetic biologists and network engineers to improve controlling and altering the form and function of cellular networks on a system level (Cameron et al. 2014).

A widely publicized development within the second wave of synthetic biology occurred at the James Craig Venter Institute (Gibson et al. 2010; Ellis et al. 2011; Elowitz and Lim 2010; Cameron et al. 2014). In 2010, the Institute announced the creation of the first synthetic cell (Gibson et al. 2010). Using a modified *Mycoplasma mycoides* genome, Venter's team demonstrated that genome design may be "constructed on a computer, chemically made in the laboratory and transplanted into a recipient cell to produce a new self-replicating cell controlled only by the synthetic genome" (JCVI 2010). In their experiment, Venter's team synthesized a version of the *M. mycoides* genome and transplanted it into an empty *Mycoplasma capricolum* bacterial shell (JCVI 2010; Cameron et al. 2014; Gibson et al. 2010; Ellis et al. 2011). This process fostered a self-replicating bacteria cell containing only the Institute's synthesized genome through transplantation of digitally synthesized genetic base pairs (Gibson et al. 2010). Within a year, a research team led by Jef Boeke at Johns Hopkins University performed a similar synthesis of *S. cerevisiae* in yeast (Dymond et al. 2011).

The Venter team's breakthrough proved the viability of constructing and editing a computerized genome for physical transplantation of a fully synthetic genome in a bacterial cell in controlled settings (JCVI 2010; Gibson et al. 2010). New developments enable scientists to cut and delete particular spots of DNA, replace portions of genes, or add entirely new genes in specific places. These techniques, collectively called "gene editing," are akin to our abilities to take pen to paper to correct typos, delete words or phrases, rearrange sentences, or add new ones. Current developments of synthetic biology applications are increasingly globalized and bring ever-expanding opportunities but more uncertainty around risk.

A team led by George Church developed the multiplex automated genome engineering (MAGE) platform to rapidly alter multiple loci in the *E. coli* genome (Wang et al. 2009; Cameron et al. 2014). This platform enabled the "proof-of-principle replacement of all TAG stop codons with the synonymous TAA codon" (Isaacs et al. 2011; Cameron et al. 2014; Wang et al. 2009). Jiang et al. (2013) and DiCarlo et al. (2013) used the clustered regularly interspaced short palindromic repeats-associated system (or CRISPR-Cas, for short) as a tool for genome editing that helped to generate genomic mutations within a cell. This increased the ability of geneticists to alter genetic structures in bacteria and yeast (Jiang et al. 2013; DiCarlo et al. 2013). CRISPR-Cas9 allows genetic engineers to mutate, swap, or add multiple genes at one time. Researchers used this approach to efficiently edit a set of 62 pig genes to produce porcine organs that harbor fewer viruses and so are safer for human transplantation (Servick 2017). Another gene-editing technique, zinc fingers, is particularly useful for engineering proteins that target specific genes (Klug 2010), allowing scientists to selectively switch specific genes on and off (Heinemann and Panke 2006; Klug 2010), and enabling more complex genetic manipulation of larger eukaryotic organisms (Khalil et al. 2012).

Animal and crop genetic engineering are heading quickly toward gene editing, not just because of its speed and creative power but also because developers recognize loopholes in oversight in the United States. Some of the new gene-editing techniques fall outside of current regulatory definitions, which are based on early genetic engineering applications. As a result, several edited crops have evaded US regulation (Kuzma 2016).

Synthetic biologists became increasingly able to alter cell DNA and produce systemic-level change to the cell's genome and behavior. However, significant challenges remain to synthetic biology, such as with the high variability of cellular part and circuit performance to overall cellular circuit construction (Nandagopal and Elowitz 2011). Smith et al. (2014) and Baltes and Voytas (2015) further note that variability within a complex intracellular environment is inherent and, at least currently, seemingly improbable to prevent or avoid.

Purnick and Weiss (2009), Andrianantoandro et al. (2006), Ellis et al. (2009), and Cheng and Lu (2012) sought to work around this problem by constructing libraries of synthesized cellular parts and rigorously quantifying the behavior and activity of these parts under certain conditions. Such libraries support assembling cellular circuits from thoroughly researched collections of parts, which would then be screened and improved as necessary for a particular function or project. The International Open Facility Advancing Biotechnology (BIOFAB) aims to construct and characterize libraries of bacterial promoters and transcription terminators (Mutalik et al. 2013; Cambray et al. 2013). Specific to this aim, BIOFAB seeks to foster a reliability score for individual cellular parts, which characterizes the potential flaws that each part may express, which can assist debugging efforts within circuit engineering exercises (Mutalik et al. 2013).

Most genetically engineered organisms (GEOs) approved for release into natural or agricultural environments are not expected to survive on their own for multiple generations because they are either less fit than the wild type or designed for human-managed systems. Confinement of the GEO and the introduced genes has been desirable for current applications of GEOs such as GE plants in agriculture or GE microorganisms for environmental pollution remediation. However, a recent advancement in gene editing enables spreading altered genes through entire populations in either a self-limited or unlimited way.

Most genes in sexually reproducing species follow the laws of Mendelian inheritance: an introduced gene is carried on one of a pair of chromosomes and is thus inherited by about half of the offspring in the first generation. If there is no selective advantage to the gene, it will dilute in the population over time. However, diverse genetic mechanisms can enable genes to occur more frequently than the expected 50% of first-generation offspring. Evolutionary biologists have been studying these naturally occurring "selfish genetic elements" for over 80 years, but only in the past decade have researchers synthesized genetic elements with these properties. Experimental new "gene drive" systems allow an edited gene on one chromosome to copy itself into its partner chromosome to ensure that nearly all offspring will inherit the engineered gene. Thus, even if just a few organisms with gene drives are released into the wild, species with short generation times and random mating could spread an engineered gene through a large population within just a season. Synthetic "gene drive" systems took a leap forward with CRISPR-Cas9 technology, which greatly increases the ease and pace at which engineered organisms with drive mechanisms can be produced (Esvelt et al. 2014, Mali et al. 2013).

Gene drives have not yet been released into the wild, but they have been demonstrated in laboratory-cage experiments with fruit flies and mosquitos. In the wild, the drives could spread killer genes to destroy unwanted pest populations, invasive species, or disease-carrying organisms. Releasing a few individuals with killer-drive systems could theoretically eradicate a whole population, like mosquitos carrying dengue, malaria, or Zika virus. Gene drives could also add beneficial genes to populations to immunize endangered species against disease or protect them from the effects of climate change. In some cases, gene drives might be the only option to save an endangered species or to protect humans from great harm.

The ecological and health risks and benefits of synthetic biology will depend on the species engineered, the type of alteration carried, the place where it is released, the strategy for release and monitoring, and the properties of the genetic alterations. Some questions for thinking about these risks of synthetic biology include the following: How might a genetic construct(s) spread to a natural population, both intentionally or not? Does the construct(s) alter the characteristics of individuals in the population (ability to transmit pathogens), decrease population size, or both? Could the construct(s) cause extinction of the population or even the species, and is that desirable? Will the construct(s) remain in a wild population or be lost with time? Is there a means for "recalling" (or eliminating) the initially released construct(s) by releasing other variants of the target species? How would changes in the target population affect the overall ecosystem? Could other more harmful species fill the ecological niches of the eradicated organisms, perhaps ones spreading even more detrimental human or ecological disease?

These are scientific questions with potentially broad social science implications. Many chapters in this volume provide new data and approaches that demonstrate the feasibility for multi-stakeholder efforts involving policymakers, regulators, industrial developers, workers, experts, and societal representatives that can together create effective and acceptable governance in the face of uncertain risk probabilities. Such participation bestows responsibility and is a partial remedy for ignorance and may provide a pathway for humanity to maximize its benefits from synthetic biology while minimizing risks.

A Brief Introduction to *Synthetic Biology 2020's* Chapters

The chapters in this book provide perspectives of historical synthetic biology developments and implications for the technology's applications in the future. Topically, they range from general background, to differing perspectives on risk assessment and management, to governance, to risk communication and ethical decision-making.

In chapter "Synthetic Biology: Research Needs for Assessing Environmental Impacts," the team of Warner, Carter, Lance, Crocker, Meeks, Adams, Magnuson, Rycroft, Pokrzywinski, and Perkins reviews multiple platforms of synthetic biology research to better understand their potential environmental implications. Such a comprehensive review of technological use scenarios as well as their hazards and exposure considerations helps to structure the research environment and identify areas where potential threats or safety challenges may be likely or unacceptable.

Chapter "Transfish: The Multiple Origins of Transgenic Salmon" documents the development of transgenic salmon and the various mechanisms that render such an innovation "safe" in the eyes of US consumers. It describes consumer suspicions and misunderstandings of synthetic biology products (will fish antifreeze make ice cream taste fishy?) and the approval process in the United States.

In chapter "The State of Synthetic Biology Scholarship: A Case Study of Comparative Metrics and Citation Analysis," Cegan applies a network/connectivity analysis to synthetic biology publications to better understand the different disciplines and actors evaluating synthetic biology and where they are headed. Using network and citation analysis, this chapter demonstrates how more advanced analytical tools can help make sense of a rapidly growing body of literature such as the physical/natural and social sciences of synthetic biology.

In chapter "Synthetic Biology, GMO, and Risk: What Is New, and What Is Different?," Trump offers a background of synthetic biology's various threats, including those related to biosafety and biosecurity. This chapter examines what types of risk may be unusual, novel, or particularly difficult for a stakeholder to assess and thereby contribute to governance challenges where no clear best practice or operating procedure has yet been identified.

Chapter "Estimating and Predicting Exposure to Products from Emerging Technologies" (Vallero) describes the risk assessment and management paradigm applied by environmental agencies in the United States and methodologies for estimating human exposures to contaminants. Vallero asserts that better understanding exposures and subsequent risks will support informed decisions involving synthetic biology and emerging technologies.

In chapter "Mosquitoes Bite: A Zika Story of Vector Management and Gene Drives," Berube examines the trade-offs of continued disease infection and disease reduction enabled by releasing varieties of engineered mosquitos, and the public reception of those options. Berube writes, "What stands between us and addressing one of the biggest public health issues in the world is not science. It's how we talk about science."

In chapter "Synthetic Biology Industry: Biosafety Risks to Workers," Murashov, Howard, and Schulte consider the risks of synthetic biology to the workers in the biotech industry. The authors describe the safety measures that can be implemented to mitigate worker risk from exposure to hazardous materials. Demonstrated containment and control of synthetic biology will support the safety of innovation processes.

Finkel (chapter "Designing a "Solution-Focused" Governance Paradigm for Synthetic Biology: Toward Improved Risk Assessment and Creative Regulatory Design") reviews challenges associated with quantitative risk assessment relative to synthetic biology and describes how a complementary approach, known as "solution-focused risk assessment," can better inform the trade-offs and implications of synthetic biology applications in areas of considerable uncertainty. As new technologies empower humans, new inventions must examine the full spectrum of trade-offs, including risks, their associated uncertainties, and their variation across different populations. Finkel advocates for transparency in hidden influential value judgments as part of risk communication.

Chapter "A Solution-Focused Comparative Risk Assessment of Conventional and Emerging Synthetic Biology Technologies for Fuel Ethanol" (Wells, Trump,

Finkel, and Linkov) compares conventional options for energy production from corn and sugarcane feedstock to biofuel produced from engineered algae. Wells et al. utilize one permutation of Finkel's proposed solution-focused risk assessment to identify areas of novel risk to weigh against the benefits of algal biofuels and incorporate the uncertainty of synthetically engineered biofuel impacts. This holistic process could be a tool for assuaging public concerns surrounding synthetic biology for fuel production.

Chapter "An Initial Framework for the Environmental Risk Assessment of Synthetic Biology-Derived Organisms with a Focus on Gene Drives" (Landis, Brown, and Eikenbary) uses a more advanced analytical approach to demonstrate how the environmental impacts of synthetic biology-derived organisms can be assessed for environmental risk implications. Whereas a major concern of the potential use of gene drives is the irreversibility and disruption posed by engineered organisms upon the natural environment, Landis et al.'s approach is one that can help inform developers and policymakers alike of the safe use requirements and best practices that synthetic biology research should incorporate.

Kuiken (chapter "Biology without Borders: Need for Collective Governance?") begins by using the competition iGEM and its Safety and Security Committee as a study of governance for synthetic biology. Kuiken expands on ideas of equity and safety as he profiles the synthetic biology community that emerged from independent informal laboratories around the United States and the world. He shares a model for collective governance that can reconcile international laws and provide a means for overseeing synthetic biology as it evolves.

Trump, Siharulidze, and Cummings (chapter "Synthetic Biology and Risk Regulation: The Case of Singapore") differ from other governance chapters by reviewing the hard and soft law activities within Singapore. As a growing developer of synthetic biology and its enabling technologies, Singaporean governance of synthetic biology differs from their Western counterparts due to its inherently diverging political and institutional frameworks and can cause it to address and govern the risks posed by such technologies in an equally divergent manner.

Novossiolova, Bakanidze, and Perkins (chapter "Effective and Comprehensive Governance of Biological Risks: A Network of Networks Approach for Sustainable Capacity Building") document the various factors spurring innovation in synthetic biology and emphasize that regulations must manage risk without stymieing innovation. Lela examines various governance structures and asserts that biological security will benefit from both top-down actions, such as regulations and inspections, and bottom-up approaches, including education and standardized procedures.

Ndoh, Cummings, and Kuzma (chapter "The Role of Expert Disciplinary Cultures in Assessing Risks and Benefits of Synthetic Biology") review not only how different expectations, norms, values, and operating requirements within various disciplinary domains shape their perception of synthetic biology and its products, but also how risk analysis, policy, and governance are crafted and executed to capture the technology. Their notion of "expert culture" is one that demonstrates the usefulness of a collaborative and mixed-method approach toward synthetic biology governance moving forward, including the combined views of natural and social scientists.

Howell et al. (chapter "Scientists' and the Publics' Views of Synthetic Biology") analyze surveys of scientific experts and members of the American public to examine their respective risk perceptions of synthetic biology. Howell recognizes the polarization that has grown around issues like genetically modified crops or stem cell research, and the path forward for synthetic biology appears to be through public engagement in decision-making.

In chapter "Dignity as a Faith-based Consideration in the Ethics of Human Genome Editing," Austriaco explores the notion of human dignity to explain divergent views on synthetic biology between religious and lay communities. The confluence of values like dignity, free will, and agency supports understanding the emotional connotations that communities voice around issues like gene editing for human babies.

In chapter "Highlights on the Risk Governance for Key Enabling Technologies: From Risk Denial to Ethics," Merad explores the confusion between science and counter science and the denial of scientific fact. This can help frame how scientific information is developed, disseminated, and consumed and the trade-offs are considered by different actors in that process.

Ultimately, synthetic biology can provide valuable benefits to humanity that likely cannot be achieved by alternate means. Such innovations will certainly also enrich the teams that create them. The incentives are prodigious and obvious, and the public deserves assurances that all potential downsides are understood and minimized accordingly. Social science may impose additional constraints on the innovation process, but its simultaneous support in improving end product acceptability to society at large is a worthy trade-off.

References

Agtmael, V., Michiel, A., Eggelte, T. A., & van Boxtel, C. J. (1999). Artemisinin drugs in the treatment of malaria: From medicinal herb to registered medication. *Trends in Pharmacological Sciences, 20*(5), 199–205.

Andrianantoandro, E., Basu, S., Karig, D. K., & Weiss, R. (2006). Synthetic biology: New engineering rules for an emerging discipline. *Molecular Systems Biology, 2*(1), 2006–0028.

Ball, P. (2004). Synthetic biology: starting from scratch. *Nature – London,* (7009), 624.

Baltes, N. J., & Voytas, D. F. (2015). Enabling plant synthetic biology through genome engineering. *Trends in Biotechnology, 33*(2), 120–131.

Bayer, T. S., & Smolke, C. D. (2005). Programmable ligand-controlled riboregulators of eukaryotic gene expression. *Nature Biotechnology, 23*(3), 337–343.

Calvert, J., & Martin, P. (2009). The role of social scientists in synthetic biology. *EMBO Reports, 10*(3), 201–204.

Cambray, G., Guimaraes, J. C., Mutalik, V. K., Lam, C., Mai, Q.-A., Thimmaiah, T., Carothers, J. M., Arkin, A. P., & Endy, D. (2013). Measurement and modeling of intrinsic transcription terminators. *Nucleic Acids Research, 41*(9), 5139–5148.

Cameron, D. E., Bashor, C. J., & Collins, J. J. (2014). A brief history of synthetic biology. *Nature Reviews Microbiology, 12*(5), 381.

Cheng, A. A., & Lu, T. K. (2012). Synthetic biology: An emerging engineering discipline. *Annual Review of Biomedical Engineering, 14*, 155–178.

Cummings, C., Frith, J., & Berube, D. (2013). Unexpected appropriations of technology and life cycle analysis: Reframing cradle to grave approaches. In N. Savage, M. E. Gorman, & A. Street (Eds.), *Emerging technologies: Socio-behavioral life cycle approaches* (pp. 251–271). CRC Press, Boca Raton, FL.

DiCarlo, J. E., Norville, J. E., Mali, P., Rios, X., Aach, J., & Church, G. M. (2013). Genome engineering in Saccharomyces cerevisiae using CRISPR-Cas systems. *Nucleic Acids Research, 41*(7), 4336–4343.

Dymond, J. S., Richardson, S. M., Coombes, C. E., Babatz, T., Muller, H., Annaluru, N., Blake, W. J., et al. (2011). Synthetic chromosome arms function in yeast and generate phenotypic diversity by design. *Nature, 477*(7365), 471.

Ellis, T., Wang, X., & Collins, J. J. (2009). Diversity-based, model-guided construction of synthetic gene networks with predicted functions. *Nature Biotechnology, 27*(5), 465.

Ellis, T., Adie, T., & Baldwin, G. S. (2011). DNA assembly for synthetic biology: From parts to pathways and beyond. *Integrative Biology, 3*(2), 109–118.

Elowitz, M. B., & Leibler, S. (2000). A synthetic oscillatory network of transcriptional regulators. *Nature, 403*(6767), 335.

Elowitz, M., & Lim, W. A. (2010). Build life to understand it. *Nature, 468*(7326), 889.

Endy, D., Deese, I., & Wadey, C. (2005). Adventures in synthetic biology. *Nature, 438*(7067), 449–453.

Engler, C., Kandzia, R., & Marillonnet, S. (2008). A one pot, one step, precision cloning method with high throughput capability. *PLoS One, 3*, e3647.

Esvelt, K. M., Smidler, A. L., Catteruccia, F., & Church, G. M. (2014). Emerging technology: Concerning RNA-guided gene drives for the alteration of wild populations. *Elife3, 3*, e03401.

Ferber, D. (2004). Time for a synthetic biology Asilomar? *Science, 303*, 159.

Ferguson, S. A., Leaf, W. A., Williams, A. F., & Preusser, D. F. (1996). Differences in young driver crash involvement in states with varying licensure practices. *Accident Analysis & Prevention, 28*(2), 171–180.

Gardner, T. S., Cantor, C. R., & Collins, J. J. (2000). Construction of a genetic toggle switch in Escherichia coli. *Nature, 403*(6767), 339.

Gibson, D. G., Young, L., Chuang, R.-Y., Craig Venter, J., Hutchison, C. A., III, & Smith, H. O. (2009). Enzymatic assembly of DNA molecules up to several hundred kilobases. *Nature Methods, 6*(5), 343.

Gibson, D. G., Glass, J. I., Lartigue, C., Noskov, V. N., Chuang, R.-Y., Algire, M. A., Benders, G. A., et al. (2010). Creation of a bacterial cell controlled by a chemically synthesized genome. *Science, 329*(5987), 52–56.

Gronvall, G. K. (2018). Safety, security, and serving the public interest in synthetic biology. *Journal of Industrial Microbiology & Biotechnology, 45*(7), 463–466.

Guan, Z.-J., Schmidt, M., Pei, L., Wei, W., & Ma, K.-P. (2013). Biosafety considerations of synthetic biology in the international genetically engineered machine (iGEM) competition. *Bioscience, 63*(1), 25–34.

Hale, V., Keasling, J. D., Renninger, N., & Diagana, T. T. (2007). Microbially derived artemisinin: A biotechnology solution to the global problem of access to affordable antimalarial drugs. *The American Journal of Tropical Medicine and Hygiene, 77*(6_Suppl), 198–202.

Heinemann, M., & Panke, S. (2006). Synthetic biology—Putting engineering into biology. *Bioinformatics, 22*(22), 2790–2799.

Hood, L., Heath, J. R., Phelps, M. E., & Lin, B. (2004). Systems biology and new technologies enable predictive and preventative medicine. *Science, 306*(5696), 640–643.

Hucka, M., Finney, A. B. B. J., Bornstein, B. J., Keating, S. M., Shapiro, B. E., Matthews, J., Kovitz, B. L., et al. (2004). Evolving a lingua franca and associated software infrastructure for computational systems biology: The Systems Biology Markup Language (SBML) project. *Systems Biology, 1*(1), 41–53.

Ideker, T., Galitski, T., & Hood, L. (2001). A new approach to decoding life: Systems biology. *Annual Review of Genomics and Human Genetics, 2*(1), 343–372.

iGEM. Team list for iGEM 2019 Championship. https://igem.org/Team_List. Last accessed 3 Apr 2019.

Ingber, D. E. (2003). Tensegrity I. Cell structure and hierarchical systems biology. *Journal of Cell Science, 116*(7), 1157–1173.

Isaacs, F. J., et al. (2004). Engineered riboregulators enable post-transcriptional control of gene expression. *Nature Biotechnology, 22*, 841–847.

Isaacs, F. J., Carr, P. A., Wang, H. H., Lajoie, M. J., Sterling, B., Kraal, L., Tolonen, A. C., et al. (2011). Precise manipulation of chromosomes in vivo enables genome-wide codon replacement. *Science, 333*(6040), 348–353.

Jasanoff, S. (2009). *The fifth branch: Science advisers as policymakers*. Cambridge: Harvard University Press.

JCVI. (2010). First self-replicating, synthetic bacterial cell constructed by J. Craig Venter Institute Researchers [Press Release]. Retrieved from https://www.jcvi.org/first-self-replicating-synthetic-bacterial-cell-constructed-j%C2%A0craig-venter-institute-researchers

Jiang, W., Bikard, D., Cox, D., Zhang, F., & Marraffini, L. A. (2013). RNA-guided editing of bacterial genomes using CRISPR-Cas systems. *Nature Biotechnology, 31*(3), 233.

Kelwick, R., Bowater, L., Yeoman, K. H., & Bowater, R. P. (2015). Promoting microbiology education through the iGEM synthetic biology competition. *FEMS Microbiology Letters, 362*(16), fnv129.

Khalil, A. S., & Collins, J. J. (2010). Synthetic biology: Applications come of age. *Nature Reviews Genetics, 11*(5), 367.

Khalil, A. S., Lu, T. K., Bashor, C. J., Ramirez, C. L., Pyenson, N. C., Joung, J. K., & Collins, J. J. (2012). A synthetic biology framework for programming eukaryotic transcription functions. *Cell, 150*(3), 647–658.

Kindermans, J.-M., Pilloy, J., Olliaro, P., & Gomes, M. (2007). Ensuring sustained ACT production and reliable artemisinin supply. *Malaria Journal, 6*(1), 125.

Kitano, H. (2002). Systems biology: A brief overview. *Science, 295*(5560), 1662–1664.

Klug, A. (2010). The discovery of zinc fingers and their applications in gene regulation and genome manipulation. *Annual Review of Biochemistry, 79*, 213–231.

Kong, L. Y., & Tan, R. X. (2015). Artemisinin, a miracle of traditional Chinese medicine. *Natural Product Reports, 32*(12), 1617–1621.

Kuzma, J. (2016). Policy: Reboot the debate on genetic engineering. *Nature, 531*, 165–167.

Levskaya, A., Chevalier, A. A., Tabor, J. J., Simpson, Z. B., Lavery, L. A., Levy, M., Davidson, E. A., et al. (2005). Synthetic biology: Engineering Escherichia coli to see light. *Nature, 438*(7067), 441.

Linkov, I., et al. (2018). Comparative, collaborative, and integrative risk governance for emerging technologies. *Environment Systems and Decisions, 38*(2), 170–176.

Liu, C., Fu, X., Liu, L., Ren, X., Chau, C. K., Li, S., et al. (2011). Sequential establishment of stripe patterns in an expanding cell population. *Science, 334*(6053), 238–241.

Mali, P., Esvelt, K. M., & Church, G. M. (2013). Cas9 as a versatile tool for engineering biology. *Nature Methods, 10*(10), 957.

Marris, C. (2015). The construction of imaginaries of the public as a threat to synthetic biology. *Science as Culture, 24*(1), 83–98.

Mercer, D. (2015). iDentity and governance in synthetic biology: Norms and counter norms in the international genetically engineered machine (iGEM) competition. *Macquarie Law Journal, 15*, 83.

Mutabingwa, T. K. (2005). Artemisinin-based combination therapies (ACTs): Best hope for malaria treatment but inaccessible to the needy! *Acta Tropica, 95*(3), 305–315.

Mutalik, V. K., Guimaraes, J. C., Cambray, G., Lam, C., Christoffersen, M. J., Mai, Q.-A., Tran, A. B., et al. (2013). Precise and reliable gene expression via standard transcription and translation initiation elements. *Nature Methods, 10*(4), 354.

Nandagopal, N., & Elowitz, M. B. (2011). Synthetic biology: Integrated gene circuits. *Science, 333*(6047), 1244–1248.

Nosten, F., & White, N. J. (2007). Artemisinin-based combination treatment of falciparum malaria. *The American Journal of Tropical Medicine and Hygiene, 77*(6_Suppl), 181–192.

Oye, K. A. (2012). *Proactive and adaptive governance of emerging risks: The case of DNA synthesis and synthetic biology*. International Risk Governance Council.

Palma-Oliveira, J. M., Trump, B. D., Wood, M. D., & Linkov, I. (2018). Community-driven hypothesis testing: A solution for the tragedy of the anticommons. *Risk Analysis, 38*(3), 620–634.

Park, S. H., Zarrinpar, A., & Lim, W. A. (2003). Rewiring MAP kinase pathways using alternative scaffold assembly mechanisms. *Science, 299*, 1061–1064.

Pauwels, E. (2009). Review of quantitative and qualitative studies on U.S. public perceptions of synthetic biology. *Systems and Synthetic Biology, 3*(1–4), 37–46.

Pauwels, E. (2013). Public understanding of synthetic biology. *Bioscience, 63*(2), 79–89.

Polizzi, K. M., Stanbrough, L., & Heap, J. T. (2018). A new lease of life: Understanding the risks of synthetic biology. An emerging risks report published by Lloyd's of London. https://www.lloyds.com/news-and-risk-insight/news/lloyds-news/2018/07/a-new-lease-of-life. Accessed 4 Sept 2018.

Porcar, M., & Peretó, J. (2014). Synthetic biology in action. In *Synthetic biology* (pp. 45–53). Dordrecht: Springer.

Purnick, P. E. M., & Weiss, R. (2009). The second wave of synthetic biology: From modules to systems. *Nature Reviews Molecular Cell Biology, 10*(6), 410.

Ro, D.-K., Paradise, E. M., Ouellet, M., Fisher, K. J., Newman, K. L., Ndungu, J. M., Ho, K. A., et al. (2006). Production of the antimalarial drug precursor artemisinic acid in engineered yeast. *Nature, 440*(7086), 940.

Servick, K.. (2017). CRISPR slices virus genes out of pigs, but will it make organ transplants to humans safer? Science Magazine. Available https://www.sciencemag.org/news/2017/08/crispr-slices-virus-genes-out-pigs-will-it-make-organ-transplants-humans-safer

Shapira, P., Youtie, J., & Li, Y. (2015). Social science contributions compared in synthetic biology and nanotechnology. *Journal of Responsible Innovation, 2*(1), 143–148.

Singh, M., & Vaidya, A. (2015). Translational synthetic biology. *Systems and Synthetic Biology, 9*(4), 191–195.

Smith, M. T., Wilding, K. M., Hunt, J. M., Bennett, A. M., & Bundy, B. C. (2014). The emerging age of cell-free synthetic biology. *FEBS Letters, 588*(17), 2755–2761.

Stemerding, D. (2015). iGEM as laboratory in responsible research and innovation. *Journal of Responsible Innovation, 2*(1), 140–142.

Tabor, J. J., Salis, H. M., Simpson, Z. B., Chevalier, A. A., Levskaya, A., Marcotte, E. M., Voigt, C. A., & Ellington, A. D. (2009). A synthetic genetic edge detection program. *Cell, 137*(7), 1272–1281.

Tocchetti, S., & Aguiton, S. A. (2015). Is an FBI agent a DIY biologist like any other? A cultural analysis of a biosecurity risk. *Science, Technology, & Human Values, 40*(5), 825–853.

Torgersen, H., & Schmidt, M. (2013). Frames and comparators: How might a debate on synthetic biology evolve? *Futures, 48*, 44–54.

Trump, B. D. (2017). Synthetic biology regulation and governance: Lessons from TAPIC for the United States, European Union, and Singapore. *Health Policy, 121*(11), 1139–1146.

Trump, B., Cummings, C., Kuzma, J., & Linkov, I. (2018). A decision analytic model to guide early-stage government regulatory action: Applications for synthetic biology. *Regulation & Governance, 12*(1), 88–100.

Trump, B. D., Cegan, J., Wells, E., Poinsatte-Jones, K., Rycroft, T., Warner, C., et al. (2019). Co-evolution of physical and social sciences in synthetic biology. *Critical Reviews in Biotechnology, 39*(3), 351–365.

Wang, H. H., Isaacs, F. J., Carr, P. A., Sun, Z. Z., Xu, G., Forest, C. R., & Church, G. M. (2009). Programming cells by multiplex genome engineering and accelerated evolution. *Nature, 460*(7257), 894.

Weiss, R., & Knight Jr, T. (2001). Engineered communications for microbial robotics. In lncs 2054.

Westfall, P. J., Pitera, D. J., Lenihan, J. R., Eng, D., Woolard, F. X., Regentin, R., Horning, T., et al. (2012). Production of amorphadiene in yeast, and its conversion to dihydroartemisinic acid, precursor to the antimalarial agent artemisinin. *Proceedings of the National Academy of Sciences, 109*(3), E111–E118.

White, N. J. (2008). The role of anti-malarial drugs in eliminating malaria. *Malaria Journal, 7*(1), S8.

Williams, A. F. (2009). Licensing age and teenage driver crashes: A review of the evidence. *Traffic Injury Prevention, 10*(1), 9–15.

Synthetic Biology: Research Needs for Assessing Environmental Impacts

Christopher M. Warner, Sarah R. Carter, Richard F. Lance, Fiona H. Crocker, Heather N. Meeks, Bryn L. Adams, Matthew L. Magnuson, Taylor Rycroft, Kaytee Pokrzywinski, and Edward J. Perkins

Introduction and Methods

Synthetic biology refers to the design and construction of new biological entities such as enzymes, genetic circuits, and cells or the redesign of existing biological systems (Keasling 2005). This capability is rooted in traditional molecular biology and engineering and incorporates newer techniques, including de novo DNA synthesis, CRISPR (clustered regularly interspaced short palindromic repeats)-based genome editing, and xenobiology. Synthetic biology, along with a wide range of emerging tools and techniques, will enable a new generation of biotechnology products of unprecedented scale and complexity over the next 5–10 years (NASEM 2017a).

Within the United States, synthetic biology and its applications are currently estimated to be a multi-billion-dollar industry and growing rapidly (Gronvall 2015) with significant private investment (see, e.g., Stevenson 2017). Many of these products are likely to have beneficial applications for military use, including new

C. M. Warner (✉) · R. F. Lance · F. H. Crocker · T. Rycroft · K. Pokrzywinski · E. J. Perkins
Environmental Laboratory, Engineer Research and Development Center, U.S. Army Corps of Engineers, Vicksburg, MS, USA
e-mail: Christopher.M.Warner@usace.army.mil

S. R. Carter
Science Policy Consulting, LLC, Arlington, VA, USA

H. N. Meeks
Defense Threat Reduction Agency, Ft. Belvoir, VA, USA

B. L. Adams
Adelphi Laboratory Center, CCDC Army Research Laboratory, Adelphi, MD, USA

M. L. Magnuson
U.S. Environmental Protection Agency, Cincinnati, OH, USA

© Springer Nature Switzerland AG 2020
B. D. Trump et al. (eds.), *Synthetic Biology 2020: Frontiers in Risk Analysis and Governance*, Risk, Systems and Decisions, https://doi.org/10.1007/978-3-030-27264-7_2

approaches to manage natural resources and ranges, produce fuels and other materials, as well as protect the warfighter. For this reason, the US government, including the Department of Defense (DoD), has made significant investments in synthetic biology (OTI 2015; Wilson Center 2015). While many of the next generation of products will be similar to existing biotechnology products, others are likely to be novel, including many with probable or intended release into the environment. These technologies may challenge our current regulatory and environmental risk assessment frameworks (Carter et al. 2014; Drinkwater et al. 2014; NASEM 2017a). The US Army Engineer Research and Development Center (ERDC) is well-positioned to address some of the critical environmental questions that these new products will raise. In doing so, ERDC can help ensure that the US DoD, regulatory agencies, broader government stakeholders, commercial entities, and others have the information and tools necessary to make informed decisions on development and potential use in the environment of synthetic biology organisms (whole organisms that have been engineered using synthetic biology) and components (constructs and circuits that may be used or deployed outside of a living organism).

In addition to posing challenges for environmental risk assessment, new biotechnologies raise broader regulatory and societal issues (Trump et al. 2019). For example, in some cases, there is uncertainty in the regulatory pathway that these products will have to traverse before they can be tested or deployed in the environment (Carter and Friedman 2016; NASEM 2017a). The potential for field testing and deployment of organisms engineered with gene drives (a class of synthetic biology organisms that have intended interactions with the natural environment; see case study below) has generated much discussion about the need for community and stakeholder engagement (NASEM 2016, 2017a), with early engagement activities already underway in some contexts (Swetlitz 2017). These regulatory and engagement activities are likely to require time and commitment, with some types of products likely to face greater scrutiny and more challenges than others (Trump et al. 2018a). A better understanding of these issues will be critical in the development and application of a wide range of biotechnology products.

In May 2017, ERDC held a case study-based workshop in Lexington, MA, that aimed to identify key challenges to the deployment of advanced biotechnologies. The 2.5-day meeting brought together 60 participants from a wide range of organizations including ERDC and other DoD entities, universities, commercial companies, federal regulatory agencies, and nongovernmental and international organizations. The agenda included context-setting plenary talks and five breakout sessions. Each breakout group focused on one of four specific case studies, including organisms engineered with gene drives to control infectious disease vectors, engineered microbes for bioremediation, cell-free applications for advanced chemical production, and engineered viruses for water treatment. The case studies (see online version for complete full case study prompts) were chosen because they represent realistic biotechnologies that together reflect the wide range of synthetic biology technologies that could be submitted to regulatory agencies for consideration in the near future. To ensure balanced and cross-disciplinary discussions of these case studies, each group included participants with backgrounds in basic and applied research, including biological, engineering, social sciences, and regulatory processes. The breakout discussions within each case study were divided into topic areas including (1) "horizon scanning" to identify the scope of technologies relevant to the case study; (2) "environmental impacts" to discuss potential environmental impacts; (3) "safety and regulation" to identify the current regulatory frameworks that apply to these technologies; (4) "community engagement" to identify broader societal issues; and (5) "challenges and opportunities" to identify the key themes and challenges to deployment.

Workshop discussions aimed to identify high-priority information, data, and capabilities needed to provide a basis for decision-making with respect to deployment of synthetic biology organisms and components in the environment. The focus of the workshop was primarily on environmental impacts, including potential hazards and risks. Throughout the workshop and this document, we defined risk as the probability of an effect on a specific endpoint or set of endpoints due to a specific stressor or set of stressors, hazard as a harmful effect, and impact as any effect which can be beneficial or harmful. Risk assessment is defined as the process by which all available evidence on the probability of effects is collected, evaluated, and interpreted to estimate the probability of the sum total of effects (NASEM 2016).

Research and development needs related to understanding and monitoring potential environmental impacts are described below. Throughout the workshop, these needs were discussed primarily in the context of regulatory assessment and approval but are also relevant in the context of nonregulatory (voluntary) assessment and mitigation measures that developers may choose to undertake (e.g., to reduce product liability). Some of these biotechnologies also have critical regulatory and societal uncertainties associated with their deployment; these information and capability needs are flagged as well. Without explicit and careful inclusion of these aspects in product development or release plans, beneficial applications of synthetic biology organisms and components could be delayed or never realized. Synthetic biology is also an important topic within the context of biosecurity and bioterrorism (NSABB 2010; NASEM 2017b). Although these are important discussion topics and were briefly considered, these types of risks were not a focus of the workshop.

Section Two of this chapter describes themes that emerged from workshop discussions, as well as research, information, and capability needs that were identified. Many of these themes and needs were common across all case studies and represent opportunities for future high-impact research and development. Section Three includes discussion summaries from each of the four case studies. Section Four describes workshop conclusions, including the need for a strategic approach across the US government for assessing the environmental impacts of synthetic biology organisms and components.

Although this chapter draws on the collected expertise of workshop participants and others, this is not a consensus report, and the conclusions are those of the authors alone and do not represent any government position. Nevertheless, despite the wide range of perspectives provided at the workshop, there was a high level of agreement on many issues.

Common Themes and Research Needs

A number of common themes and research needs emerged from the consideration of the workshop case studies. This section describes the research, information, and capability needs that were identified, including many that were common across all case studies. These needs represent opportunities for high-impact research and development.

Modeling

Because biological systems and their interactions with dynamic ecosystems are complex, development, refinement, and ongoing evaluation of models will be critical to understanding the characteristics and interactions of synthetic biology organisms and components, as well as their potential risks and benefits. The need to populate models with useful data will also provide an important basis for many of the research needs listed below. These needs include:

- Modeling of how synthetic biology organisms and components will interact with native populations and ecosystems, including scenarios of intentional release and escape.
- Modeling of how synthetic biology organisms and components may change or evolve over time in different contexts and environments.
- Experimental or observational evaluation of models, including the ability to ensure that relevant data are reliably generated and incorporated into models.

Fate and Transport of Synthetic Biology Organisms and Components

The fate and transport of biological components, engineered or otherwise, have long been identified as a research need, but much work remains to be done. Needs include:

- Understanding of gene transfer for different types of nucleic acids (e.g., naked oligonucleotides, viral-encapsulated DNA and RNA, microbial plasmid and genomic DNA, and eukaryotic DNA) in both natural contexts and with synthetic biology organisms and components. This includes studies of the potential for hybridization of synthetic biology organisms with nontargeted, natural populations.
- Modeling (including evaluation of models) and measurement of the distance synthetic biology organisms and components are likely to travel within specific

environments and the length of time they are likely to persist in different contexts.

Control and Stability of Synthetic Biology Organisms and Components

A key challenge for many synthetic biology organisms and components is ensuring that they are controllable, stable, and predictable in the environment. Needs include:

- Improved control of synthetic biology organisms and components. For example, organisms engineered with gene drives that only survive where and when they are wanted with the characteristics that are intended and microbial or viral systems that contain improved and stable intrinsic biocontainment mechanisms (e.g., kill switches and auxotrophic metabolism) to limit the spread of synthetic biology organisms and components in the environment
- Development of predictive tools and methodologies to identify potential novel outcomes, such as genetic rearrangements, unintended enzymatic or metabolic activity, or unwanted reproduction

Monitoring and Surveillance

Discussions for each case study identified the need for improved monitoring and surveillance of environmental systems, both for improved baseline understanding of the naïve environment prerelease and for tracking synthetic biology organisms and components following deployment. Needs include:

- Monitoring and surveillance tools for synthetic biology organisms and components in the environment, including development of metrics to track their spread, stability, and persistence
- Baseline characterization of native environments into which synthetic biology organisms and components are likely to be deployed to detect and contextualize post-deployment changes

Oversight, Regulation, and Community Engagement

Several common themes emerged in discussions about regulatory oversight and community engagement for the four case studies. For regulatory decision-making, there was an awareness in each group of the need for case-by-case evaluation of synthetic biology organisms and components and potential environmental releases due to the wide variety of uses and contexts. There was also a desire for clarification of the regulatory process, including timelines and data requirements. Early and frequent engagement with regulators was identified as critical to successful navigation of the regulatory process. Another theme that arose was the need to evaluate the impacts of synthetic biology organisms and components against the impacts of alternative actions, including no action. Phased testing and evaluation of synthetic biology technologies were identified as a way to improve products and generate the data necessary to make decisions on eventual deployment in the environment.

The need for effective community engagement also emerged as a common theme in discussions of the four case studies, with one case study (gene drives) identifying it as fundamental to successful testing and deployment in the environment. When pursuing community engagement activities, product development teams should provide a means for community members and other stakeholders to impact decision-making. Such a process should include a well-defined intention, a thoughtful analysis of who should be included, what information needs to be shared by the product development team, and how discussions with community participants can best inform decision-making. By establishing engagement and building trust in the community early in the development and deployment process, decisions can be made with clarity and mutual respect. Throughout the workshop, there was discussion about whether and how engagement processes and deployment of the "first" synthetic biology technologies may impact perceptions and potential deployment of those that come later. Needs for regulatory and community engagement include:

- Development of processes to determine environmental endpoints of interest and clarity on how these should be measured/assessed. Regulators may already have some guidance, but it may be appropriate to clarify stakeholder roles and include community input in some cases.
- Characterization of and guidance for successful community engagement processes. This includes lessons learned from other types of community engagement and best practices developed for similar types of products and technologies. It also includes an understanding of how community perceptions and engagement processes are affected by previous and ongoing engagement on related technologies. Understanding whether successful deployment in the environment of one synthetic biology technology affects how the next is perceived and what factors influence these perceptions (e.g., type of technology, environment, or developer) is essential to successful community engagement.
- Improved communication tools, along with access to and awareness of potential collaborations with those experienced and skilled in community engagement, will empower scientists in fostering successful community engagement activities. This includes the development of more effective approaches for inclusion of stakeholder needs and perceptions throughout the development cycle.

Case Studies

This section includes discussion summaries for each of the four synthetic biology case studies: organisms engineered with gene drives to control infectious disease vectors, microbial engineering for bioremediation, cell-free technologies for advanced chemical production, and viral systems for water treatment. As mentioned in Section One, each discussion group included technical experts, as well as those familiar with environmental, regulatory, policy, and other societal implications of biotechnologies. The groups met independently (with opportunities to report conclusions to the larger group), and the written reports below represent these separate discussions. Common themes and research priorities identified in discussions across case studies are outlined in Section Two.

Case Study: Organisms Engineered with Gene Drives to Control Infectious Disease Vectors

Introduction

Gene drives are "systems of biased inheritance that enhance the ability of a genetic element to pass from an organism to its offspring through sexual reproduction" (NASEM 2016). Throughout this document, we use the terms "gene drive," "gene drive constructs," and "organism engineered with gene drives." Gene drive is the system of biased inheritance that enhances the ability of a genetic element to pass from an organism to its offspring through sexual reproduction (NASEM 2016); gene drive construct refers to the engineered genetic construct that contains elements that are preferentially inherited by the progeny of an organism; and organism engineered with gene drives refers to an organism that contains in its genome a gene drive construct. Naturally occurring gene drives have been studied for decades (Burt and Trivers 2009). However, in recent years, genome editing using CRISPR (clustered regularly interspaced short palindromic repeats) has overcome technical challenges involved in engineering gene drives. CRISPR allows insertion of genetic material targeted to a specific DNA sequence; some types of CRISPR-based gene drive constructs can ensure that nearly 100% of offspring inherit that genetic material (NASEM 2016). A wide range of gene drive constructs and applications are currently being considered and developed in laboratory settings but will require multiple stages of confined testing before being approved for field testing and deployment in the environment. It has been estimated that the first organisms engineered with gene drives are likely to be ready for field testing and regulatory consideration in 5–10 years. Oversight and assessment of this process will largely be guided using frameworks already in place for genetically engineered organisms (WHO 2014; EFSA 2013; CBD 2012; FDA 2017a), though frameworks specific to organisms engineered with gene drives are under development to address some of their specific governance challenges (NASEM 2016). The workshop case study involved a *Aedes aegypti* mosquito engineered with gene drives for suppression of wild populations to reduce disease.

Horizon Scanning

Organisms engineered with gene drives can be developed for a wide range of purposes and applications (Esvelt et al. 2014). Most anticipated applications of organisms engineered with gene drives at this time are for population suppression (i.e., decreasing numbers of an undesirable species). These include suppression of disease vectors (e.g., mosquitoes), invasive species (e.g., mice, rats, other mammals, cane toads, some invasive plant species), and agricultural weeds and pests (e.g., pigweed, screwworm, desert locust). Because gene drives require sexual reproduction, asexual or facultatively sexual species, such as many plants and fungi, may not be candidates for gene drives. To date, CRISPR-based gene drives have been demonstrated in fruit flies (*Drosophila*) and in mosquito species that are significant disease vectors (Gantz and Bier 2015; Hammond et al. 2016). Organisms engineered with gene drives are also under development for management of invasive populations of the house mouse (*Mus musculus*) on islands where this species devastates native fauna, particularly birds (Lewis 2017). A wide range of gene drive constructs for a variety of potential applications are currently under development (DARPA 2017; Target Malaria 2017).

In addition to population suppression, organisms engineered with gene drives can also be used to replace existing populations with those that are composed of individuals that carry (and pass on) genetic constructs that express one or more desired traits. Such traits could include enhanced resistance (or susceptibility) to pesticides, enhanced immunity to disease, reduced capacity to harbor disease-bearing parasites or pests, capacities for environmental remediation of pollutants or contaminants, or others. The diversity of potential uses for organisms engineered with gene drives is only beginning to be realized.

Regardless of application, organisms engineered with gene drives can be categorized based on the way that they are predicted to function. They can be self-sustaining (i.e., designed to spread in a population unless and until the population generates resistance) or self-exhausting (i.e., designed to spread in a limited way in time and space). If a organism engineered with gene drives is threshold-independent, then only a small number of individuals may allow the gene drive construct to spread in a population (unless and until the population generates resistance). In contrast, for a threshold-dependent gene drive, the engineered organism must be present in sufficient numbers relative to the wild-type individuals in a population (i.e., at or above a certain threshold) before it is likely to spread in that population; below that threshold, it will die out. By definition, releases of threshold-dependent organisms engineered with gene drives are reversible; organisms engineered with gene drives can be outcompeted by releasing sufficient numbers of wild-type organisms. They are also dispersal-limited; if a organism engineered with gene drives migrates into a largely wild-type population, the organism engineered with gene drives would be at below threshold levels and should be extirpated in that population.

The simplest CRISPR-based gene drive constructs (e.g., insertion of a single CRISPR construct targeted to one sequence within that genome) are self-sustaining and threshold-independent, which, modeling indicates, may allow them to spread and persist in the environment even when only a few individual organisms are released. When an area-wide application is intended, this type of gene drive construct could provide significant benefits. However, genetic stability is a major challenge for these types of gene drive constructs. A single nucleotide mutation (or naturally occurring polymorphism in the release site population) in the targeted sequence has the potential to reduce or eliminate the functionality of the gene drive construct and prevent it from "driving." This could in some cases also confer a selective advantage on individuals with the mutation, which could give rise to population-level resistance to the gene drive construct and bring about the extirpation of the organism engineered with gene drives in the population over time, thus decreasing the benefits of the product (Noble et al. 2017). This challenge may be addressed by using multiplexed CRISPR-based gene drive constructs, where multiple DNA sequences are targeted for insertion of the gene drive construct; this is an active area of research.

A major focus of current research in gene drive laboratories is on designing and developing gene drives with limited spatiotemporal spread. For example, as described above, self-exhausting organisms engineered with gene drives are designed to spread in a wild population for a limited time. Another approach is to target specific subpopulations that have unique DNA sequences so that the gene drive construct will only spread among those subpopulations. These gene drive constructs may be more appropriate for limited applications (e.g., when only a localized pest population or subset of a pest population is targeted).

All organisms engineered with gene drives face a significant challenge when testing at scale. It is difficult to collect meaningful data from contained cage trials that would be applicable to populations at the ecosystem scale where the gene drive construct is designed to function. Such experimental systems cannot capture the full ecological and environmental complexity that will be experienced by organisms engineered with gene drives upon release during field testing or deployment in the environment. Also, failure modes for many types of gene drive constructs (e.g., multiple mutations giving rise to resistance in a multiplexed system) are anticipated to be very rare events; thus, having a sufficient number of individuals in a cage trial to reliably detect the mutation rate and rate of drive failure is a significant challenge. Furthermore, an understanding of the population dynamics of target organisms (e.g., short-term dispersal and gene flow patterns) is often lacking, limiting the reliability of modeling efforts. To address these challenges, additional data on target organisms and their population dynamics, as well as environmental factors, are needed. Researchers can also draw on information generated and lessons learned from the release of previous generations of genetically engineered organisms, as well as non-engineered biological control organisms and pesticide applications. As the field progresses, data generated from these sources as well as contained laboratory and field trials of organisms engineered with gene drives should feed back into models to improve their predictive power.

Environmental Impacts

The potential environmental impacts of release of a organisms engineered with gene drives must be considered prior to release. Critically, effects associated with the release of a organism engineered with gene drives must be understood in relation to the alternatives (e.g., the use of pesticides) and/or no action (e.g., continuing human disease burden). A key issue in environmental risk assessment is problem formulation: identifying the environmental endpoints (protection goals) that we care most about. Because it is impossible to monitor all parts of an ecosystem, even at a small scale, there must be some prioritization of endpoints in order to evaluate risks and benefits. Although regulators can help define key environmental endpoints, engagement with those publics who might be (or perceive that they might be) affected by testing and deployment of a organism engineered with gene drives in the environment should be involved in setting these priorities (e.g., Roberts et al. 2017; Linkov et al. 2018).

For all organisms engineered with gene drives, monitoring of the environmental endpoints of interest and of the organisms themselves (e.g., spread of the organisms, gene drive phenotype, effectiveness, and stability) will be critical. Prior to release of the organism engineered with gene drives, some baseline monitoring of key features of the ecosystem will be needed to increase the likelihood that the impacts of release (if any) can be detected and measured. Effective monitoring also requires access to sites and potentially affected habitats, as well as financial support for a sustained effort. The goals of any monitoring program (including endpoints of interest and types of analyses to be performed) should be explicit to best ensure that the program generates data and information that informs decision-making. Tools for detecting and tracking the spread of the organisms engineered with gene drives will also be required, both for understanding the impact of the gene drive construct on the ecosystem and its effectiveness at spreading in the target population. Since there are a wide variety of gene drive constructs with different characteristics under development, specific environmental considerations for each potential organisms engineered with gene drives will have to be evaluated on a case-by-case basis.

For the specific workshop case study on suppression of *Aedes aegypti* mosquito populations in the United States, a number of potential environmental factors would need to be explored in greater detail prior to release. These include trophic interactions, potential for interbreeding with other species of mosquitoes, impacts on vector competence, and potential for niche effects on *A. albopictus* and other mosquitoes (e.g., the suggestion that *A. albopictus* may spread more quickly in the absence of *A. aegypti*). Additionally, there may need to be a better understanding of how the specific gene drive construct might spread within the local mosquito population through modeling efforts that take local conditions into account. Research on wild-type populations of *Aedes* (and other pest control programs such as pesticide applications and sterile insect techniques) could provide data related to population dynamics, gene flow, and genetic diversity. Such information would strengthen existing models and assessment of impacts from release of organisms engineered with gene drives and help developers improve product design and effectiveness.

Studies on mosquito populations and on potential environmental impacts must be considered in the context of the specific ecosystem into which these mosquitoes might be deployed (Finkel et al. 2018). For example, *A. aegypti* are adapted to living with humans, so their dispersal rates and methods are influenced by human movement in the area. Some populations of *A. aegypti* are invasive and have arrived relatively recently (e.g., those in the United States); eradication of those populations may be seen as restoring the native ecosystem rather than a perturbation in the ecosystem.

Safety and Regulation

Laboratory biosafety and containment in the United States are overseen primarily at the institutional level by institutional biosafety committees (IBCs). IBCs are not required under any regulation, but are a term and condition of funding from the US National Institutes of Health (NIH) and most other federal agencies. However, the NIH Guidelines for biosafety, which provide guidance for IBCs, are primarily focused on human pathogens and potential impacts on human health. As such, membership and expertise on many IBCs may be inadequate to evaluate and address potential environmental impacts that may arise from organisms engineered with gene drives. Laboratories that work on insects (especially insect vectors of disease) also follow Arthropod Containment Guidelines developed by the American Society of Tropical Medicine and Hygiene (ACG 2004); these, too, are widely applied, though voluntary. Internationally, there is no consensus on appropriate biosafety precautions for working with organisms engineered with gene drives, with different countries taking very different approaches.

Within the United States, organisms engineered with gene drives will be regulated based on their intended use (OSTP 2017). An organism intended for pest suppression may be regulated by the Environmental Protection Agency (EPA) under their rules for pesticides. If it is intended to decrease human disease burden, then it is likely to be regulated by the Food and Drug Administration (FDA) under the Federal Food, Drug, and Cosmetic Act (FDCA). If it is a plant or animal pest, then it may be regulated by the US Department of Agriculture (USDA) under its plant and animal protection rules. In some cases, the organism engineered with gene drives will be regulated by multiple agencies with these three primary agencies working together to coordinate testing, approval, and oversight. Regardless of its regulatory pathway, all organisms engineered with gene drives intended to be marketed in the United States must undergo some environmental risk assessment to comply with a federal regulatory agency. Outside of the United States, many countries have a "process-based" regulatory system whereby genetically engineered organisms are regulated under laws specifically designed to regulate products derived using biotechnology. These regulations will also apply to organisms engineered with gene drives that are being registered for in-country use.

For the specific workshop case of an *Aedes aegypti* mosquito release intended to prevent the spread of diseases including Zika and dengue, regulatory oversight in the United States would be provided by FDA (Trump 2017). This case would be regulated by FDA because the product is intended to reduce disease burden. FDA and EPA recently finalized guidance on mosquito products that clarified that those products with health claims will be regulated by FDA while those with pesticide claims will be regulated by EPA under their pesticide provisions (FDA 2017b). Under FDA regulations, a mandatory pre-market approval would be required, and a product would be approved only if it is shown to be "safe" (i.e., causes no greater harm to humans, other animals, and the environment as compared to non-engineered *A. aegypti*) and "effective." Effectiveness is determined based on the claim that the applicant intends to make, which should be specific and supported by data (provided by the developer or publicly available).

In addition to meeting FDA requirements, product developers would also need to develop an environmental assessment (EA) in compliance with the National Environmental Policy Act (NEPA) as part of the product approval process. The FDA would evaluate the EA for investigational use and issue either a Finding of No Significant Impact (FONSI) (allowing product development and testing to move forward) or require that a full environmental impact statement (EIS) be conducted. The EIS is typically a broader and more rigorous analysis than the EA. Once the EIS is published, along with a record of decision, product development can proceed. When all other FDA requirements are met, including the publication of a final EA/FONSI or EIS/record of decision, the developer can file for an approval. The NEPA process requires publication of the draft EA or EIS, including opportunities for public comment, when the agency action is without precedent (i.e., if the type of product has never been approved by FDA in the past, which would likely include organisms engineered with gene drives). If the organism engineered with gene drives is expected to spread near the range of a federally listed endangered species or critical habitat, then the product developer may also be required to provide data and information for an assessment under the Endangered Species Act.

Regulators from the United States and other nations are likely to have (and are working to develop) some common requirements for organisms engineered with gene drives. These might include information on the organism's molecular biology and resulting phenotype, quality control, construct and trait stability, and safety, along with tools and assays for detecting and monitoring the organism once released. A major unmet need for the regulation of organisms engineered with gene drives is an understanding of their phenotypic and genotypic stability over generations. The regulatory agency will need to have some confidence that the product will remain stable over time because approval is given for a specific product with specific characteristics. Gene drive researchers will have to work with regulatory agencies to help define the requirements for stability and product quality control in this context.

There has been some guidance specific to performing environmental risk assessments for testing and deploying organisms engineered with gene drives in the environment (NASEM 2016), and extensive guidance has been published for earlier generations of genetically engineered organisms (EFSA 2013; FDA 2017a), including mosquitoes (WHO 2014; CBD 2012). Much of this guidance emphasizes a phased approach, with Phase 1 focused on laboratory studies, Phase 2 on physically and/or ecologically confined field trials, Phase 3 on unconfined release, and Phase 4 on post-release surveillance. However, risk assessment for organisms engineered with gene drives may pose some challenges beyond those posed by earlier generations of genetically engineered organisms. For example, for some types of gene drives (e.g., those that are threshold-independent), even a small number of escapees from a confined field test could have a significant impact on native populations; best practices at this stage are not yet clear. Several groups, including the WHO, are working to develop guidance to address risk assessment challenges associated with mosquitoes engineered with gene drives. The first organism engineered with gene drives to be developed will likely help define the regulatory pathway and the appropriate milestones and precautions to incorporate into this phased approach.

Community Engagement

Organisms engineered with gene drives are typically designed to persist in the environment and impact wild populations, often at large scales. Although they hold great promise, they also hold some uncertainty about potential environmental impacts. These factors raise important issues about the responsibilities of product development teams beyond just environmental risk assessment and regulatory approvals. Because organisms engineered with gene drives have the potential for broad impact, decision-making for their deployment in the environment should also be broad and include community and stakeholder engagement from the early stages of development (NASEM 2016; Carter and Friedman 2016). These efforts will require significant dedication and commitment from funders and product development teams.

A well-organized engagement process should be designed by product development teams with the intention of involving local communities throughout the phased development process to help guide product design, site preparation, early testing, product development and deployment in the environment, post-deployment monitoring, reporting and communication, etc. The product development team itself should include social scientists alongside researchers and other transition partners (such as companies or nonprofit entities). There are multiple other contexts (such as public health and agriculture) within which community engagement processes have been used to successfully guide decision-making, and these may provide some lessons for releases of organisms engineered with gene drives. Examples include the Eliminate Dengue Program in Australia (Kolopack et al. 2015), efforts in support of field trials of genetically engineered mosquitoes in Mexico (Ramsey et al. 2014; Lavery et al. 2010), and the "mosquito-free Hawai'i" initiative, which has brought together community members and scientists to evaluate options for eliminating invasive mosquitoes from the islands; this process has included discussion of mosquitoes engineered with gene drives as a far-future possibility (Revive and Restore 2017).

Case Study: Microbial Engineering for Bioremediation

Introduction

Although genetically engineered microbes have been used for decades in laboratories and for commercial purposes, genetic and metabolic engineering of microbes has become both much easier and more complex in recent years (Chari and Church 2017; also see, e.g., Temme et al. 2012). Domesticated microbes are regularly genetically modified and utilized in high-throughput commercial services (e.g., Ginkgo Bioworks, Zymergen); however, these microbes function in precisely maintained and optimized bioreactors. Engineering microbes that can survive and function as designed in the environment remains a major challenge. Even so, an increasingly wide variety of engineered microbes with intended uses in the environment are under development, including microbes used for bioremediation, biomining, and chemical production (Bates et al. 2015). The workshop case study involved a microbe engineered for bioremediation.

Horizon Scanning

Much of the current work on engineered microbes for environmental applications has been focused on designs to overcome challenges and limitations related to release. Most successful microbial engineering endeavors have used well-characterized microbes that have been cultured for generations in the laboratory (such as *Escherichia coli*, *Saccharomyces* spp., and *Bacillus* spp.). Ensuring their survival in the natural environment will be an additional challenge. Furthermore, the engineered genetic constructs and tools developed in laboratory strains like *E. coli* are not universally functional in other microbes, and the extent to which they can be adapted and transferred to other chassis is not yet clear (Adams 2016; Kushwaha and Salis 2015). This issue is difficult to study because there has been limited research on how to quantify functional fitness in the field. There is also a lack of understanding of how genetic and phenotypic traits are correlated with an organism's fitness in the environment. Survival and growth are also related to the variation of microbiome diversity and complexity. Adjacent ecosystems may also contain variable nutrients and toxicants (especially relevant to microbes developed for bioremediation applications), which may impact survival and growth.

   To address these challenges, more studies are needed on natural microbial communities, including survival factors, interactions between microbes, microbial evolution, and transfer of genetic information among different microbial strains and species. Such data will allow better prediction and monitoring of the broader impacts of engineered microbes in the environment. These studies will also allow more effective and predictable outcomes in engineering microbes to express products that penetrate into natural systems (e.g., mobile genetic elements that can be passed to multiple types of microbes for enhanced effectiveness), which have thus far been difficult.

Another significant challenge in developing novel functional systems is the lack of predictive tools. In particular, bioinformatics capabilities are needed that will enable researchers to discover functional components from unexplored genomes in order to expand the range of tools that can be utilized in the future. The development of machine learning and artificial intelligence will likely lead to more rapid advances in the future, but these approaches will require reliable structured datasets and significant improvements in our underlying understanding of relationships between primary sequence, macromolecular structure, and function.

One major theme for engineered microbes is engineering systems for biocontrol and biocontainment. A variety of methods are being pursued by researchers. For example, codons can be reassigned so that only the intended microbial host is capable of reading the engineered DNA, or novel promoters can be inserted into engineered organisms so metabolic activity is controlled through addition of a chemical typically not found, or quickly degraded, in the environment. Additionally, engineered microorganisms can be designed to exclusively utilize nonnatural amino acids or nucleic acids that do not exist in nature (Mandell et al. 2015). Such organisms are orthogonal to natural systems and may therefore appear "invisible" to native organisms (Schmidt 2010).

Traditional biocontainment methods can also be incorporated into advanced engineered microbes, but these methods require additional development in order to be effective. There have been significant research efforts in the development of auxotrophic systems, where the microbe is engineered to be dependent on a specific chemical or nutrient. While highly effective in controlled laboratory settings, microbes in complex natural environments are often able to find the required nutrients or suitable alternatives directly in the environment. There is the additional challenge that genetic constructs conferring a growth and survival disadvantage place a selective pressure on the organism to evolve away from those constraints in order to increase environmental fitness. For auxotrophic systems, there is the potential for engineered microbes to overcome the nutrient dependence through natural genetic mutations or by acquiring the necessary genes by foreign genetic material uptake from the environment (Moe-Behrens et al. 2013). Similarly, kill switches face the same challenges. A typical kill switch system contains a continuously expressed toxin that is lethal to the host cell. By linking an external signal (chemical) to neutralization of the toxic protein or genetic repression, the engineered microbe will only survive in the presence of the specific signal or chemical. However, this also provides a strong selective pressure against the kill switch and can escape (Moe-Behrens et al. 2013). Incorporating multiple biocontainment mechanisms will likely have a greater chance of success in limiting survival and propagation of an engineered microorganism in the environment.

Environmental Impact

The potential environmental impacts of engineered microbes will depend on the particular engineered function and will have to be evaluated on a case-by-case basis. In some instances, advances in synthetic biology may reduce potential hazards, for example, the use of xenobiology or orthogonal genetic systems to prevent the transfer of genetic material to native organisms. However, the complexity of synthetic biology technologies may, in some cases, increase uncertainty. Risk assessors have little experience with proteins composed of nonnatural amino acids, and potential impacts on the environment will have to be determined. Also, it can be difficult to identify the secondary and tertiary metabolites in complex metabolic pathways and to understand how these pathways interact with the natural environment. In all cases, engineered microbes that closely resemble previously evaluated microbes will be easiest to assess for safety and environmental impacts.

A critical challenge in determining the environmental impact of engineered microbes is that current understanding of natural microbial communities is lacking. The undisturbed (baseline) state of microbial ecosystems is often unknown, and indicators of "healthy" or "pristine" microbial ecosystems are not defined. The most relevant timescale for detecting impacts from engineered microbes is also unclear, and it will be difficult to determine the cause of observed perturbations in a microbial ecosystem. Environmental applications of engineered microbes are designed to have a measurable effect, and it may be difficult to understand if observed changes in the microbial communities are beneficial or detrimental to the environment. This dilemma is particularly clear for the workshop case study of an engineered microbe developed for bioremediation. Sites where such microbes would be deployed are likely to be highly polluted, so environmental changes in this context are likely desirable. Furthermore, polluted sites will naturally give rise to unusual microbial ecology, thus complicating what might be considered a baseline state. Research on natural microbial ecosystems in a variety of contexts will help to define healthy ecosystems, thus providing critical context for understanding the desirability of perturbations in those systems, and indicators for gauging microbial community resilience.

Due to the diversity of microbial ecosystems where engineered microbes may be released, each release should be evaluated within its ecological context. For example, soil microbes have developed competitive strategies such as production of antibacterial metabolites. Native microbial products at a specific site would affect survival and activity of the introduced engineered microbes, as well as the natural population. Knowledge of these metabolites both will improve the design of the microbes and may also provide a better understanding of changes in the ecosystem. Critical information on these effects can be obtained from microcosm or small-scale field experiments, and data should be collected in phases from the lab to the field in order to evaluate predictive models. Development of models will be especially critical when engineered microbes are intended for use in multiple areas or in a broad area that may include multiple microbial-scale ecosystems.

There are tools available for monitoring and detection of intentionally released engineered microbes. DNA sequencing of environmental samples using next-generation sequencing can provide knowledge of the existing genes at a site. These methods, combined with increased annotation of genetic information and detection of DNA markers/barcodes, can provide useful information on the survival of the introduced microbes and can provide data on changes in the microbial ecosystem. Although these methods detect DNA, including DNA that is part of an engineered pathway, they cannot determine if that DNA remains in an engineered microbe or has been incorporated into a native organism (or persists outside of a cell). As discussed above, genetic containment methods are under development but require more research to become reliable. The possibility of gene transfer has been studied for many years, but key questions still remain. These include the probability of chromosomal integration of introduced DNA and factors that affect this probability, such as nutrient levels in the environment and competence factors for different strains of bacteria. Genomic differences and cellular factors affecting gene transfer among native bacteria are difficult to study because most microbes are largely unknown or not sequenced and cannot be cultured in the lab. Furthermore, the impact of gene transfer on the recipient microbes is not clear. It is believed that, in most cases, the engineered genes are likely to confer a selective disadvantage for the recipient and will be purged from the population over time; however, potential impacts will have to be evaluated on a case-by-case basis. As discussed above, changes in microbial ecology at highly polluted sites may be desirable.

Advances in monitoring and detection of accidental releases of engineered microbes will also be necessary. Currently, engineered microbes are predominantly located in governmental and academic research laboratories or commercial production facilities and are securely maintained in physical containment systems (e.g., bioreactors) with safeguards in place. An accidental release at such an institution is not likely, but could be significant. Also, engineered microbes, albeit with less complexity and at smaller scales, are increasingly produced in shared community laboratories and other small-scale facilities by the DIYBio (do-it-yourself biology) community that have variable oversight (Grushkin et al. 2013). DNA sequencing of environmental samples can be used to track accidental releases, but the lack of environmental baselining may make it difficult to detect an engineered DNA sequence. More data on natural microbial ecosystems would aid monitoring efforts.

Safety and Regulation

Engineered microbes will be regulated in the United States depending on their intended use. If they are developed as a therapeutic (e.g., an engineered gut microbe, Synlogic 2017; Garber 2015), then it will be regulated by FDA. Microbial pesticides will be regulated by EPA under its pesticide authorities (the Federal Insecticide, Fungicide, and Rodenticide Act). Other types of engineered microbes developed for commercial use, including microbes developed for bioremediation, will be regulated by EPA under the Toxic Substances Control Act (TSCA; OSTP 2017). If the microbe produces a new chemical, then EPA will separately regulate that chemical as well. In its current review process, EPA considers all potential uses of a microbe and, if necessary, issues a "significant new use rule" under which it can impose new safeguards or restrictions on the developer for uses not initially proposed.

One challenge the regulatory system may face in the near future is its ability to keep pace with the speed at which new bioengineered microbial strains and compounds are developed (Trump et al. 2017). The current regulatory framework has been adequate to date; the numbers of applications and products have increased at a manageable rate for EPA. As high-throughput synthetic biology approaches become more widely used and development times become faster, EPA and other regulatory agencies could be overwhelmed, and the review process could slow considerably (NASEM 2017a; Carter et al. 2014).

The lack of comparators for risk assessment presents another challenge for the regulation of engineered microbes. To date, EPA has been able to conduct risk assessments on engineered microbes by comparing them to naturally occurring microbes and previous technologies. However, future engineered microbes may incorporate increasingly novel traits, such as synthetic genetic elements, unusual chassis, and nonnatural nucleic acids or amino acids. For previous generations of engineered microbes, the "host" organism for the engineered DNA construct has been clear, but newer engineered microbes may combine critical components from many species. Data requirements for such products may be more rigorous than for previous technologies. Early engagement with regulators at EPA will help identify critical questions and knowledge gaps for specific engineered microbes.

The value of current biocontainment measures, such as auxotrophic systems, kill switches, and (even further into the future) orthogonal biology, remains unclear as they are still under development. However, if and when they are fully successful (i.e., are shown to adequately reduce the microbe's persistence in the environment and/or horizontal gene transfer), EPA may consider them as an appropriate containment measure. Under TSCA, there are some exemptions for well-characterized microbes that contain well-understood DNA constructs and are used in contained systems. Similar exemptions could be considered, far in the future, for microbes with adequate biocontainment constructs, should such technologies prove operational. Such exemptions would, however, require new formal rule making, which can be a time-consuming process.

Community Engagement

The decision to deploy an engineered microbe in the environment should be transparent and should incorporate actionable input from community members. The focus of this engagement should be on defined applications and uses of engineered microbes and issues of concern such that specific risks, benefits, and concerns can be articulated and discussed. This process will require interdisciplinary teams that include experts in community engagement, as well as early and frequent collaboration and communication with stakeholders outside of the development team. The range of stakeholders will be broad and will include local communities, funders, regulators, policy makers, and others.

Case Study: Cell-Free Technologies for Advanced Chemical Production

Introduction

For the purposes of the workshop, synthetic cell-free technologies were defined as the suite of synthetic biology technologies that allow for the exploitation of transcription and translation systems outside of the cell. This definition excludes more general enzymatic reactions that typically occur within a cell but can be leveraged outside of the cell, as well as technologies like DNA storage, DNA barcoding, and DNA forensics. Many cell-free technologies are under development, including paper-based cell-free systems for detecting chemical threat agents or pollutants (e.g., Ma 2013) and for on-demand chemical synthesis using cell-free protein production systems (Carlson et al. 2012). These tools can be compact in size, fitting on a small piece of paper, and can rapidly analyze the environment for specific target molecules or produce chemicals of interest. The workshop case study included several possible applications of cell-free systems to best generate discussion.

Cell-free systems may present a unique opportunity to serve as a proving ground to identify and answer foundational questions around hazard exposure, risk, and public perception of synthetic biology. Many of the safety and efficacy questions that stakeholders may have about complex synthetic biology technologies, such as engineered microorganisms and insects engineered with gene drives, can be addressed empirically using simpler cell-free systems. Cell-free tools may enable researchers to make certain determinations much more quickly than they could in a living cell, although direct comparisons require evaluation. Cell-free systems are likely to face fewer regulatory restrictions because they are not living. Additionally, engagement with the public and other stakeholders for deployment of cell-free systems may be simpler than for other synthetic biology organisms and components.

Horizon Scanning

State-of-the-art cell-free technologies include paper-based gene circuits and cell-free manufacturing reactors. In general, paper-based gene circuit tools incorporate cell-free extracts that power a gene-based indicator. The cellular components for a detect-and-report system are freeze-dried on a porous medium (paper) and, once hydrated, "boot up" the genetic circuitry. After a few hours, the circuitry has created enough detectable product (i.e., protein, RNA, or other macromolecule) to indicate the presence or absence of a specific target. Current gene circuits are relatively simplistic, employing "if this, then that" logic with simple colorimetric reporters (e.g., green fluorescent protein). For example, Pardee et al. (2016) describe a paper-based Zika virus diagnostic tool that detects 24 different Zika RNA sequences, turning the paper from yellow to purple. In the future, paper-based diagnostic gene circuits will likely be more elaborate with multistep logic circuits that are more sensitive to the target molecules and that have more rapid reporting times and more vivid indicators. In addition, other matrix materials are currently being studied, including cloth, silk, hydrogel beads, plastics, and wax-printed channels on glass. Further into the future, living cells might be included in the hydrating agent to make the diagnostic tool capable of more complex detection, thus providing improved readouts.

Paper-based systems are based on either whole-cell extracts or pure cellular components. Systems utilizing pure cellular components are very stable but are very expensive due to the process required to obtain the purified components. Whole-extract systems, on the other hand, are inexpensive to make but are less stable, in part, because proteases present in cellular extracts may degrade important machinery. The stability of whole extracts may be increased by using alternate matrices, listed above. For example, a 3D assay may incorporate channels that move material selectively and allow for components to be added in a multistage fashion.

Cell-free manufacturing reactors are currently in use and under development for a variety of products, including targeted biopesticides, drugs, and systems for complex chemical synthesis (e.g., GreenLight Biosciences 2017). This type of manufacturing is conducive to both macroscale production in a manufacturing facility (i.e., bioreactors) and microscale production in a field-deployed situation. In both cases, very little DNA is required to make these systems functional. Cell-free systems also have the potential to be used in other consumer products, such as customized face creams and bioluminescent lip balm.

One key advantage of cell-free systems is that they are lightweight and more conducive to transport compared to other technologies. Active pharmaceuticals or chemicals needed in a remote location may be easier and more cost-effective to create on-site via cell-free methods than to transport, avoiding the complex logistics required for equipment security, component stability (e.g., refrigeration), and reliable power supply. Cell-free systems also offer shorter development timelines and greater modularity, which is likely to enable a wide range of applications to be developed by a wide range of actors, including the DIYBio community. These advantages may also, however, provide opportunities for nefarious uses, such as small-scale production of toxins or narcotics or the transport of benign components across borders for later incorporation into or manufacturing of harmful products. It is also possible that cell-free systems may have security gaps, such as components that can be sabotaged by viruses or other exogenously applied DNA.

There are many limitations for paper-based diagnostics that must be resolved, including target diversity, reliability, sensitivity, stability, and speed. Development of recognition elements is still quite cumbersome, and all sensors must be designed for specific, known chemicals or biomolecules. Cell-free sensors are best suited for screening (i.e., environmental analyses or other high-throughput applications with many samples) and are not currently reliable or sensitive enough to use as confirmatory diagnostic tools for human health. Another challenge is that cell-free diagnostic readouts are largely qualitative, with precise quantification requiring sophisticated techniques such as mass spectrometry analysis of well-defined and purified extracts. Lastly, current cell-free gene circuits take 90–120 minutes for optimal readouts, and although there are techniques to reduce this time slightly, it is still too slow for many health or sensor applications. It remains to be seen if cell-free systems mimic the functionality of the organism from which they are derived or if differences exist in biological activity. Cell-free systems are a new and emerging technology, with highly active research and development efforts underway. These efforts will likely address many of the challenges outlined above and bring this technology into wider use in the near future.

Environmental Impact

There are likely to be many cell-free systems that are developed and deployed because of their ease of use and their potential for relatively low regulatory hurdles (particularly for environmental applications, as discussed below). The environmental impact of cell-free systems is likely to be less than that of other synthetic biology technologies (e.g., engineered viruses, bacteria, or organisms engineered with gene drives) because of the lack of self-replication, the minute amount of cell-free material used in each product, and the low likelihood of components interacting with living systems.

Despite the anticipated low environmental impacts of cell-free components, potential impacts still require investigation. Many of the environmental risks posed by cell-free systems are also posed by other synthetic biology technologies. For example, the DNA component of a cell-free system could transform native organisms via horizontal gene transfer (transformation or transduction) and could provide new functionality to the unintended host, potentially disrupting an ecological balance. The impact of such an event would depend on the function encoded by the cell-free system DNA (e.g., DNA encoding antimicrobial or cell lysing proteins may kill the transformed host). Engineered controls embedded in the DNA, such as irregular codons or artificial promoters not found in natural organisms, may limit genetic transfer, integration, and expression. Other strategies include tightly binding DNA to the matrix to prevent uptake by other organisms, engineering designs that result in the rapid degradation of system components if released from the matrix, or using DNA constructed using nonnatural nucleic acids that cannot be easily incorporated into or replicated by native flora.

Cell-free manufacturing applications may pose some environmental hazard if they are designed to produce a hazardous end product, although these pose similar concerns to traditional chemical manufacturing. If the cell-free components escape containment in a form that remains functional, a potential would exist for those toxic substances to continue to be manufactured and released directly into the environment. The waste from producing cell-free extracts – rather than the extracts themselves – may also be an environmental hazard (again, this hazard may be similar to hazards posed by traditional chemical manufacturing).

Multiple studies are needed to qualify and quantify these risks. More lab-scale and field-scale studies are needed to characterize (1) the likelihood of horizontal gene transfer; (2) the efficacy of built-in biocontainment mechanisms as safety controls (e.g., programmed cell lysis if native organisms acquire genetic components from cell-free systems); and (3) the quantification of viral replication in cell-free extracts. Field trials could be conducted in facilities with controlled experimental chambers (e.g., mesocosms).

Many of the questions about environmental impacts have been identified previously (such as those pertaining to stability and horizontal gene transfer of genetic material) but have not been answered empirically due to limited funding and a lack of prioritization by funding agencies. There are models for tracking fate and transport of genetic material in the environment, but evaluation of these methods and empirical data is limited (Furlan et al. 2016). Cell-free systems may serve as excellent tools for measuring environmental impacts related to synthetic biology organisms and components before deploying more complex technologies in the environment. For example, cell-free systems could be used to study gene flow and the uptake rate of free DNA in the environment. Other data derived from cell-free studies could be used to inform models developed to assess the impacts of synthetic biology organisms.

Safety and Regulation

Laboratory biosafety requirements for cell-free systems are similar to those for other biochemical and biotechnological facilities and protocols. Within academia, research in cell-free synthetic biology is typically overseen by Institutional Biosafety Committees (IBCs) under the NIH Guidelines. For commercial products developed with cell-free systems (e.g., paper-based diagnostics, manufactured specialty chemicals, etc.), current regulations in the United States are likely to provide adequate oversight. For diagnostics and other health-related applications, regulatory oversight will be provided by the FDA, with any necessary environmental assessment performed in compliance with NEPA (see gene drive discussion summary). Other cell-free systems and components with novel genetic arrangements, including those intended for use in the environment, are likely to be regulated by the EPA under the Toxic Substances Control Act as new chemicals. Internationally, restrictions on transporting "living modified organisms" across borders are not likely to apply to cell-free systems, allowing easier deployment of these systems compared to living synthetic biology organisms.

Community Engagement

Relative to other synthetic biology organisms and components, cell-free systems might not pose as great a challenge for community engagement as these systems can be seen as incremental advances over currently used technologies (such as home chemistry sets or pregnancy tests). Furthermore, because these systems do not contain living organisms, public concerns about environmental impacts and the uncertainties around those impacts may be reduced. Community engagement can help ensure that cell-free systems are pursued in ways that are welcomed and embraced by the public, but the level of engagement required for cell-free systems may not be as in-depth as that required for living synthetic biology organisms, such as organisms engineered with gene drives.

Case Study: Viral Systems for Water Treatment

Introduction

Viruses are "semi-living" entities composed of single- or double-stranded DNA or RNA surrounded by a protein capsid. They function by infecting a host, harnessing host cellular machinery for replication, and then releasing new viruses. Although there are many different types of viruses that infect a variety of host cells, the focus of the workshop was on bacterial viruses, also called bacteriophages (phages). Phages contain highly compact genomes ranging from 10^4 to 10^5 nucleotide base pairs and constitute the most diverse genetic entities on Earth. There can be as many as 10^7 viral particles in 1 mL of sea water, an order of magnitude larger than marine microbes (Parsons et al. 2012).

Viruses are an appealing system for engineering, as they are relatively easy to work with and well-studied and can transfer genetic material into a host genome with varying degrees of specificity. Viruses have been used for many industrial purposes, including medical, agricultural, and veterinary, as well as the production of novel materials. The workshop case study focused on the deliberate release of engineered phages through wastewater or as a result of wastewater treatment to inactivate high-consequence pathogens. In this context, "wastewater" includes sewage, storm water, precipitation runoff, firefighting runoff, and other ways that an aqueous solution of virus can be generated and potentially enter the environment.

Horizon Scanning

Phage therapy (the use of phages to kill harmful bacteria) predates the use of antibiotics. In Western countries, including the United States, there are some approved phage-based agricultural products in addition to ongoing human clinical trials using phages (Vandenheuvel et al. 2015; Parracho et al. 2012). All of these products use cocktails of natural phage isolates cultured in traditional large-scale fermenters. Advances in technology over the last decade, including next-generation sequencing, droplet microfluidics, single-cell omics, protein design, receptor docking, and biogeochemical modeling/bio-cycling, have revolutionized the study and understanding of phages. DNA synthesis, genetic editing, cell culture systems, and transformation protocols have all advanced to the point that phage engineering has become a common laboratory practice.

Despite these technological gains, there are several challenges remaining for those that work with and engineer phages. A better understanding of the interactions between phages with their microbial hosts is needed, including the interaction of phage genetic material with the bacteria. Another need is a greater characterization of viral and bacterial communities (i.e., viral and microbiomes) and studies of community dynamics. In order to gain this level of understanding, phage propagation is critical, but that is itself a challenge. Not only is there limited knowledge as to what comprises the microbial communities in wastewater, but the ability to isolate and culture non-model organisms under laboratory conditions is also lacking. Beyond simple aquatic, terrestrial, or aerosolized environments, complicated biofilms and microbial mats with three-dimensional and asymmetrical variation present unique challenges to studying viral dynamics. Research has focused on ways to improve or circumvent host cell culturing, including the development of cell-free systems to produce phages. Engineering bacterial hosts for expanded range, developing mixed cell culture systems that more closely mimic natural environments, and increasing the number of microbial host strains that can be grown in the laboratory are ongoing efforts.

Phages have a complex life cycle that is poorly understood. They infect bacteria through interactions between viral capsid proteins and bacterial cell surfaces. Capsid proteins often target conserved cell surface regions (e.g., receptors or lipid rafts) of target bacterial hosts. Many current research efforts are focused on modifying these receptors, either to expand the phage host range or to more precisely target a cell type. Once inside the cell, questions remain about both the efficiency of incorporation of viral genetic material into host genomes and the process of phage-mediated cell lysis. By understanding, engineering, and optimizing these factors, a variety of commercial phage-based applications may be developed in the future.

The ability to monitor and contain engineered phages after release into the environment is critical. One approach currently under investigation is the incorporation of kill switches into engineered phages. Like bacterial kill switches, these genetic components can trigger cell death or halt reproduction (or other metabolic activity) in the presence of an external stimulus, such as pH, temperature, or the addition of an enzyme or chemical. There are many ongoing research efforts to identify pathways to improve kill switch efficiency and to develop alternative biocontainment methods for phages. The same challenges face biocontainment of viral systems as do bacterial systems discussed in the engineered microbes for bioremediation case study, as above.

Environmental Impacts

Phages impact the environment in a variety of ways, each of which is habitat-specific and should be evaluated on a case-by-case basis. Prior to the release of engineered phages, characterization of the complex and dynamic ecological interactions in the natural environment is critical. Endpoints of concern should be established so that monitoring efforts can be directed toward meaningful data. Phased testing may provide a better understanding of how introduced phages may interact with native wild-type organisms and ecological processes and provide relevant data for modeling efforts.

The treatment of wastewater (the case study discussed here) is a likely use of engineered phages, although such an application would still require substantial development and controlled testing. For example, testing could be conducted on a laboratory or pilot scale and follow the typical protocol for wastewater treatment:

1. Large particles are settled out naturally.
2. Microbes degrade contaminants.
3. Advanced treatment including chemicals, filtration membranes, disinfection, or elimination of microbes. This approach allows phages to be investigated in a contained setting, and critical comparisons could be made, such as determining if modified organisms become more resistant to typical disinfection processes.

Many environmental concerns surrounding the use of phages in the environment mirror those of industrial use and release of chemicals or use of pesticides. Essential considerations for understanding environmental impacts include the persistence of phages over time, the likelihoods of phage infection in new or unexpected bacterial strains or species, and the potential for unintended toxicity (due to, e.g., endotoxin release from cell lysis). When using phages for wastewater treatment, assessments should be made of the potential for phage transport into and persistence in downstream water bodies (e.g., irrigation systems and wastewater by-products). Aerosolization and transfer of phage to aquatic organisms should be evaluated (Withey et al. 2005). Potential exposure of humans to engineered phages should also be considered (although phages do not infect human cells, they may find suitable hosts in the human microbiome). In all cases, it will be important to develop models and evaluate them by gathering meaningful data from laboratory experiments and phased releases.

While some environmental concerns surrounding the use of engineered phages are similar to those associated with industrial chemicals, engineered phages present unique challenges, particularly in regard to their ability to both replicate and mutate. Mutations are usually deleterious and lead to nonviable viruses yet occasionally can result in novel properties. Some of these mutations have been shown to expand the host range, introduce novel virulence factors, or decrease phage susceptibility to neutralization (e.g., via UV light, cold, or desiccation). Mutation rates vary both among viruses and among host bacterial strains. RNA viruses, in particular, benefit from high mutation rates that promote rapid adaptation. Mutations and the emergence of fitness-enhancing traits in phages can be extremely difficult to detect and model. Gene transfer mechanisms, including horizontal gene transfer (movement and incorporation of DNA segments among viruses and bacteria), can also confer new capabilities on phages and surrounding microbiomes. One particular concern, among others, is the introduction of DNA encoding pathogenic traits into a new host through horizontal gene transfer, followed by increased virulence in a previously non-virulent species. Methods for studying horizontal gene transfer have not been standardized, due in part to the lack of understanding of how it occurs. Basic research in this area is badly needed.

Accidental and unforeseen risks (e.g., natural disasters that damage containment infrastructure) should also be considered when evaluating potential environmental impacts. Product developers should be encouraged to develop worst-case scenario plans and include genetically engineered biocontainment strategies when possible. Such strategies could, for example, focus on impaired replication and/or reproduction. Disaster mitigation plans may also include materials specifically designed to remove viruses from the water system, such as selective absorption filtration systems that use membrane-bound bacteria or specialized nanomaterials.

Safety and Regulation

Engineered phage products have yet to be addressed by US regulatory agencies but will be regulated based on their intended use. There are examples of non-engineered phage products that have been approved for various applications in the United States. For example, a cocktail of non-engineered phage isolates is used to treat *Staphylococcus aureus* and *Pseudomonas aeruginosa* infections and is regulated by the FDA (Kingwell 2015). Future engineered phage therapeutics for humans will be similarly regulated by the FDA under its human drug provisions. Likewise, engineered phages used as pesticides will be regulated by EPA under its pesticide provisions (e.g., AgriPhage, a non-engineered pesticidal phage cocktail). Engineered phages for wastewater treatment (the case study presented here) would be regulated by EPA under the Toxic Substances Control Act (TSCA), which requires a premarket review for engineered microbes (including those intended for environmental release). Researchers should engage with regulators early on to ensure that the regulatory agencies can anticipate upcoming products and that planning and experimentation are aligned with current regulatory standards. Development of standardized forms and questions for environmental risk assessment that are more relevant to a wider range of synthetic biology technologies would also help researchers anticipate regulatory needs; e.g., see (EPA 2017).

A variety of nonregulatory mechanisms also contribute to biosafety and appropriate use of engineered phages. Trainings for researchers on safe use of engineered phages (above and beyond that required by the NIH Guidelines) could also contribute to overall biosafety. Regulations are legal requirements, but companies also comply with environmental risk mitigation measures aimed at reducing legal liabilities, as well as maintaining trade secrets. For example, insurance for wastewater facilities using engineered phages will likely be simpler for those including well-characterized containment mechanisms.

In order to enhance safety and regulation for deployment of engineered phages, there is a need to advance the science that guides regulation. For example, better detection capabilities are needed for phages, including rapid-result, field-deployable platforms. Ecologically relevant animal models are also needed in order to determine potential impacts of engineered phages on native animal species and their microbiomes. The potential role of animal vectors in uncontrolled phage dispersal is also an important research gap. In addition, studies should evaluate the limitations of laboratory or caged trials in fully capturing phenomena that occur at larger scales and aim to develop more powerful methodologies for test chamber or mesocosm studies. Models that integrate data and information from multiple sources will also be critical.

Community Engagement

As with any new and unfamiliar technology, engaging with the public early and providing information in an understandable way will help ensure that the decision to deploy engineered phages (or not) is made responsibly. Key challenges in community engagement include identifying the stakeholders that should be involved and finding the right communicators. Frequently, scientists are not trained in communications or stakeholder engagements with nontechnical audiences; therefore, product developers should team with appropriate experts. In all cases, the benefits and risks of deployment of synthetic biology technologies, including the uncertainties of both, should be clearly articulated.

Conclusions

One theme that provided a key foundation for the workshop and resonated in each of the discussions about interactions between synthetic biology organisms and components and the environment was the need for research to support environmental risk assessment and regulatory decision-making. Such research will be critical to the development, testing, and deployment of synthetic biology technologies. Given the level of investment in synthetic biology and its applications by the DoD and the broader US government, commensurate investments in environmental and regulatory science may be warranted. Indeed, one of the recommendations from the NASEM report on Future Products of Biotechnology (NASEM 2017a) is that those US government agencies that fund advanced biotechnology should also allocate funds for advancing regulatory science. By addressing some of the research needs described in Section Two of this chapter, ERDC and DoD can set an example for responsible development of synthetic biology technologies.

Many technical hurdles identified at the workshop, such as the need for improved control and modeling of engineered organisms and surveillance and monitoring tools, have been identified previously (NASEM 2016, 2017a; Drinkwater et al. 2014; Carter and Friedman 2016). DARPA's Safe Genes program is working to overcome some of these hurdles for gene drives (DARPA 2017), and the IARPA FELIX program aims to develop tools for the identification of genetically engineered organisms in the environment (IARPA 2017). The need for more information on natural environmental processes, such as horizontal gene transfer in microbial communities, has also been highlighted in earlier reports (Drinkwater et al. 2014). Many of these research needs, particularly for environmental baselining and potential environmental interactions of synthetic biology organisms and components, have not been prioritized and remain underfunded.

There is a strategic opportunity to meet these technical needs not only for development and deployment of synthetic biology organisms and components but also for a wide range of US government goals. For example, more effective and efficient methods for monitoring and surveying the environment for engineered organisms or components, or for characterizing ecological communities pre- and post-deployment, will be important for fielding synthetic biology technologies. These efforts could also support basic research (e.g., NSF's National Ecological Observatory Network), public health efforts for tracking the spread of vector-borne diseases (e.g., the Global Emerging Infections Surveillance System) and/or antibiotic resistance (PCAST 2014), detection of inadvertent or nefarious biosecurity threats (PCAST 2016), efforts to address invasive species (NISC 2017), and characterization of ecosystems and ecosystem services (PCAST 2011; USGEO 2016). Monitoring efforts already underway within the US government could be adapted and coordinated to better serve these multiple purposes. Improved control of the persistence and spread of genetically engineered organisms and components will yield benefits not only for potential environmental applications but also for medical advances and countermeasures (DARPA 2017; PCAST 2016).

Many of the common themes identified at the workshop and in other venues address the critical need for regulatory and community engagement before, during, and after development and deployment of synthetic biology organisms and components. Case-by-case evaluation of products and environmental risk assessment have been hallmarks of the US Coordinated Framework for the Regulation of Biotechnology since it was established in 1986 (OSTP 2017). Even so, these newer products raise well-described challenges for the US regulatory system, with options and recommendations available (NASEM 2017a; Carter et al. 2014). The need for broader community and stakeholder engagement has also been identified repeatedly, especially for more complex engineered organisms, such as those that contain gene drive constructs (NASEM 2016; Carter and Friedman 2016). Establishing research priorities for synthetic biology organisms and components, including research providing the basis for environmental risk assessments, should be done in a coordinated way that best supports regulatory and community engagement needs (Trump et al. 2018b). International perspectives should also be included where international deployment of synthetic biology technologies either is intended or may be possible. The US State Department is active in multilateral fora where synthetic biology is a topic of interest, including the Convention on Biological Diversity and the Biological Weapons Convention.

One challenge that the US government faces with the development of synthetic biology organisms and components is the dual-use nature of the technology. Biosecurity risks were not the focus of the workshop, but have been topics for other meetings, workshops, and publications (PCAST 2016; Regalado 2016; NSABB

2010; NASEM 2017b). However, there is a repeated emphasis, even within the biosecurity and biodefense communities, on promoting innovation and ensuring that the benefits of synthetic biology can be harnessed (including development of countermeasures and other applications that may improve security). As these technologies are developed and research is prioritized, it will be important to include biosecurity and biodefense perspectives.

The wide-ranging applications and the promise of synthetic biology organisms and components will require a strategic and cross-disciplinary US government approach to ensure that they are developed in a way that meets multiple objectives. Such an approach should include prioritizing research that underpins environmental risk assessment of the technologies, integrates with other relevant research and surveillance efforts, supports regulatory decision-making, limits the potential for unintended and nefarious use, and is guided by community and stakeholder needs.

References

ACG (Arthropod Containment Guidelines). (2004). Arthropod containment guidelines version 3.1. *Vector Borne and Zoonotic Diseases, 3*(2), 57–98.

Adams, B. L. (2016). The next generation of synthetic biology chassis: Moving synthetic biology from the laboratory to the field. *ACS Synthetic Biology, 5*(12), 1328–1330.

Bates, M. E., Grieger, K. D., Trump, B. D., Keisler, J. M., Plourde, K. J., & Linkov, I. (2015). Emerging technologies for environmental remediation: Integrating data and judgment. *Environmental Science & Technology, 50*(1), 349–358.

Blake, W. J., Cunningham, D. S., & GreenLight Biosciences (2017) Engineered proteins with a protease cleavage site. US Patent 20170159058, 20170096692, 9688977.

Burt, A., & Trivers, R. (2009). *Genes in conflict: The biology of selfish genetic elements.* Cambridge: Harvard University Press.

Carlson, E. D., Gan, R., Hodgman, C. E., & Jewett, M. C. (2012). Cell-free protein synthesis: Applications come of age. *Biotechnology Advances, 30*(5), 1185–1194.

Carter, S. R., & Friedman, R. M. (2016). *Policy and regulatory issues for gene drives in insects, gene drives to control insect-borne human disease and agricultural pests: A workshop to examine regulatory and policy issues.* UC San Diego: J. Craig Venter Institute.

Carter, S. R., Rodemeyer, M., Garfinkel, M. S., & Friedman, R. M. (2014). *Synthetic biology and the US biotechnology regulatory system: Challenges and options.* J. Craig Venter Institute. http://www.jcvi.org/cms/fileadmin/site/research/projects/synthetic-biology-and-the-us-regulatory-system/full-report.pdf. Accessed 31 Oct 2017.

CBD (Convention on Biological Diversity). (2012). *Guidance on risk assessment of living modified organisms. Open-ended online forum and the Ad Hoc Technical Expert Group.* Hyderabad: United Nations Environment Programme.

Chari, R., & Church, G. (2017). Beyond editing to writing large genomes. *Nature Reviews Genetics, 18*, 749. https://doi.org/10.1038/nrg.2017.59.

DARPA (Defense Advanced Research Projects Agency). (2017). Building the safe genes toolkit. https://www.darpa.mil/news-events/2017-07-19. Accessed 31 Oct 2017.

Drinkwater, D., Kuiken, T., Lightfoot, S., McNamara, J., & Oye, K. (2014). Creating a research agenda for the ecological implications of synthetic biology. MIT Program on Emerging Technologies and the Woodrow Wilson International Center for Scholars Synthetic Biology Project.

EFSA (European Food Safety Authority). (2013). Guidance on the environmental risk assessment of genetically modified animals. *EFSA Journal, 11*(5), 3200, 190pp.

EPA. (2017). Points to consider: Engineered microbial applications. https://www.epa.gov/regu-lation-biotechnology-under-tsca-and-fifra/tsca-biotechnology-regulatory-and-policy-related. Accessed 31 Oct 2017.

Esvelt, K. M., Smidler, A. L., Catteruccia, F., & Church, G. M. (2014). Concerning RNA-guided gene drives for the alteration of wild populations. *eLife, 3*, e03401.

FDA (Food and Drug Administration). (2017a). Guidance for industry regulation of intention-ally altered genomic DNA in animals. Draft Guidance, FDA. https://www.fda.gov/downloads/AnimalVeterinary/GuidanceComplianceEnforcement/GuidanceforIndustry/ucm113903.pdf. Accessed 31 Oct 2017.

FDA (Food and Drug Administration). (2017b). FDA issues final guidance clarifying FDA and EPA jurisdiction over mosquito-related products. https://www.fda.gov/AnimalVeterinary/NewsEvents/CVMUpdates/ucm578420.htm. Accessed 31 Oct 2017.

Finkel, A. M., Trump, B. D., Bowman, D., & Maynard, A. (2018). A "solution-focused" compara-tive risk assessment of conventional and synthetic biology approaches to control mosquitoes carrying the dengue fever virus. *Environment Systems and Decisions, 38*(2), 177–197.

Furlan, E. M., Gleeson, D., Hardy, C. M., & Duncan, R. P. (2016). A framework for estimating the sensitivity of eDNA surveys. *Molecular Ecology Resources, 16*(3), 641–654.

Gantz, V. M., & Bier, E. (2015). The mutagenic chain reaction: A method for converting heterozy-gous to homozygous mutations. *Science, 348*(6233), 442–444.

Garber, K. (2015). Drugging the gut microbiome. *Nature Biotechnology, 33*, 228–231.

Gronvall, G. K. (2015). US competitiveness in synthetic biology. *Health Security, 13*(6), 378–389.

Grushkin, D., Kuiken, T., & Millet, P. (2013). Seven myths and realities about Do-It-Yourself biol-ogy. Woodrow Wilson International Center for Scholars Synthetic Biology Project. http://www.synbioproject.org/site/assets/files/1292/7_myths_final-1.pdf. Accessed 31 Aug 2017.

Hammond, A., et al. (2016). A CRISPR-Cas9 gene drive system targeting female reproduction in the malaria mosquito vector *Anopheles gambiae*. *Nature Biotechnology, 34*, 78–83.

IARPA (Intelligence Advanced Research Projects Agency). (2017). Finding Engineering-Linked Indicators (FELIX). https://www.iarpa.gov/index.php/researchprograms/felix. Accessed 31 Oct 2017.

Keasling, J. (2005). The promise of synthetic biology. *The Bridge, 35*(4), 18–21.

Kingwell, K. (2015). Bacteriophage therapies re-enter clinical trials. Nat Rev Drug Discov. 14, 515. doi: 10.1038/nrd4695.

Kolopack, P. A., Parsons, J. A., & Lavery, J. V. (2015). What makes community engagement effec-tive?: Lessons from the eliminate dengue program in Queensland Australia. *PLoS Neglected Tropical Diseases, 9*(4), e0003713.

Kushwaha, M., & Salis, H. M. (2015). A portable expression resource for engineering cross-species genetic circuits and pathways. *Nature Communications, 6*, 7832.

Lavery, J. V., et al. (2010). Towards a framework for community engagement in global health research. *Trends in Parasitology, 26*, 279–283.

Lewis, A. (2017). Driving down pests. *The Scientist.* http://www.the-scientist.com/?articles.view/articleNo/50180/title/Driving-Down-Pests. Accessed 31 Oct 2017.

Linkov, I., Trump, B. D., Anklam, E., Berube, D., Boisseasu, P., Cummings, C., et al. (2018). Comparative, collaborative, and integrative risk governance for emerging technologies. *Environment Systems and Decisions, 38*(2), 170–176.

Ma, M. (2013). UW awarded $10 million to design paper-based diagnostic medical device. UW News. http://www.washington.edu/news/2013/06/25/uw-awarded-10-million-to-design-paper-based-diagnostic-medical-device. Accessed 31 Oct 2017.

Mandell, D. J., et al. (2015). Biocontainment of genetically modified organisms by synthetic pro-tein design. *Nature, 518*(7537), 55–60.

Moe-Behrens, G. H., Davis, R., & Haynes, K. A. (2013). Preparing synthetic biology for the world. *Frontiers in Microbiology, 4*, 5.

NASEM (National Academies of Sciences, Engineering, and Medicine). (2016). *Gene drives on the horizon: Advancing science, navigating uncertainty, and aligning research with public val-ues*. Washington, DC: The National Academies Press, 230pp.

NASEM (National Academies of Sciences, Engineering, and Medicine). (2017a). *Preparing for future products of biotechnology*. Washington, DC: The National Academies Press, 220pp.

NASEM (National Academies of Sciences, Engineering, and Medicine). (2017b). *A proposed framework for identifying potential biodefense vulnerabilities posed by synthetic biology: Interim report*. Washington, DC: The National Academies Press, 51pp.

NISC (National Invasive Species Council). (2017). National Invasive Species Council. https://www.doi.gov/invasivespecies. Accessed 31 Oct 2017.

Noble, C., Olejarz, J., Esvelt, K. M., Church, G. M., & Nowak, M. A. (2017). Evolutionary dynamics of CRISPR gene drives. *Science Advances, 3*, e1601964.

NSABB (National Science Advisory Board for Biosecurity). (2010). Addressing biosecurity concerns related to synthetic biology. https://osp.od.nih.gov/wp-content/uploads/NSABB_SynBio_DRAFT_Report-FINAL-2_6-7-10.pdf. Accessed 31 Oct 2017.

OSTP (Office of Science and Technology Policy). (2017). Modernizing the regulatory system for biotechnology products: Final version of the 2017 update to the coordinated framework for the regulation of biotechnology. US Environmental Protection Agency, US Food and Drug Administration, and US Department of Agriculture. https://www.epa.gov/sites/production/files/2017-01/documents/2017_coordinated_framework_update.pdf. Accessed 31 Oct 2017.

OTI (Office of Technical Intelligence). (2015). Technical assessment: Synthetic biology. US Department of Defense. http://defenseinnovationmarketplace.mil/resources/OTI-SyntheticBiologyTechnicalAssessment.pdf. Accessed 31 Oct 2017.

Pardee, K., et al. (2016). Rapid, low-cost detection of Zika virus using programmable biomolecular components. *Cell, 165*, 1255–1266.

Parsons, R. J., Breitbart, M., Lomas, M. W., Carlson, C. A. (2012). Ocean time-series reveals recurring seasonal patterns of virioplankton dynamics in the northwestern Sargasso Sea. ISME J. 6, 273–284

Parracho, H. M., Burrowes, B. H., Enright, M. C., McConville, M. L., & Harper, D. R. (2012). The role of regulated clinical trials in the development of bacteriophage therapeutics. *Journal of Molecular and Genetic Medicine: An International Journal of Biomedical Research, 6*, 279–286.

PCAST. (2011). Report on the intersection of the nation's ecosystems and the economy. Office of the President. https://obamawhitehouse.archives.gov/sites/default/files/microsites/ostp/pcast_sustaining_environmental_capital_report.pdf. Accessed 31 Oct 2017.

PCAST. (2014). Report on antibiotic resistance. Office of the President. https://obamawhitehouse.archives.gov/sites/default/files/microsites/ostp/PCAST/pcast_amr_jan2015.pdf Accessed 31 Oct 2017. https://obamawhitehouse.archives.gov/administration/eop/ostp/pcast/docsreports

PCAST (President's Council of Advisors on Science and Technology). (2016). Letter report on action needed to protect against biological attack. Office of the President. https://obamawhitehouse.archives.gov/sites/default/files/microsites/ostp/PCAST/pcast_biodefense_letter_report_final.pdf. Accessed 31 Oct 2017.

Ramsey, J. M., et al. (2014). A regulatory structure for working with genetically modified mosquitoes: Lessons from Mexico. *PLoS Neglected Tropical Diseases, 8*(3), e2623.

Regalado, A. (2016). Top US intelligence official calls gene editing a WMD threat. *MIT Technology Review*. https://www.technologyreview.com/s/600774/top-us-intelligence-official-calls-gene-editing-a-wmd-threat/. Accessed 31 Oct 2017.

Revive and Restore. (2017). The plan to restore a mosquito-free Hawaii. http://reviverestore.org/the-plan-to-restore-a-mosquito-free-hawaii/. Accessed 31 Oct 2017.

Roberts, A., et al. (2017). Results from the workshop "problem formulation for the use of gene drive in mosquitoes". *The American Journal of Tropical Medicine and Hygiene, 96*(3), 530–533.

Schmidt, M. (2010). Xenobiology: A new form of life as the ultimate biosafety tool. *BioEssays, 32*(4), 322–331.

Stevenson, C. (2017). Three new mega-raises put SynBio on track for record-breaking year of funding. *SynBioBeta*. https://synbiobeta.com/three-new-mega-raises-put-synbio-track-record-breaking-year-funding/. Accessed 31 Oct 2017.

Swetlitz. (2017). In a remote West African village, a revolutionary genetic experiment is on its way – If residents agree to it. *STAT*. https://www.statnews.com/2017/03/14/malaria-mosquitoes-burkina-faso/ Accessed 31 Oct 2017.

Synlogic. (2017). Synlogic™ doses first subject in phase 1 trial of novel class of Synthetic Biotic™ medicines. https://www.synlogictx.com/news/press-releases/synlogic-doses-first-subject-phase-1-trial-novel-class-synthetic-biotic-medicines/. Accessed 31 Oct 2017.

Target Malaria. (2017). Target Malaria. http://targetmalaria.org/. Accessed 31 Oct 2017.

Temme, K., Zhao, D., & Voigt, C. A. (2012). Refactoring the nitrogen fixation gene cluster from *Klebsiella oxytoca*. *PNAS, 109*(18), 7085–7090.

Trump, B. D. (2017). Synthetic biology regulation and governance: Lessons from TAPIC for the United States, European Union, and Singapore. *Health Policy, 121*(11), 1139–1146.

Trump, B., Cummings, C., Kuzma, J., & Linkov, I. (2017). A decision analytic model to guide early-stage government regulatory action: Applications for synthetic biology. *Regulation & Governance, 12*(1), 88–100.

Trump, B. D., Cegan, J. C., Wells, E., Keisler, J., & Linkov, I. (2018a). A critical juncture for synthetic biology: Lessons from nanotechnology could inform public discourse and further development of synthetic biology. *EMBO Reports, 19*(7), e46153.

Trump, B. D., Foran, C., Rycroft, T., Wood, M. D., Bandolin, N., Cains, M., et al. (2018b). Development of community of practice to support quantitative risk assessment for synthetic biology products: Contaminant bioremediation and invasive carp control as cases. *Environment Systems and Decisions, 38*(4), 517–527.

Trump, B. D., Cegan, J., Wells, E., Poinsatte-Jones, K., Rycroft, T., Warner, C., Martin, D., Perkins, E., Wood, M., & Linkov, I. (2019). Co-evolution of physical and social sciences in synthetic biology. *Critical Reviews in Biotechnology, 39*, 351. https://doi.org/10.1080/07388551.2019.1566203.

USGEO (US Group on Earth Observations). (2016) US Group on earth observations. Office of Science and Technology Policy. https://obamawhitehouse.archives.gov/administration/eop/ostp/nstc/committees/cenrs/usgeo. Accessed 31 Oct 2017.

Vandenheuvel, D., Lavigne, R., & Brüssow, H. (2015). Bacteriophage therapy: Advances in formulation strategies and human clinical trials. *Annual Review of Virology, 2*, 599–618.

Wilson Center. (2015). US trends in synthetic biology research funding. Woodrow Wilson International Center for Scholars Synthetic Biology Project. http://www.synbioproject.org/site/assets/files/1386/final_web_print_sept2015.pdf. Accessed 31 Oct 2017.

Withey, S., Cartmell, E., Avery, L. M., & Stephenson, T. (2005). Bacteriophages—Potential for application in wastewater treatment processes. *Science of the Total Environment, 339*, 1–18. https://doi.org/10.1016/j.scitotenv.2004.09.021. ISSN 0048-9697.

World Health Organization (WHO). (2014). Guidance framework for testing of genetically modified mosquitoes. WHO Special Programme for Research and Training in Tropical Diseases. http://apps.who.int/iris/bitstream/10665/127889/1/9789241507486_eng.pdf. Accessed 31 Oct 2017.

Transfish: The Multiple Origins of Transgenic Salmon

Hallam Stevens

Introduction

This chapter traces the development of AquAdvantage (AA) salmon from the initial scientific insights in the 1970s that led to its invention to the eventual approval of the fish for human consumption by the FDA in 2015. Since AA salmon was the first genetically engineered animal approved for human consumption, its story is significant on its own terms. In particular, it is important to understand how AA salmon came to be approved in the face of substantial opposition to – and public concern with – genetically modified foods (GMFs).

By examining how risks were managed and framed around AA salmon, this chapter aims to provide insight into how genetically modified and synthetic organisms are perceived and understood by GMF opponents and regulators. I argue that in this case the perception of risk was successfully managed through a series of displacements; by (literally and figuratively) moving the fish into different categories and zones, its creators were able to convince regulators that the fish posed no substantial risk for human consumption. This success reveals a great deal about how GM organisms are perceived. But it also raises significant questions about who is risked and what is risked in the making of GMFs and synthetic organisms. The construction of AA salmon as a *transnational* animal suggests how risks and rewards are increasingly unevenly distributed within global scientific and capital-commercial flows.

Importantly, my aim here is not to argue for or against AA salmon in terms of health, safety, environmental impact, or any other factor; rather, my objective is to

H. Stevens (✉)
History Programme, School of Humanities, Nanyang Technological University, Singapore, Singapore
e-mail: hstevens@ntu.edu.sg

© Springer Nature Switzerland AG 2020
B. D. Trump et al. (eds.), *Synthetic Biology 2020: Frontiers in Risk Analysis and Governance*, Risk, Systems and Decisions, https://doi.org/10.1007/978-3-030-27264-7_3

explore *how* risks and benefits were framed and counter-framed by those on both sides.

After some background on AA salmon, this chapter examines the historical development of the fish from Newfoundland, Canada, to Boston to Washington, DC, to Panama. The main sections of the paper explore how AA successfully constructed the fish as both a "foreign" and "safe" object.

Background

AA salmon is a genetically modified Atlantic salmon with several types of genetic modifications. First, it has a growth hormone gene taken from a Chinook salmon so that it can grow (and reach maturity) faster. Second, it has a promoter borrowed from an ocean pout that causes the growth hormone to be expressed all year round, not just in spring and summer. In the pout, this promoter – known as an anti-freeze promoter (AFP) – turns on an "anti-freeze protein" that prevents blood from freezing and allows the fish to survive in arctic waters. This promoter is repurposed in AA salmon for turning on growth genes in winter months. And third, AA salmon also has some genetic modifications for increased disease resistance.

The most important and novel element of this cassette of modifications was the AFP. The development of this element dates back to work undertaken by Singaporean biologist Hew Choy Leong working in Canada in the 1970s. Hew and Garth Fletcher isolated anti-freeze proteins in flounder and began to study them in detail. In 1991, Hew, Fletcher, and Elliot Entis founded A/F Proteins Inc. to explore possibilities for commercialization of anti-freeze proteins. At this time, they had in mind applications in frozen foods, cosmetics, and cryogenic surgery.

In 1995, A/F Proteins acquired the intellectual property for transgenic salmon, giving them the right to produce and sell the modified fish. In the same year, they submitted an application for a "New Animal Drug" to the US Food and Drug Administration (FDA). This marked the beginning of a 20-year investigation by the FDA to establish the safety of AA salmon. During this time, anti-freeze proteins were also adapted and developed for use in other foods (e.g., ice cream) and cosmetics. News of "anti-freeze" in food made consumers more aware and warier of such genetically engineered products.

Perhaps most importantly for the FDA's review, however, AA salmon were not farmed within the United States. Juvenile transgenic salmon were transferred directly from facilities in Canada to inland fish breeding facilities in Panama. There, distant from the United States, AA salmon could be portrayed as non-threatening to US wild salmon stocks and rendered "safe" in a variety of ways.

Biotechnology and genetic engineering have long been discussed in terms of hybridity and boundary crossing. Donna Haraway's account of genetically modified laboratory "oncomouse," for example, described it as a gender-bending, category crossing, radically unstable, monstrous vampire of a creation (Haraway 1997). Newer biotechnological creations – like AA salmon – are perhaps even more radically "crossed" and hybridized. This ability to challenge and bend biological categories is perhaps now so strong that we need new theoretical tools to account for it and understand it. Sophia Roosth outlines ways in which we might draw on queer theory for understanding the new and category crossing formations of synthetic biology: "Now queer kinship theory may be applied to objects of synthetic biology – and perhaps all biotechnologically-made transgenic organisms – to make sense of how synthetic biologists arrange and define biological relatedness" (Roosth 2017).

Queer theory can also help us understand how objects such as transgenic fish stabilize themselves through the performances of "trans"-identity. How are such "trans" objects constructed as "same" or "other" by different parties (both advocates and opponents)? How are such objects policed? How do they become "deviant" or "normal"? The tools of queer theory – in challenging binaries and exposing the ways in which identities are performatively constructed – may be of help in understanding the kinds of cultural roles transgenic organisms are occupying. In so doing, they can also help us understand how regulators and GMF opponents construct and perceive risks of particular kinds of "foreign" objects.

Cori Hayden's work has pointed to the connections between kinship and intellectual property, especially in the regimes of bioprospecting. As she notes, both are, at root, about genealogy of people, ideas, and things: "How do new property relations rearrange genealogical grids, not only turning natural kinds into brands but also creating novel, 'propertized' kinds in the first place?" (Hayden 2007, p. 342). AA salmon emerges from exactly such a rearrangement of "genealogical grids" into new forms of property and profit. But it also trades on the rearrangement of "geographical grids," re-organizing space and place to accommodate the emergence of new kinds of crosses, objects, and brands.

AA salmon is "trans" in multiple dimensions: transgenic, transgender, and transnational. This "trans"-ness is critical to how this salmon is perceived as alternatively risky, safe, familiar, foreign, animal, monster, edible, and inedible.

Methods

This account is based largely on historical methods. It uses a range of primary sources to reconstruct the origins and development of AA salmon from the 1970s to 2015. These sources include oral history interviews with scientists, published scientific and technical literature, as well as documentary sources drawn from the FDA and the civil society organizations.

The Making of a Transgenic Fish

Singapore to Newfoundland

Hew Choy Leong was born in British Malaya and educated in Singapore at Nanyang University, graduating in 1963. After obtaining a Ph.D. in Vancouver and completing postdoctoral fellowships at Yale and Toronto, in 1974 Hew was appointed assistant professor at Memorial University in St. John's, Newfoundland. Hew's field was insulin precursor pathways. Since insulin from cod was exceptionally easy to purify, his work took him to the Ocean Science Center in Logy Bay, a remote and frigid part of the Canadian Maritimes (Hew 2017).

As one of the earliest areas of settlement for Europeans in North America, the Maritimes have a long history of economic activity centered on shipping and fishing. The fishing of cod, whale, haddock, herring, lobster, scallops, and other marine animals began in the sixteenth century and remained a prosperous activity into the twentieth century.

As such, Logy Bay was a natural location at which to study these fish. Hew's cod were imported into his tanks from further south in Nova Scotia. The tanks themselves were fed with seawater from the ocean nearby and heated to ensure that the fish, which were not accustomed to the colder water of Newfoundland, did not freeze. One night in February 1975, however, disaster struck! Just as Hew's experiments on insulin were getting underway, the power in the Ocean Science Center failed overnight; the heaters went off and Hew's 200 cod died in their tanks (Hew 2017).

As Hew recalls, "While I was devastated, I was at the same time the most popular guy in the Center that day because everyone wanted cod for dinner!" But Hew also noticed something peculiar. He shared his lab and the tanks with a fish physiologist, Garth Fletcher, who was studying flounder. Although the flounder were in the same freezing water as the cod, they did not die. Why, Hew wondered? Fletcher immediately suggested that the flounder must produce some kind of anti-freeze into their bodies to keep them alive in the cold waters of the North Atlantic (Hew 2017).

Seizing the opportunity, Hew put aside his work on insulin and began to collaborate with Fletcher to attempt to figure out exactly how the flounder stayed alive – what was this anti-freeze and how did it work? Answering this question consumed the next decade of Hew's life. With Fletcher, Hew built the world's leading laboratory for the study of the molecular biology of anti-freeze proteins.

During the 1970s, Hew and his co-workers gave little thought to the application of their work, let alone the possibility of commercial fish farming. It was only in 1982, in what Hew attributes to another "serendipitous" turn, that his attention was turned to salmon. During a coffee break at a conference, a salmon aquaculturalist mentioned that some salmon farms had had problems with freezing fish. Was it possible to put the anti-freeze in the salmon? (Hew 2017).

Motivated by this question, over the next several years, Hew and Fletcher aimed to do exactly this. They were also inspired by the work of Ralph Brinster and Richard Palmiter. In 1982, Brinster and Palmiter managed to inject a rat growth hormone gene into the embryo of a mouse; the gene was not only taken up by the mouse but also passed on to subsequent generations to create "Mighty Mice" (Palmiter et al. 1982). If such a gene transfer could be done for a mouse, it could be done for a fish, Hew thought (Hew 2017).

In fact, transferring genes from one species of fish to another proved difficult – fish eggs are very delicate and it proved tricky to get foreign DNA to be taken up into the nucleus of the egg. Hew and Fletcher eventually came up with a way of achieving the transfer using a hollow needle inserted through the micropyle of the fish egg (Fletcher et al. 1988). They filed a patent on this method and on transgenic anti-freeze fish in November 1988 (Health and Social Care Research Development Corporation 1988). Although their salmon stably took up and expressed the flounder anti-freeze gene, the absence of other crucial proteins to interact with the flounder anti-freeze meant that it remained in a relatively ineffective form. The salmon still froze in sub-zero water.

Undeterred, Hew and Fletcher saw that there might be different uses for their anti-freeze gene and for their transgenic technique. In particular, their idea was that the anti-freeze *promoter* could be spliced to a fish growth hormone and used to speed up fish growth. The anti-freeze promoter would ensure that the growth hormone gene would be turned on even in cold weather and fish could grow throughout the year. By 1988, having put the idea of cold-resistant fish aside, the Canadian team was able to rapidly grow big salmon using their technique.

Toronto to Boston

By the late 1980s, the long-prosperous trade in cod the Canadian Maritimes had begun to enter a serious crisis. Although a substantial 1.4 million tons of cod were fished in Atlantic Canada in 1988 (with a total value of over $1 billion), stocks were already under pressure. By 1992, Canada's Fisheries and Oceans Minister called for a moratorium on cod fishing in the area. Bans on other fish such as haddock and other groundfish soon followed. This was a major economic blow to the region – 40,000 individuals in Quebec and the Atlantic provinces lost their jobs (Gough 2013).

Hew and Fletcher's work in Newfoundland was developing at a time when other economic opportunities in the area were in rapid decline. The National Sciences and Engineering Research Council of Canada, which had funded their work, was keen to see their new discovery commercialized. The crossing of fish with biotechnology was just the sort of high-tech product that might help to revive the region by bringing new jobs and new industries.

The opportunity for such commercialization presented itself in the form of an entrepreneurial fish salesman, Elliot Entis. In 1990, Entis was based in New England and specialized in exporting fish from Atlantic fisheries to the Caribbean. Entis was searching for better methods of preserving frozen fish in transit and on the shelf. This interest led him to the work of Hew and Fletcher. A few conversations with them led Entis to the realization that fish anti-freeze proteins had multiple potential uses – not just for fish or other foods but also potential medical applications such as freezing tumors or human organs during surgeries.

It was with this goal in mind – the preservation of cells, tissues, and organs – that Entis, Fletcher, and Hew established A/F Protein Inc. in 1991 to pursue possibilities for commercialization of anti-freeze technologies (Powell 2006). The new company was established in Waltham, Massachusetts, a suburb of Boston (Entis's hometown) with a subsidiary in St. John's, Canada. Hew, in particular, was keen to make this move – he saw Boston as a more dynamic place for biotech and start-up companies. Targeting the US market involved one substantial factor: the innovation's success or failure ultimately depended on the approval of the US Food and Drug Administration (FDA).

Since A/F was focused on these other commercialization opportunities, the development of fish as a technology in its own right remained on hold. It was only 2 years into A/F Protein's operations that one of its scientists mentioned to Entis their discovery of fast-growing salmon. Entis quickly seized on this as another possible lucrative business opportunity. By 1995, A/F Protein had acquired the intellectual property rights for transgenic salmon from the University of Toronto and Memorial University and began the process of gaining regulatory approval. In the same year, A/F Protein established an Investigational New Animal Drug file for transgenic salmon with the Center for Veterinary Medicine (CVM) at the FDA.

Fish to Ice Cream

During the 20-year FDA process, anti-freeze proteins emerged in several different contexts. These came to play an important role in how the debates around AA salmon played out. The identity of the first GM animal for human consumption was shaped by broader debates and fears around GM foods.

As it turns out, flounder are not the only organisms that contain anti-freeze proteins. Also occurring in other fish such as smelt, herring, and sea raven, anti-freeze proteins exist in plants, insects, and various microorganisms that live in sea ice. In 1999, Unilever, the Dutch food giant, filed for a patent on utilizing such anti-freezes to an "unaerated ice confection" (Unilever 1999). By 2004, the company had applied to the FDA for approval of an ice cream product containing genetically modified anti-freeze components.

Unilever had in fact taken an anti-freeze protein from an ocean pout and modified it for insertion into yeast. These GM yeast, when cultivated, would produce a version of the fish anti-freeze. Appropriately utilized in ice cream, the protein can cause the ice cream to melt more slowly and provide better "mouthfeel" (the anti-freeze causes smaller ice crystals to be formed, yielding a smoother product). This proved particularly useful for the production of "low-fat" varieties of ice cream. By 2006, anti-freeze proteins produced from yeast were approved for sale and were appearing on the shelves in products such as "Breyers Light Double Churned" ice cream bars.

The introduction of fish proteins into ice cream resulted in immediate opposition. The idea of fish (or anti-freeze) in one's ice cream seemed intuitively distasteful to many consumers. This was used by opponents of GMOs to portray the new ice cream as scary and unsafe (Reidhead 2006). One Unilever representative reported that he had repeatedly had to explain that the new product would not taste "fishy" (Moskin 2006).

Such arguments are influential because they rely on cultural ideas about the kinship of objects – ice cream and fish don't go together! Assumptions about categories and kinship are deeply related to (and structured by) all sorts of other divisions, hierarchies, and power relations – between men and women, homo and hetero, for instance. Crosses are threatening not just on their own terms but because they are deeply imbricated with one another. The incorporation of fish products into ice cream established public awareness of GM fish as an "unnatural" product and set the stage for the battle over regulatory approval of genetically modified salmon.

Boston to Washington

During the long approval process, A/F Protein and Entis continued to advocate forcefully for their product. In 2000, the company changed its name to AquaBounty Farms and spun off A/F Protein as a separate company. After submitting its first regulatory studies to the FDA in 2003, the company again changed its name (in 2004) to AquaBounty Technologies. In 2006, AquaBounty was successfully listed on the London Stock Exchange Alternative Investment Market, raising $30 million in an initial public offering of stock. By 2009, the FDA had completed its site visits to AquaBounty's facilities, and the company had submitted all its studies for approval (AquaBounty Technologies 2014).

The length of this process was partly due to the fact that this was uncharted territory – the FDA had never approved a genetically modified *animal* for human consumption. Transgenic crops were regulated under the Coordinated Framework for the Regulation of Biotechnology – an agreement between the FDA, the EPA, and the US Department of Agriculture – but the process for animals was unclear. In fact, the FDA gave A/F Protein a choice between two regulatory pathways: a New Animal Drug Application or a Food Additive Application (Juma 2016, pp. 266–268).

The FDA considered a large number of factors in attempting to decide whether the fish was "safe": efficacy (did it actually grow faster?), stable inheritance over multiple generations, stable base sequence and copy number of modified genes, equivalence to standard Atlantic salmon (did it have similar tissue composition, hormone levels, and levels of vitamins, minerals, and amino acids?), salmon health (did it have similar rates of disease?), and environmental safety (FDA 2015).

The key to AquaBounty's (and the FDA's) argument that the fish was safe for human consumption was the construction of a "standard of identity" of the fish that was at once technically novel but also "normal" in the sense of behavior and appearance. This was central to the methodology of the AA's testing, showing that the GM salmon could "pass" as regular Atlantic salmon in as many respects as possible: in DNA testing, in appearance, in general health, and so on. The following quotation is typical:

> There are no significant adverse outcomes associated with the introduction of the *opAFP-GHc2* construct… Most of the adverse outcomes that are observed (e.g. morphological changes) were present in comparators or have been described in the peer-reviewed literature… their consequences are likely to be small and within the range of abnormalities affecting rapid growth phenotypes of Atlantic Salmon. (FDA 2015, p. 18)

The FDA used language such as this ("no meaningful differences," "no biologically relevant differences") to argue that the GM salmon, while showing some differences, was within "normal" ranges expected for fish and that such differences are unlikely to pose any risks.

The data from these tests was presented to the Veterinary Medicine Advisory Committee at a public meeting in Washington, DC, in 2010. Subsequently, in December 2012, the FDA released a draft "Environmental Assessment" and "Finding of No Significant Impact" (FONSI) based on their preliminary analysis. There was an outpouring of opposition. During the early 2013, the FDA received 38,000 public comments in response to these documents. Many of these came from civil society organizations including the Center for Food Safety, Food & Water Watch, Friends of the Earth, Organic Consumers Association, and Food Democracy Now (Juma 2016).

These objections and comments can be divided into four broad classes. First, opponents contended that the process of assessment was not transparent enough and that the FDA had not released sufficient data to the public. Second, critics argued that the FDA's review, and especially its Environmental Assessment, was not sufficiently wide in scope. In particular, commenters argued that it should have covered overseas facilities, that it should have taken into account the effects on endangered species, that it should have sought approval from the Environmental Protection Agency and the Fisheries and Wildlife Service, that it should have considered "economic, social, and cultural impacts," and that it should have considered the effects on minority populations and on "Indian Tribal rights." Third, comments suggested that the FDA was not in a position to monitor fish-growing facilities and would therefore not be able to enforce any conditions under which approval would be granted (FDA 2017). Fourth, many worried that the "containment" of the fish would be insufficient and that AA salmon would escape and potentially outcompete wild populations. Since this became the most controversial issue, it will be discussed separately in the next section.

Significantly, these first three concerns centered on what remained obscured or left out by the FDA's investigation. It was not that the public or advocacy groups could point to known and specific dangers or risks associated with AA salmon, but rather the concern was based on what remained unknown (with respect to existing data, with respect to broader impacts, and with respect to future enforcement). In other words, AA salmon was dangerous because it remained foreign and unfamiliar.

The fact that the fish straddled existing categories (including institutional and regulatory categories) made it especially dangerous to GMF opponents.

Washington to Panama

As the debates around AA salmon continued, *place* became more and more important: the geography of the fish became critical to its identity (and, ultimately, its approval). As noted above, this was framed as an issue of "containment" – whether salmon could or would escape into wild, survive there, and potentially destabilize wild salmon populations or other ecosystems. Even if GM salmon could not survive themselves in the wild, critics suggested that "Trojan" genes from AA salmon could spread to wild salmon or that intensively farmed fish could be breeding waters for diseases that could then spread to wild populations.[1]

In order to anticipate and mitigate these risks, AquaBounty developed three related "containment" strategies for their salmon. First, AquaBounty developed a strategy of "biological containment" in which all fish shipped outside Canada would be female and triploid (have three copies of their chromosomes instead of two). This rendered the populations doubly impotent – females could not breed with females, and triploid fish are usually sterile. If AA salmon ever escaped into the wild, AquaBounty argued that they would be unable to propagate any offspring. Second, a system of "physical containment" was developed through which breeding and growth would take place in tanks (not nets or pens) in well-secured facilities far from any ocean or waterway. AA salmon would find it difficult to "escape" into the environment. Third, salmon were isolated from their wild counterparts through "geographical containment": broodstock were raised on Prince Edward Island in Canada and the fish would be grown to market size in an inland facility in Panama. Even if fish did manage to escape from their tanks and swim to the ocean, Panama is surrounded by tropical water – salmon, AquaBounty argued, are cold water fish and would not be able to grow or reproduce in this climate.

Through these techniques, AquaBounty and the FDA could argue that the fish was sufficiently "contained" not to pose any realistic threat of spreading in ways that would threaten wild populations. Perhaps more importantly, however, the FDA could credibly claim that Canada and Panama were outside the jurisdiction of their Environmental Assessment:

> because these facilities are outside the United States, and because NEPA [National Environmental Policy Act] does not require analysis of impacts in foreign sovereign countries, the EA considered environmental impacts in Canada and Panama only to the

---

[1] The "Trojan gene hypothesis" emerged from a widely cited 1999 paper by ecologists William Muir and Richard Howard (Muir and Howard 1999). It argues that genes could spread from genetically modified fish to wild relatives and eventually decimate both populations.

extent necessary to determine whether there would be significant effects on the environment in the United States due to exposure pathways originating from the facilities in Canada and Panama. (FDA 2017)

The FDA also undertook an "Analysis of Potential Impacts on the Environments of the Global Commons and Foreign Nations" but concluded there would be "no significant impacts" due to the low probability of escape and survival of AA salmon.

The successful mobilization of biological, physical, and geographical containment suggests, first, how the transnationalism of AA salmon was critical to their approval and regulation. Only fish grown in Canada and Panama in the specified AquaBounty facilities, slaughtered outside the United States, and then imported were granted approval by the FDA.[2] This mobility not only contributed to the perceived "safety" of the fish (far away from US salmon stocks, in tropical waters) but limited the scope of FDA's investigation.

This geographical isolation also conforms to a familiar pattern of bio-neo-colonialism in which colonies and former colonies are utilized as "testing grounds" for technologies that may not be politically or socially acceptable within the borders of the United States. For example, in the late 1950s, clinical trials for the contraceptive pill were conducted in Puerto Rico by Massachusetts-based scientists and clinicians (Tone 2002). The FDA's abjuring of any detailed investigation into Panama implies that such "foreign" zones can tolerate greater risks for the potential benefit of US consumers. This uneven distribution of risk and reward became critical to AA salmon's identity.

Second, the "containment" of AA salmon mixes geography with gender: it is instituted not merely territorially but also through the sexing of the fish. Control over the fish is exercised at once by policing space but also by policing sex and reproductive capacity. The fish become safe by "othering" them in multiple ways, both sexually and territorially. This pairing is not coincidental; there is an enduring association between queerness and geographical foreignness (see, e.g., Chávez 2015). Marking the fish as doubly "foreign" reassures us that it cannot compete with or mix with US wild salmon and interfere with domestic species or markets.

This issue of mixing and competition became salient when, in 2011, Representatives Don Young (R – Alaska) and Lynn Woolsey (D – California) proposed axing the FDA's budget if it approved AA salmon. The ostensible aim was protecting the Alaskan wild salmon-fishing industry (with 78,000 jobs and worth $5.8 billion) (Juma 2016, pp. 271–272). AA salmon, Young assumed, would undercut the prices of Alaskan salmon. But Alaskan salmon remained a premium product, serving a top-end section of the market; AA and other farmed salmon aimed at a lower price segment. The othering of AA salmon served to reinforce this sense of inferiority.

---

[2] More recently, AquaBounty has sought approval for a facility in the United States in Albany, Indiana (Slabaugh 2018).

Discussion and Conclusion

In one final twist in this story, AA salmon finally became entangled with skin care. Even the entrepreneur, in 2010, Entis founded another company (again in Waltham, Massachusetts): Liftlab, a proprietor of cosmeceuticals. In a video advertisement on their website, set to images of glaciers and icy mountainscapes, Liftlab describes its unique products:

> In the unforgiving arctic, bio-tech scientists first uncovered an evolutionary secret: proteins that allow sea life to survive in sub-zero temperatures. Naturally protecting; intensely moisturizing; damage reversing. Cell Protection Proteins also deliver softer, smoother, more sumptuous skin. (http://www.liftlabskincare.com/about/story/)

Through Liftlab, anti-freeze proteins have found another life, beyond ice cream and fish, in skin care and anti-aging remedies. Significantly, Liftlab's marketing mobilizes not only biotech science but also the specific nature and geography of the arctic in order to sell its products. Our skin, Liftlab suggests, can be protected by the "proteins that protect the health and survival of plant and marine life that thrive in extreme cold, dryness, and unrelenting UVA/UVB of Arctic regions" (http://www. liftlabskincare.com/about/story/). Not only fish, but the environs of "arctic" Canada, become bound to the circuits of global capitalism. As Entis tells viewers during another promotional video, it was his father's Boston wholesale seafood business that initially led him to searching the North Atlantic for novel products.

In their essay "Capitalism and the Commodification of Salmon," Stefano Longo, Rebecca Clausen, and Brett Clark argue that AA salmon is just the latest episode in increasing industrial and commercial control and capitalist exploitation of oceans and fisheries. The commodification of salmon as AA salmon will inevitably lead to further capitalist exploitation; they contend: "This long process has resulted in the privatization of commons, concentration of ownership, loss of subsistence livelihoods, exploitation of natural resources, and disintegration of local knowledge" (Longo et al. 2014). But "commodification" in and of itself is not sufficient to describe how AA salmon came to be approved (if not widely accepted) as fit for human consumption. I have described this as a process of "othering" and displacement that constructed the fish as foreign, alien, different, and mixed in various degrees. The perspective of queer theory is particularly useful here in showing how constructions of similarity/difference, like/not alike, and domestic/ foreign pervade our understanding of biotechnologies and their risks. Thinking with Hayden, this suggests that "commodification" is accompanied by attempts to remake the genealogies of salmon in ways that would cast it as either threatening or safe. The making of the "property" of AA salmon went alongside the making of the "family" of AA salmon. The "success" of AA salmon can be partly attributed to the construction of a particular version of its relations: sufficiently close to wild salmon to be safe for eating but also sufficiently distant to pose no direct threats to US territories.

This territorial dimension further suggests that the "commodification" that Longo, Clausen, and Clark refer to is likely to have a very uneven distribution of risks and rewards. The history of AA salmon demonstrates how its ultimate approval by the FDA in 2015 depended on the usage of a range of territories and spaces outside the United States. Both "arctic" Canada and tropical Panama were mobilized in order to realize the value of the GM salmon as a novel invention and a commodity product. Kaushik Sunder Rajan has argued in the case of pharmaceutical trials in India that this phenomenon represents another way in which the global South is "risked," by the global North (Rajan 2010). Indian subjects, who are unlikely to be able to afford the drugs tested in the trials, stand to gain few "rewards." AA salmon suggests that other biotechnologies may be following similar trajectories – rich countries and zones stand to gain (economically, environmentally, medically, and so on), while poorer nations and zones are put at risk.

Ultimately, any assessment of risks of new biotechnologies needs to take a more transnational view. As the case of AA salmon demonstrates, this "transnationalism" and its significance may not be immediately obvious without careful attention to the origins and development of new synthetic organisms – it is through the history of this fish that we can see the importance of geography in its development and approval.

References

AquaBounty Technologies. (2014). *Chronology of AquAdvantage Salmon and AquaBounty Technologies*. Retrieved from https://www.aquabounty.com/wp-content/uploads/2014/01/Chronology-of-AquAdvantage-Salmon-F1.pdf

Chávez, K. R. (2015). The precariousness of homonationalism: The queer agency of terrorism in post-9/11 rhetoric. *QED: A Journal in GLBTQ Worldmaking, 2*(3), 32–58.

Fletcher, G. L., Shears, M. A., King, M. J., Davies, P. L., & Hew, C. L. (1988). Evidence for antifreeze protein gene transfer in Atlantic salmon. *Canadian Journal of Fisheries and Aquatic Sciences, 45*(2), 352–357. https://doi.org/10.1139/f88-042.

Food and Drug Administration. (2015). Freedom of information summary/original new drug application. Retrieved from https://www.fda.gov/AnimalVeterinary/DevelopmentApprovalProcess/GeneticEngineering/GeneticallyEngineeredAnimals/UCM466215.pdf

Food and Drug Administration. (2017). AquAdvantage Salmon – response to public comments on the environmental assessment. Retrieved from https://www.fda.gov/AnimalVeterinary/DevelopmentApprovalProcess/GeneticEngineering/GeneticallyEngineeredAnimals/ucm466220.htm

Gough, J. (2013, Aug 12). History of commercial fisheries. In *The Canadian Encyclopedia*. Retrieved from https://www.thecanadianencyclopedia.ca/en/article/history-of-commercial-fisheries/

Haraway, D. J. (1997). *Modest_Witness@Second.Millennium.FemaleMan_Meets_Oncomouse: Feminism and technoscience*. New York: Routledge.

Hayden, C. (2007). Kinship theory, property, and the politics of inclusion: From lesbian families to bioprospecting in a few short steps. *Signs: Journal of Women in Culture and Society, 32*(2), 337–345. https://doi.org/10.1086/508220.

Health and Social Care Research Development Corporation. (1988). Microinjection procedure for gene transfer in fish. Patent. Retrieved from https://patents.google.com/patent/CA1341553C/en

Hew, C. L. (2017). *Interview with Hallam Stevens*. Singapore: National University of Singapore.

Juma, C. (2016). *Innovation and its enemies: Why people resist new technologies*. Oxford: Oxford University Press.

Longo, S. B., Clausen, R., & Clark, B. (2014). Capitalism and the commodification of salmon: From wild fish to a genetically modified species. *Monthly Review, 66*(7). Retrieved from https://monthlyreview.org/2014/12/01/capitalism-and-the-commodification-of-salmon/), 35.

Moskin, J. (2006, July 26). Creamy, healthier ice cream? What's the catch? *New York Times*. Retrieved from https://www.nytimes.com/2006/07/26/dining/26cream.html

Muir, W. M., & Howard, R. D. (1999). Possible ecological risks of transgenic organism release when transgenes affect mating success: Sexual selection and the Trojan gene hypothesis. *Proceedings of the National Academy of Sciences, 96*(24), 13853–13856. https://doi.org/10.1073/pnas.96.24.13853.

Palmiter, R. D., Brinster, R. L., Hammer, R. E., Trumbauer, M. E., Rosenfeld, M. G., Birnberg, N. C., & Evans, R. M. (1982). Dramatic growth of mice that develop from eggs microinjected with metallothionein-growth hormone fusion genes. *Nature, 300*(5893), 611–615.

Powell, K. (2006). Profile: Elliot Entis. *Nature Biotechnology, 24*, 735. https://doi.org/10.1038/nbt0706-735.

Rajan, K. S. (2010). The experimental machinery of global clinical trials: Case studies from India. In A. Ong & N. N. Chen (Eds.), *Asian Biotech: Ethics and communities of fate* (pp. 55–80). Durham: Duke University Press.

Reidhead, P. (2006). Unilver (Breyer's & Good Humor) using genetically modified fish "antifreeze" protein in ice creams. *The Milkweed*, 6–7. Accessed at https://www.saynotogmos.org/ud2006/antifreeze.pdf

Roosth, S. (2017). *Synthetic: How life got made*. Chicago, IL: University of Chicago Press.

Slabaugh, S. (2018). Indiana fish farm will grow genetically engineering AquAdvantage salmon in 2018. Genetic Literacy Project. Star Press. Retrieved from https://geneticliteracyproject.org/2018/04/04/indiana-fish-farm-will-grow-genetically-engineered-aquadvantage-salmon-in-2018/

Tone, A. (2002). *Devices and desires: A history of contraceptives in America*. New York: Hill & Wang.

Unilever. (1999). Ice confection with antifreeze protein. Retrieved from https://patents.google.com/patent/EP1158865B8/

The State of Synthetic Biology Scholarship: A Case Study of Comparative Metrics and Citation Analysis

Jeffrey C. Cegan

Introduction

Synthetic biology is a growing field of study within scholarly literature. It is the deliberate engineering of existing biological systems or the creation of novel biological systems. Advancements in synthetic biology have been driven by key technologies that increase efficiency and reduce resource costs for DNA synthesis and sequencing (Raimbault et al. 2016). Such research generally seeks to empower greater design capacity and control over biological systems and phenotypic expression – the results of which have been posited to offer benefits in fields ranging from medicine to energy to environmental remediation (Khalil and Collins 2010; Church and Regis 2014).

While motivations of synthetic biology were described in the early twentieth century, the modern field has evolved through breakthroughs such as synthetic circuit engineering (Elowitz and Leibler 2000) and the creation of a bacterial cell with a chemically synthesized genome (Gibson et al. 2010). These and other related pieces of scholarship represent the iterative and evolutionary nature of synthetic biology, where scholars build from seminal works in the field to improve cellular design and control (Cameron et al. 2014). Such research was described in existing scholarship such as Raimbault et al. (2016), Cameron et al. (2014), and Oldham et al. (2012), yet these efforts focused upon reviewing the state of biological and computational sciences relevant to synthetic biology and generally did not describe narratives and developments within social sciences literature (Trump 2017). A more thorough and holistic quantification of synthetic biology literature will help visualize and describe the growth and development of the larger synthetic biology

J. C. Cegan (✉)
Environmental Laboratory, Engineer Research and Development Center,
US Army Corps of Engineers, Washington, DC, USA
e-mail: Jeffrey.C.Cegan@usace.army.mil

© Springer Nature Switzerland AG 2020
B. D. Trump et al. (eds.), *Synthetic Biology 2020: Frontiers in Risk Analysis and Governance*, Risk, Systems and Decisions, https://doi.org/10.1007/978-3-030-27264-7_4

65

community over time and will respond to foundational or seminal discussions noted within key publications (Trump et al. 2019).

With this inspiration, this chapter constructs and analyzes a citation network of synthetic biology scholarship in both technical and social sciences. The initial search of articles is described in Trump et al. (2019). Described herein as "communities of practice," synthetic biology literature is categorized into various paths of inquiry including (i) "state-of-science" literature detailing biological experimentation or computational models, (ii) "products" literature detailing the application of synthetic biology research to a potential product application, (iii) "risk" detailing literature that characterizes synthetic biology risks or provides metrics for a potential risk assessment, (iv) "governance" including literature that describes regulatory and governance needs, and (v) "ELSI" literature pertaining to ethical, legal, and moral implications that synthetic biology may incur. Using these five communities of practice, this chapter explored how a compilation of 712 publications related to synthetic biology acknowledge and discuss developments within and between their respective communities. We used network science to identify how articles within each community of practice cite one another. The connections among these communities of practice are quantitatively assessed through the construction of a citation network. Overall, such a citation network provides a holistic quantitative measure of the importance of different communities of practice within synthetic biology and identifies key performers with the greatest degree of influence upon the larger synthetic biology community. This chapter utilized multiple quantitative metrics to identify those publications with the greatest amount of network centrality or those that displayed a high degree of motivational influence upon synthetic biology's communities of practice from 2000 to 2017. Through such quantification, it may be possible to identify which publications and communities of practice have spurred the evolutionary development of synthetic biology's enabling technologies in particular and the larger discussion of its use and impacts in general.

The synthetic biology citation data were used to create a network, which is a graph that contains a number of nodes, a.k.a. vertices, connected by links, a.k.a. edges (Jacob et al. 2017). Each publication is represented by a vertex, and each citation is shown as an arrow into or out of the vertex, depending on whether the publication is being cited or citing another publication, i.e., the citation network is directed. This network is also disconnected because all nodes cannot be reached by all other nodes.

Developed by Euler in 1736, network science studies pairwise relations within a network. McPherson et al. (2001) introduced the concept of homophily, which suggests that similarity breeds connection in a network. Synthetic biology is an evolving field in which much of the discussion and research is steered by a few communities of practice. Their research is then picked up and recreated or furthered by universities, NGOs, and other labs on the periphery. Since the field is primarily driven by those few communities, the network analysis explored whether or not similar types of publications are more likely to cite within their own groupings.

Applying graph theory to citation networks was pioneered by Garfield (1955) and De Solla Price (1965) and has since been applied to several fields to analyze

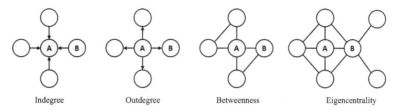

Indegree   Outdegree   Betweenness   Eigencentrality

Fig. 1 Network centrality measures

networks and attempt to quantify the importance of a publication. Citation networks represent publications and their citations using nodes that are connected through links, or edges. Centrality measures are one method to quantify node analysis and seek the node(s) most central, or critical, to a network. The simplest measure is degree centrality, which is found by counting the edges connected to each node. Edges are further divided directionally, into and out of each vertex, called indegree and outdegree, respectively (Weisstein 2017). Betweenness centrality measures the extent that a vertex is positioned on the shortest path between other pairs of vertices in the same network (Leydesdorff 2007). Eigenvector centrality, or eigencentrality, uses the eigenvector of the largest eigenvalue and standardizes to the length of the eigenvalue (Bonacich 1972), giving greater weight to vertices connected to other highly connected vertices. In each of the networks shown in Fig. 1, node A has a greater centrality than node B according to the particular measure referenced. Such quantitative measures offer objective viewpoints regarding which publications might be driving discussion within larger synthetic biology literature and ultimately detail how the field has grown over time to incorporate various elements of biological and social sciences discussion.

Methodology

To populate the synthetic biology citation network, a topic search was conducted through the ISI Web of Knowledge. Specifically, 16 search terms were used to generate a curated list of 880 publications. Relevant criteria for inclusion into our citation network required that each publication (i) primarily focus upon the implications or research of synthetic biology in general or an enabling technology in particular, (ii) were indexed in a journal, book, or formal government document about synthetic biology, and (iii) were framed between the years 2000 and 2017 (see Cameron et al. 2014 for a brief description of biological developments within synthetic biology research). Each publication is classified into one of five "communities of practice," including: (1) state of science, (2) products, (3) risk, (4) governance, or (5) ethical, legal, and social implications (ELSI).

State of science included publications that described the evolutionary or iterative advancement of a synthetic biology enabling technology within a laboratory setting or via computational modeling. Product publications focused upon applying synthetic biology scientific development into a commercial product. Risk publications sought to characterize or assess risks associated with synthetic biology research and development and governance publications identified and described strategies for the regulation and governance of processes or products of synthetic biology (Linkov et al. 2018). Finally, ELSI included publications examining social impact considerations (Merad and Trump 2019), legal challenges (Marchant et al. 2013), ethical concerns, and general economic impact discussion resulting from synthetic biology development and commercialization. Table 1 shows the number of publications in each community of practice.

Communities of practice were assigned manually to each publication consistent with Trump et al. (2019). Each publication is assigned to only one community of practice, meaning that the community of practice groupings are mutually exclusive. For the case where a publication is related to multiple community of practice areas, the author chose the most dominant community of practice.

Of the 880 returns, 168 were not linked to any other publication in the network for either having no references to other review publications or their files were not readable for text mining. This left a remaining 712 publications for analysis. A brief discussion of the removed isolated publications can be found in Appendix A. Publication dates range from 2000 to 2017, with every year represented in the ISI Web of Knowledge search. A histogram of publication years for the 712 publications in the network is shown in Fig. 2.

Another important methodological issue involves the use of text mining, an approach for extracting information from a body of text. The "tm" package in R was

Table 1 Number of publications in each community of practice

	State of science	Products	Risk	Governance	ELSI
Publications	445	122	33	42	70

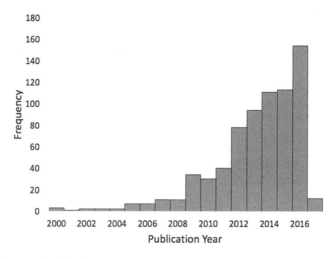

Fig. 2 Histogram of publication years

utilized to scan and identify patterns and frequencies of symbols within the PDF format of each publication. Using publication meta-data about the lead author's last name, publication year, and title information, we determined whether or not each publication was being referenced within another publication. This process produced an adjacency matrix of citations among all reviewed publications, and this quantitative data was converted into a citation network.

Results and Discussion

This section is divided into three parts. First, we show a graph of the network and cross pollination by community of practice. Second, we show top publications for each centrality and discuss pairwise correlations between measures. Finally, we examine the temporal changes in the network.

Network by Communities of Practice

In order to visualize the network, the graph shown in Fig. 3 was created to represent each of the 712 publications as a node, colored according to its respective community of practices. Links connecting one node to another are colored according to the community of practice of the terminal node, i.e., the publication being cited.

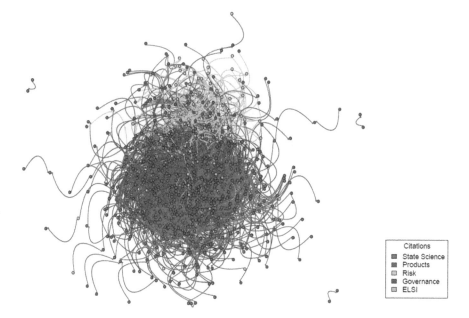

Fig. 3 Citation network of select synthetic biology literature (2000–2017)

Table 2 Cross pollination (citation counts)

	State of science	Products	Risk	Governance	ELSI	Total
State of science	1322	291	10	14	39	1676
Products	343	175	6	10	11	545
Risk	64	15	16	15	19	129
Governance	51	20	18	32	19	140
ELSI	113	39	10	19	72	253

Table 3 Cross pollination (citation percentages)

	State of science (%)	Products (%)	Risk (%)	Governance (%)	ELSI (%)	Total (%)
State of science	79	17	1	1	2	100
Products	63	32	1	2	2	100
Risk	50	12	12	12	15	100
Governance	36	14	13	23	14	100
ELSI	45	15	4	8	28	100

We analyze the interrelation of citations among communities of practice in Tables 2 and 3. The "Total" column sums to 2743, which represents the total number of links in the network.

The rows in Table 2 enumerate the citations that each community of practice made to each community column: of the 1676 network references identified in "state-of-science" publications, 1322 referenced other "state-of-science" publications, 291 referenced "product" publications, 10 referenced "risk" publications, and so on.

Table 3 shows that "state of science" is the largest share of cited publications, regardless of the publication category. This means that there is relatively little homophily among these categorical groupings because individual communities are not citing primarily within their own group. "State of science" seems to be at the forefront of this emerging field, and other synthetic biology communities are seemingly looking to "state-of-science" publications for guidance in their own field.

Network Centrality Measures

Measures of network centrality indicate the more important nodes and provide a greater understanding to key network structures. In this section, centrality concepts of degree, betweenness, and eigencentrality are used to understand the structure of the synthetic biology literature network.

Table 4 Sorted by decreasing indegree

Publication	Indegree	Outdegree	Betweenness	Eigen
Elowitz and Leibler (2000)	142	0	0	1
Gardner et al. (2000)	123	0	0	0.92766808
Endy (2005)	113	2	187.362	0.72251627
Ro et al. (2006)	95	0	0	0.47347666
Gibson et al. (2010)	91	0	0	0.38069174
Khalil and Collins (2010)	71	10	1657.26	0.63949588

Indegree

In a directed network, the indegree of a node is the number of links leading to that node. For publication citations, this means how often a publication is cited by other publications in the network. Indegree citations are the most common measure of a paper's relevance to the field, i.e., how many times it has been cited. The top six publications according to the indegree centrality measure are shown in Table 4. Table 4 also shows the profile publications' scores for the remaining three measures (outdegree, betweenness, and Eigen), which will be described in more detail in the following text.

Indicating the top-cited publications by indegree, several publications noted in Table 4 are described as seminal synthetic biology scholarship by Cameron et al. (2014). For example, Elowitz and Leibler (2000) and Gardner et al. (2000) published work describing the first synthetic circuits that were engineered to perform logical functions and increase control within and between cells. Further, Endy (2005) outlined the general principles of engineering biology, which served as a significant philosophical and scientific underpinning behind synthetic biology research moving forward. Finally, Gibson et al. (2010) described the creation of the first cell with a "synthetic genome," where synthesized DNA cassettes were assembled to recreate *M. mycoides* and inserted into a cell that would contain only the synthesized genome. While many other important scientific papers may be described here, this output signifies that key motivational or scientifically foundational works are among the top citation performers within synthetic biology scholarship.

Further, the citation network reveals that four of the six most cited publications do not cite any network publications themselves. For the top two publications (Elowitz and Leibler 2000; Gardner et al. 2000), this is due to the fact that they were both published in 2000 (the beginning year of the Web of Science search queries).

Outdegree

The outdegree of a node is the number of links leading away from that node. For publication citations, it shows how often a publication cites other publications in the network. Publications with high outdegree measures have a strong awareness of

Table 5 Sorted by decreasing outdegree

Publication	Indegree	Outdegree	Betweenness	Eigencentrality
MacDonald and Deans (2016)	0	31	0	0.36462184
Trosset and Carbonell (2015)	1	31	113.983	0.16192745
Brophy and Voigt (2014)	32	24	1669.1	0.43277875
Cameron et al. (2014)	16	21	493.538	0.40452803
Bradley et al. (2016)	1	21	66.2056	0.22901943
Xie and Fussenegger (2015)	1	18	26.7721	0.15101327

other publications in the field. The top six publications according to the outdegree centrality measure are shown in Table 5.

Each of the top six publications by outdegree was published in 2014 or later. Such articles reflect heavily upon engineering principles via enabling technologies of synthetic biology (Brophy and Voigt 2014; Bradley et al. 2016), applications of such enabling technologies to potential products (Trosset and Carbonell 2015; Xie and Fussenegger 2015), or review general progress within computational and engineering biology (MacDonald and Deans 2016; Cameron et al. 2014). Specifically, Brophy and Voigt (2014) describe principles of genetic circuit engineering, which Bradley et al. (2016) expand upon for microbial gene circuit engineering. Next, Trosset and Carbonell (2015) discuss the use of synthetic biology for pharmaceutical drug discovery, and Xie and Fussenegger (2015) discuss engineering principles for mammalian designer cells in biomedical applications. Finally, Cameron et al. (2014) offers a timeline of synthetic biology biological and computation sciences through 2014, while MacDonald and Deans (2016) further unpack the various tools, applications, and enabling technologies of synthetic biology. In general, these papers served as reviews or aggregations of synthetic biology research and heavily reference developments within the larger synthetic biology world.

Figure 2 shows that these publications were published after the majority of the network and therefore also had more literature to potentially cite. Four of these publications are cited one or fewer times by other network publications, likely due to the fact that all four were published after 2014 and have had little time to be cited. Of note, Brophy and Voigt (2014) is cited close to twice as often as the other five combined, suggesting that it both had a broad understanding of the field and quickly became an important publication.

Betweenness

Betweenness centrality measures the extent that a vertex is positioned on the shortest path between other pairs of vertices in the same network (Leydesdorff 2007). For publications within the network, a high betweenness measure indicates that the publication is playing a critical role to link publications to one another. In general, publications with high betweenness can be viewed as more interdisciplinary. The top six publications according to the betweenness centrality measure are shown in Table 6.

Table 6 Sorted by decreasing betweenness

Publication	Indegree	Outdegree	Betweenness	Eigencentrality
Brophy and Voigt (2014)	32	24	1669.1	0.43277875
Khalil and Collins (2010)	71	10	1657.26	0.63949588
Carr and Church (2009)	59	5	1437.99	0.36316962
Lu and Collins (2007)	30	9	1137.87	0.45722903
Mutalik et al. (2013)	31	7	1031.59	0.3009861
Kahl and Endy (2013)	9	7	902.154	0.12177155

Each publication in Table 6 was published between 2009 and 2014 and has non-zero indegree and outdegree centrality. This list excludes the majority of the publications from Tables 4 and 5. Brophy and Voigt (2014) repeats as a top performer here, which may derive from its review of circuit engineering techniques that are foundational to synthetic biology research. Similarly, Carr and Church (2009) discuss principles of genome engineering, while Kahl and Endy (2013) surveyed a general body of synthetic biology enabling technologies to report on the state of biological and computational research. Khalil and Collins (2010) more generally describe ongoing and potential future developments of both synthetic biology science and potential product applications. Mutalik et al. (2013) sought to address ongoing difficulties with synthetic biology research to reliably model and predict quantitative behaviors for novel genetic combinations and posed an approach to improve modeling and prediction capabilities.

High betweenness nodes have the potential to disconnect networks if they are removed. A publication with high betweenness and relatively low eigencentrality (e.g., Kahl and Endy 2013) may be an important gatekeeper between key clusters of publications. Kahl and Endy's survey of enabling technology serves as a link between the clusters surrounding the various enabling technologies of synthetic biology. Since it is a recent publication, it is not connected to many other highly connected publications. Because it brings together several technological clusters, many shortest paths pass through it and it would disrupt the network if removed, making it an important gatekeeper between clusters. The combination of many shortest paths passing through and a lack of connection to other highly connected publications give this survey a high betweenness centrality and a relatively low eigencentrality.

Eigencentrality

Eigencentrality measures the influence of a node in a network, giving greater weight to a node with more connections to other highly connected nodes. For publications within the network, a high eigencentrality measure indicates that the publication is playing an influential role to link important publications to one another while also being important itself. The top six publications according to the eigencentrality measure are shown in Table 7.

Table 7 Sorted by decreasing eigencentrality

Publication	Indegree	Outdegree	Betweenness	Eigencentrality
Elowitz and Leibler (2000)	142	0	0	1
Gardner et al. (2000)	123	0	0	0.92766808
Endy (2005)	113	2	187.362	0.72251627
Khalil and Collins (2010)	71	10	1657.26	0.63949588
Basu et al. (2005)	67	0	0	0.51157712
Ro et al. (2006)	95	0	0	0.47347666

The publications in Table 7 have a much greater range of publication years and centrality measures than the previous three tables. Four of the top six cite zero other publications in the network and have a betweenness centrality of zero, indicating that they are terminal nodes in the network. The other two, Endy 2005 and Khalil and Collins 2010, have a relatively low and high betweenness, respectively.

Within this range of papers, Elowitz and Leibler (2000) and Gardner et al. (2000) engineered the first synthetic genetic circuits that carried out specific design functions, which were described by Cameron et al. (2014) as foundational to synthetic biology research. Further, Endy (2005) describes engineering principles for biological systems, which are described by Raimbault et al. (2016) as seminal to synthetic biology's philosophical and scientific development. On a different note, Ro et al. (2006) describe how engineered yeast can be used to generate semi-synthetic artemisinic acid – an antimalarial precursor that was frequently discussed in future literature as a potential product application of synthetic biology in developing pharmaceutical applications (see also Paddon et al. 2013 for an update on this work).

Of the publications, five show up in the top six of multiple centrality measures. Elowitz and Leibler (2000), Gardner et al. (2000), Endy (2005), Khalil and Collins (2010), and Brophy and Voigt (2014) show up multiple times. In particular, Khalil and Collins (2010) shows up three times. These publications describe scientific research or perspectives on synthetic biology (Cameron et al. 2014) or have been further described as seminal or important publications in the field (Raimbault et al. 2016).

Pairwise Correlations

Table 8 shows the correlations between the centrality measures to determine how closely they are interrelated. This table compares how publications' different centrality measures generally relate to one another. Indegree and outdegree are not closely related, i.e., publications that are highly cited do not cite others in the network and publications that are cited frequently do not tend to cite others in the network. Furthermore, indegree is highly correlated with eigencentrality, demonstrating that publications which are cited the most are also the ones that are connected to other highly cited publications. This indicates high-density groupings around these publications.

Table 8 Pairwise correlations

	Indegree	Outdegree	Betweenness	Eigencentrality
Indegree	1.000	−0.025	0.392	0.774
Outdegree	−0.025	1.000	0.345	0.468
Betweenness	0.392	0.345	1.000	0.485
Eigencentrality	0.774	0.468	0.485	1.000

Evolution of the Network

In this section, we track the development of modern synthetic biology, where Cameron et al. (2014) defines 2000 as a critical starting point due to the first publications on synthetic circuit engineering (Elowitz and Leibler 2000; Gardner et al. 2000). We choose 34-year demarcations to view the growth in citation connectedness within the field and further allow observers to view growth during and immediately after important moments in modern synthetic biology history. These include the first synthetic circuit and toggle switch (Elowitz and Leibler 2000; Gardner et al. 2000), the first meeting of the Biobricks Conference Series in 2004, the description of an engineered bacteriophage for biofilm disposal, the creation of the first bacterial cell with an artificial genome (Gibson et al. 2010), and the commercial production of semi-synthetic artemisinic acid (Paddon et al. 2013). Data from each year includes all publications through that year (i.e., the data from 2004 includes publications from 2000, 2001, 2002, 2003, and 2004). Figure 4 shows what the network looked like in 2000, 2004, 2007, 2010, 2013, and 2016.

In 2000, the citation network contained only three disconnected nodes (Elowitz and Leibler 2000; Gardner et al. 2000; Ostergaard et al. 2000). Between 2007 and 2016, the number of publications in the network grows from 31 to 699 nodes, and the number of links grows from 75 to 2697. The number of links per node grows from just over two to almost four, showing that the network gets more connected as time goes on. The number of nodes in the center continues to increase, while the number of nodes on the outside remains relatively constant. The network becoming more connected suggests that synthetic biology is a unified field expanding in all directions.

As the network grows, its most central publications change over time. The top three publications for each year according to indegree, outdegree, betweenness, and eigencentrality are presented in the following figures. The evolution of top publications based on indegree centrality is shown in Fig. 5.

From the first year of the network onward, Elowitz and Leibler (2000) and Gardner et al. (2000) are the most cited publications. Their staying power and recurring citation in synthetic biology literature likely stem from (i) their description of the first successful synthetic genetic circuits and (ii) their description with reviews of synthetic biology research and progress such as Raimbault et al. (2016) or Cameron et al. (2014). In this way, Cameron et al. (2014) describes research within these publications as foundational to the larger synthetic biology field, and their importance is reflected due to their status as the most cited publications within our

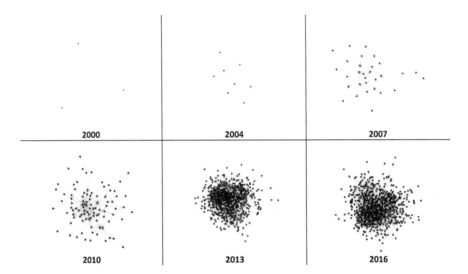

Fig. 4 Network evolution over key years

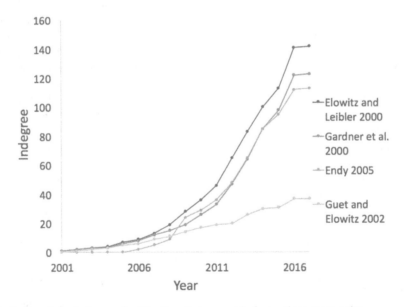

Fig. 5 Cumulative indegree of publications that appears in the top three at any point

dataset. Guet et al. (2002) was initially a high indegree performer, but did not experience the exponential growth seen in other papers circa 2008. The exponential growth of Elowitz and Leibler (2000), Gardner et al. (2000), and Endy (2005) show that these publications have had impact in shaping discourse within synthetic biology's biological, computational, and social sciences.

Unlike indegree, outdegree does not change over time, because a paper cannot cite future publications. Figure 6 shows the top publication by outdegree in each year.

In general, later publications cite more publications within the network. This is likely driven by the growth in scientific capacity and number of performers in the field (Raimbault et al. 2016; Oldham et al. 2012).

Like indegree, betweenness changes over time. The top three publications by betweenness were calculated for each year, and their full progressions are shown in Fig. 7.

Brophy and Voigt (2014), Khalil and Collins (2010), Carr and Church (2009), and Lu and Collins (2007) have significantly greater betweenness values than other publications. Each of them, particularly Brophy, became very central very quickly. Of note, Brophy and Voigt (2014) was an important "connector" in the field as soon as it was published, due to its pioneering work in genetic circuits.

Finally, the top three publications by eigencentrality were calculated for each year. Their respective eigencentralities were found over time and are shown in Fig. 8.

Elowitz and Leibler (2000) and Gardner et al. (2000) are at or near the top for the entire evolution of the network, indicating that their description and operation of synthetic genetic circuits had (and continues to have) substantial impact into various veins of synthetic biology research. Likewise, Endy (2005) maintains higher eigencentrality from 2005 to 2017, which is consistent with Raimbault et al.'s (2016) description of how the Endy lab has influenced the "programmatic discourses" of

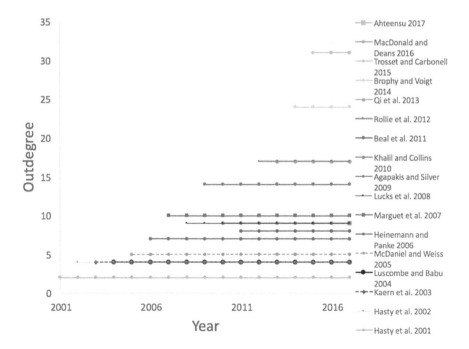

Fig. 6 Outdegree of top publication each year

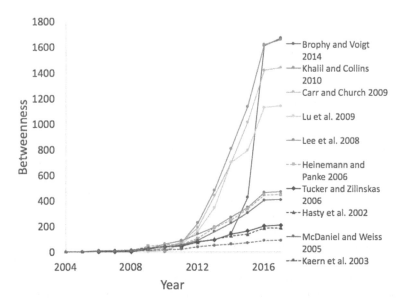

Fig. 7 Cumulative betweenness of publications to appear in the top three at any point

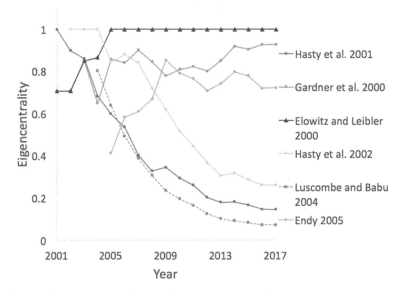

Fig. 8 Cumulative eigencentrality of publications to appear in the top three at any point

synthetic biology research by suggesting principles of engineering biology that have shaped inquiries into modeling, parts creation, and techniques for genomic assembly. These concepts are further noted as foundational to driving modern synthetic biology science by Cameron et al. (2014) and Trump (2016).

Other publications initially expressed higher eigencentrality yet declined in this metric over time. Hasty et al. (2001) and Hasty et al. (2002) build from work described in Gardner et al. (2000) regarding the design of synthetic circuits and thereby may be described as iterative rather than evolutionary developments in synthetic biology science.

Although excluded from the analysis of the network, isolates are still peer-reviewed publications and part of the synthetic biology network as a whole. These publications do not connect to any of the other publications in this particular network for a variety of reasons including corrupted or unreadable files, which do not allow the publication to be properly formatted for text mining. The numbers of publications and isolates over time are shown in Fig. 9, and the percentages of the publications that are isolates are shown in Fig. 10.

The number of publications in the network has been growing exponentially over the last 17 years. The number of isolates is as well, but not nearly as quickly. Figure 10 shows that over the last decade between 20% and 30% of publications have been isolates in any given year. Due to the small size of the network in the beginning, the percent varied greatly, particularly when there were three unconnected publications in 2000. Over the last 5 years, the percent of isolated publications has stabilized at roughly 20%, showing that the vast majority of new literature is adding to the existing network each year.

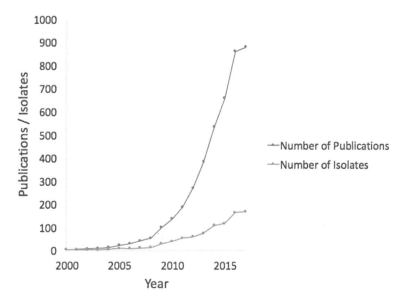

Fig. 9 Number of publications and isolates over time

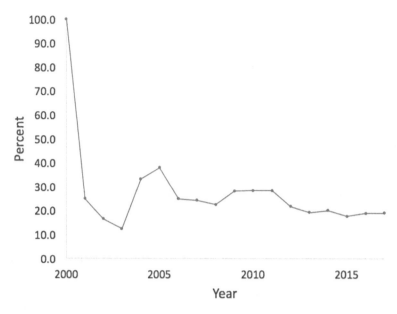

Fig. 10 Percent of publications that are isolates over time

Conclusion

The emerging field of synthetic biology is strongly linked and has significant publications that stand out among their peers, playing a critical role in connecting the network. The field is strongly linked because 80% of the 880 publications reviewed are a part of one large connected subgraph. Several publications appear multiple times in the final centrality measures and are then shown to rise to prominence rather quickly. These publications are either early seminal publications (Elowitz and Leibler 2000; Gardner et al. 2000), connecting publications (Brophy and Voigt 2014), or made important discoveries (Gibson et al. 2010; Paddon et al. 2013), showing that there are in fact several key publications connecting the field of synthetic biology. From a social sciences–oriented perspective, other publications have significant outdegree connection with state-of-sciences work (Seager et al. 2017; Cummings and Kuzma 2017) or use a mixture of natural and social sciences perspectives to inform regulatory gaps and opportunities (Oye et al. 2014; Bates et al. 2015).

As one of the first efforts to holistically characterize both biological and social sciences literature, this chapter sought to provide quantitative metrics to indicate how five individual communities of practice have developed over 2000–2017 and thereby identify key performers which influenced discourse in the field. Based upon quantitative metrics of cumulative indegree and eigencentrality, these top performers included breakthroughs in circuit engineering (Gardner et al. 2000; Elowitz and Leibler 2000; Brophy and Voigt 2014; Bradley et al. 2016), guidelines and philosophies for engineering biology (Endy 2005), reviews and progress reports of the field (Cameron et al. 2014; MacDonald and Deans 2016), and descriptions of product applications of synthetic biology (Trosset and Carbonell 2015; Xie and

Fussenegger 2015). Other important publications by cumulative indegree include important discoveries, such as Gibson et al.'s (2010) description of work to create the first cell with a synthetic genome and Paddon et al. (2013) which described the semi-synthetic production of artemisinic acid for antimalarial applications.

Furthermore, the broader synthetic biology community is not populating into various disparate groups but instead coalescing and commenting upon iterative developments and improvements to the field. More specifically, state of science and products literature appear to build directly from recent developments in synthetic biology science, in the form of either theoretical or applied research in biological or modeling research (Trump et al. 2018a; b). Risk, governance, and ELSI discussion *also* comment upon developments in "state of science" and "product" development, where they seek to respond to developments in the field and describe emerging benefits and challenges that may arise from synthetic biology's growing maturation and worldwide access (see examples such as with Tucker and Zilinskas 2006; Trump et al. 2017; Oye et al. 2014; Malloy et al. 2016; Schmidt et al. 2009). Rather than a collection of several unconnected and separated communities, synthetic biology's various communities of practice in physical and social sciences have at least acknowledged a small number of key developments in the field and may be using such works to develop a shared understanding of synthetic biology's physical and social sciences.

Such a citation network can also be helpful to identify those publications with the greatest degree of impact upon the synthetic biology community. Specifically, this chapter reviewed publication citation counts by indegree, where top performers here were generally described as motivational or seminal papers within reviews such as with Cameron et al. (2014) or Church et al. (2014). Such a citation network will help new entrants to the synthetic biology community identify key streams of research to follow (see Raimbault et al. 2016) or review possible motivations for future biological, computational, and social sciences research in the field. Such networks should be continually updated over time to account for adaptations in the synthetic biology literature, such as new trends in risk assessment protocols and testing procedures (Finkel et al. 2018; Linkov et al. 2017). Overall, network analysis and citation networks can help quantify and illustrate the growth trajectories, key performers, and disruptive events that spur growth in scholarly literature and may help the larger synthetic biology community to respond to such widely read and cited papers within and between their individual communities of practice.

References

Basu, S., Gerchman, Y., Collins, C. H., Arnold, F. H., & Weiss, R. (2005). A synthetic multicellular system for programmed pattern formation. *Nature, 434*, 1130–1134.
Bates, M. E., Grieger, K. D., Trump, B. D., Keisler, J. M., Plourde, K. J., & Linkov, I. (2015). Emerging technologies for environmental remediation: Integrating data and judgment. *Environmental Science & Technology, 50*(1), 349–358.
Bonacich, P. (1972). Factoring and weighting approaches to status scores and clique identification. *The Journal of Mathematical Sociology, 2*(1), 113–120.

Bradley, R. W., Buck, M., & Wang, B. (2016). Tools and principles for microbial gene circuit engineering. *Journal of Molecular Biology, 428*(5), 862–888.

Brophy, J. A. N., & Voigt, C. A. (2014). Principles of genetic circuit design. *Nature Methods, 11*(5), 508–520.

Cameron, D. E., Bashor, C. J., & Collins, J. J. (2014). A brief history of synthetic biology. *Nature Reviews Microbiology, 12*(5), 381–390.

Carr, P. A., & Church, G. M. (2009). Genome engineering. *Nature Biotechnology, 27*(12), 1151–1162.

Church, G. M., & Regis, E. (2014). *Regenesis: How synthetic biology will reinvent nature and ourselves.* Basic Books.

Church, G. M., Elowitz, M. B., Smolke, C. D., Voigt, C. A., & Weiss, R. (2014). Realizing the potential of synthetic biology. *Nature Reviews Molecular Cell Biology, 15*(4), 289–294.

Cummings, C. L., & Kuzma, J. (2017). Societal risk evaluation scheme (SRES): Scenario-based multi-criteria evaluation of synthetic biology applications. *PLoS One, 12*(1), e0168564.

De Solla Price, D. J. (1965). Networks of scientific papers. *Science, 149*(3683), 510–515.

Elowitz, M. B., & Leibler, S. (2000). A synthetic oscillatory network of transcriptional regulators. *Nature, 403*(6767), 335–338.

Endy, D. (2005). Foundations for engineering biology. *Nature, 438*(7067), 449–453.

Finkel, A. M., Trump, B. D., Bowman, D., & Maynard, A. (2018). A "solution-focused" comparative risk assessment of conventional and synthetic biology approaches to control mosquitoes carrying the dengue fever virus. *Environment Systems and Decisions, 38*(2), 177–197.

Gardner, T. S., Cantor, C. R., & Collins, J. J. (2000). Construction of a genetic toggle switch in Escherichia coli. *Nature, 403*(6767), 339–342.

Garfield, E. (1955). Citation indexes for science: A new dimension in documentation through Association of Ideas. *Science, 122*(3159), 108–101.

Gibson, D. G., Glass, J. I., Lartigue, C., Noskov, V. N., Chuang, R. Y., Algire, M. A., et al. (2010). Creation of a bacterial cell controlled by a chemically synthesized genome. *Science, 329*(5987), 52–56.

Guet, C. C., Elowitz, M. B., Hsing, W., & Leibler, S. (2002). Combinatorial synthesis of genetic networks. *Science, 296*(5572), 1466–1470.

Hasty, J., Mcmillen, D., Isaacs, F., & Collins, J. J. (2001). Computational studies of gene regulatory networks: In numero molecular biology. *Nature Reviews Genetics, 2*(4), 268–279.

Hasty, J., Mcmillen, D., & Collins, J. J. (2002). Engineered gene circuits. *Nature, 420*(6912), 224–230.

Jacob, R., Harikrishnan, K. P., Misra, R., & Ambika, G. (2017). Measure for degree heterogeneity in complex networks and its application to recurrence network analysis. *Royal Society Open Science, 4*, 160757. https://doi.org/10.1098/rsos.160757.

Kahl, L. J., & Endy, D. (2013). A survey of enabling technologies in synthetic biology. *Journal of Biological Engineering, 7*(1), 13.

Khalil, A. S., & Collins, J. J. (2010). Synthetic biology: Applications come of age. *Nature Reviews Genetics, 11*(5), 367–379.

Leydesdorff, L. (2007). Betweenness centrality as an indicator of the interdisciplinary of scientific journals. *Journal of the American Society for Information Science and Technology, 58*(9), 1303–1319.

Linkov, I., Trump, B. D., Wender, B. A., Seager, T. P., Kennedy, A. J., & Keisler, J. M. (2017). Integrate life-cycle assessment and risk analysis results, not methods. *Nature Nanotechnology, 12*(8), 740.

Linkov, I., Trump, B. D., Anklam, E., Berube, D., Boisseasu, P., Cummings, C., et al. (2018). Comparative, collaborative, and integrative risk governance for emerging technologies. *Environment Systems and Decisions, 38*(2), 170–176.

Lu, T. K., & Collins, J. J. (2007). Dispersing biofilms with engineered enzymatic bacteriophage. *Proceedings of the National Academy of Sciences, 104*(27), 11197–11202.

MacDonald, I. C., & Deans, T. L. (2016). Tools and applications in synthetic biology. *Advanced Drug Delivery Reviews, 105*, 20–34.

Malloy, T., Trump, B. D., & Linkov, I. (2016). Risk-based and prevention-based governance for emerging materials. *Environmental Science and Technology., 50*, 6822.

Marchant, G. E., Abbot, K. W., & Allenby, B. (Eds.). (2013). *Innovative governance models for emerging technologies*. Cheltenham: Edward Elgar Publishing.

McPherson, M., Smith-Lovin, L., & Cook, J. (2001). Birds of a feather: Homophily in social networks. *Annual Review of Sociology, 27*, 415–444.

Merad, M., & Trump, B. D. (2019). *Expertise under scrutiny: 21st century decision making for environmental health and safety*. Cham: Springer International Publishing. https://doi.org/10.1007/978-3-030-20532-4.

Mutalik, V. K., Guimaraes, J. C., Cambray, G., Lam, C., Christoffersen, M. J., Mai, Q.-A., Tran, A. B., Paull, M., Keasling, J. D., Arkin, A. P., & Endy, D. (2013). Precise and reliable gene expression via standard transcription and translation initiation elements. *Nature Methods, 10*(4), 354–360.

Oldham, P., Hall, S., & Burton, G. (2012). Synthetic biology: Mapping the scientific landscape. *PLoS One, 7*(4), e34368.

Ostergaard, S., Olsson, L., & Nielsen, J. (2000). Metabolic engineering of Saccharomyces cerevisiae. *Microbiology and Molecular Biology Reviews, 64*(1), 34–50.

Oye, K. A., Esvelt, K., Appleton, E., Catteruccia, F., Church, G., Kuiken, T., et al. (2014). Regulating gene drives. *Science, 345*(6197), 626–628.

Paddon, C. J., Westfall, P. J., Pitera, D. J., Benjamin, K., Fisher, K., McPhee, D., et al. (2013). High-level semi-synthetic production of the potent antimalarial artemisinin. *Nature, 496*(7446), 528–532.

Raimbault, B., Cointet, J. P., & Joly, P. B. (2016). Mapping the emergence of synthetic biology. *PLoS One, 11*(9), e0161522.

Ro, D.-K., Paradise, E. M., Ouellet, M., Fisher, K. J., Newman, K. L., Ndungu, J. M., Ho, K. A., Eachus, R. A., Ham, T. S., Kirby, J., Chang, M. C. Y., Withers, S. T., Shiba, Y., Sarpong, R., & Keasling, J. D. (2006). Production of the antimalarial drug precursor artemisinic acid in engineered yeast. *Nature, 440*(7086), 940–943.

Schmidt, M., Kelle, A., Ganguli-Mitra, A., & de Vriend, H. (Eds.). (2009). *Synthetic biology: The technoscience and its societal consequences*. Dordrecht: Springer Science and Business Media.

Seager, T. P., Trump, B. D., Poinsatte-Jones, K., & Linkov, I. (2017). Why life cycle assessment does not work for synthetic biology. *Environmental Science and Technology, 51*, 5861.

Trosset, J.-Y., & Carbonell, P. (2015). Synthetic biology for pharmaceutical drug discovery. *Drug Design, Development and Therapy, 9*, 6285.

Trump, B. (2016). *A comparative analysis of variations in synthetic biology regulation*. University of Michigan.

Trump, B. D. (2017). Synthetic biology regulation and governance: Lessons from TAPIC for the United States, European Union, and Singapore. *Health Policy, 121*(11), 1139–1146.

Trump, B., Cummings, C., Kuzma, J., & Linkov, I. (2017). A decision analytic model to guide early-stage government regulatory action: Applications for synthetic biology. Regulation and Governance.

Trump, B. D., Cegan, J., Wells, E., Keisler, J., & Linkov, I. (2018a). A critical juncture for for synthetic biology. *EMBO Reports., 19*(7), e46153.

Trump, B. D., Foran, C., Rycroft, T., Wood, M. D., Bandolin, N., Cains, M., et al. (2018b). Development of community of practice to support quantitative risk assessment for synthetic biology products: Contaminant bioremediation and invasive carp control as cases. *Environment Systems and Decisions, 38*(4), 517–527.

Trump, B. D., Cegan, J., Wells, E., Poinsatte-Jones, K., Rycroft, T., Martin, D., Warner, C., Perkins, E., Warner, C., Wood, M., & Linkov, I. (2019). Co-evolution of physical and social sciences in synthetic biology. *Critical Reviews in Biotechnology, 39*, 351.

Tucker, J. B., & Zilinskas, R. A. (2006). The promise and perils of synthetic biology. *The New Atlantis, 12*, 25–45.

Weisstein, E. W. (2017) Outdegree. From MathWorld – A Wolfram Web Resource.

Xie, M., & Fussenegger, M. (2015). Mammalian designer cells: Engineering principles and biomedical applications. *Biotechnology Journal, 10*(7), 1005–1018.

Synthetic Biology, GMO, and Risk: What Is New, and What Is Different?

Benjamin D. Trump

The term *synthetic biology* is a point of significant contention in both academic and policy circles. What comprises synthetic biology, as well as what does not, has implications ranging from regulatory requirements and risk-based testing challenges to communication with an often concerned and skeptical public. Synthetic biology has been defined as a unique field to the latest iteration along the spectrum of genetic engineering. Complicating matters is the notion that, simultaneously, both are likely true.

Pertaining to the design and construction of biological modules, biological systems, and biological machines or re-design of existing biological systems for useful purposes (Nakano et al. 2013), synthetic biology is a field with tremendous promise yet also possesses significant uncertainty related to its risks. Even 20 years since the development of the first synthetic circuits and genetic switches (Gardner et al. 2000; Cameron et al. 2014), the field's hazards remain broadly underexplored and uncharacterized (Trump et al. 2018a). Likewise, exposure assessment for the field remains in a relative infancy. This makes the job of a risk assessor, such as those within the US Environmental Protection Agency, Department of Agriculture, or Food and Drug Administration, quite difficult (Carter et al. 2014). Thankfully, however, this task is one that will ease over time and can draw from decades of thought and policy from genetically modified organisms (i.e., microbial risk assessment) to inform feasible best practices and minimum operating procedures within a given country and product domain.

In order to progress our risk assessment tools and capabilities for synthetic biology, an essential task is to understand, from a risk-based perspective, what its novelties are relative to other established, emerging, or enabling technologies. This

B. D. Trump (✉)
US Army Corps of Engineers, Engineer Research and Development Center,
Vicksburg, MS, USA
e-mail: bdtrump@umich.edu

© Springer Nature Switzerland AG 2020   85
B. D. Trump et al. (eds.), *Synthetic Biology 2020: Frontiers in Risk Analysis and Governance*, Risk, Systems and Decisions, https://doi.org/10.1007/978-3-030-27264-7_5

chapter[1] provides some insight via scholarly literature and policy debate regarding where the novelty of synthetic biology lies and some early thoughts regarding how it may be addressed in the near future.

General Background of Emerging Technologies and Synthetic Biology

Emerging technologies (or those technologies with novel characteristics or components that differ from conventional options) challenge the understanding of policy-makers and regulators to understand the technology's potential risks and benefits due to the uncertainty that such technologies inherently possess (Ludlow et al. 2015). With regard to their scope, a report issued by the United States' Presidential Commission for the Study of Bioethical Issues describes emerging technologies as revolutionary and/or evolutionary technological and scientific advances that are geared to improve various aspects of human life (PCSBI 2010). This report specifically sought to review the risks and regulatory concerns of novel biotechnologies and noted on behalf of the US government that the technologies' uncertainties make regulatory reform for such technologies difficult to accomplish without further experimentation and research (PCSBI 2010). Further, the report also acknowledges that many such technologies have "dual use" concerns, where the perceived benefits from a particular technology or innovative product may also be coupled with risks driven by an intentional misuse of technological innovation for deliberately hazardous purposes (PCSBI 2010).

For such emerging technology enterprises with the potential for uncertain levels of risk to human and/or environmental health, understanding the differences between novel and generally unconventional health risks and well-understood conventional health risks is of high importance to regulators and decision-makers (Bates et al. 2015). This is driven by concerns where novel risks may arise from the emerging materials or engineering processes that traditional regulatory paradigms may or may not be able to properly cover (Carter et al. 2014).

However, not all emerging technologies possess the uncertainty and novel health risks that regulators and research stakeholders must consider when reviewing an emerging technology's regulation. All innovation poses some degree of uncertainty and risk, yet a contrasting point for those emerging technologies containing uncertain risks includes their novel physical characteristics that could drive them to act in an unpredictable and irreversible manner (Moe-Behrens et al. 2013; Linkov et al. 2017). Current examples of this include nanotechnology and synthetic biology. For the former, Maynard (2007) notes that "some purposely made nanomaterials will present hazards based on their structure—as well as their chemistry—thus

---

[1] This chapter includes information previously discussed in the dissertation: Trump, B. (2016). A Comparative Analysis of Variations in Synthetic Biology Regulation. University of Michigan.

challenging many conventional approaches to risk assessment and management," indicating that the chemical and structural novelties of such nanomaterials pose possible novel yet uncertain threats to human and environmental health. Likewise for synthetic biology, the substantial modification of an organism's DNA may contribute to the transfer of novel genetic information to the natural environment that could yield uncertain and irreversible risks to plant, animal, and human biology (Schmidt et al. 2008; Cardinale and Arkin 2012; Dana et al. 2012).

It is tempting to lump all emerging technologies into a common basket of risk assessment and governance. However, doing so neglects the unique chemical and biological factors associated with the experimentation, generation, commodification, and disposal of organisms engineered via one of the several enabling technologies of synthetic biology. Further, such rough grouping also neglects the unique assessment, testing, and regulatory environments that genetic engineering research must comply within a given country and for a specific product. For synthetic biology, this means that any risk assessment or governance process must simultaneously account for the unique physical properties associated with genetic modification of various organisms, as well as the unique institutional challenges and requirements associated with genetic engineering. The remainder of this chapter discusses the unique risk concerns associated with synthetic biology and its enabling technologies and describes what areas of risk or governance may be novel or particularly difficult and complex for a risk assessment and regulatory audience to grapple with.

Synthetic Biology and Risk: What Is the Same, and What Is Different?

Synthetic biology is one of the more recent cases of emerging technology development and uncertain risks and benefits, where the technology is purported to contain significant potential benefits to a variety of industries (Tucker and Zilinskas 2006; Neumann and Neumann-Staubitz 2010; d'Espaux et al. 2015). The "novelty" expressed within synthetic biology research includes several different factors but generally includes the ability of synthetic biology research to conduct greater control of genetic systems and the enabling of novel gene expression through the application of standardized engineering techniques to biology and thereby create organisms or biological systems with novel or specialized functions (Cameron et al. 2014; Tabor et al. 2009; PCSBI 2010).

The ability to alter, manipulate, and control cell expression has driven many scholars to hypothesize the technology's potential benefits within fields ranging from medicine (Dormitzer et al. 2013; Paddon and Keasling 2014) to ethanol production to insect population control (Georgianna and Mayfield 2012; Harris et al. 2012). One specific area of this includes pharmaceutical development, where synthetic biology has been purported to provide several benefits to this field (Weber and

Fussenegger 2012). Such discussed benefits include the ability to speed up the rate of drug and vaccine production (Dormitzer et al. 2013; Rojahn 2013), facilitate the production of pharmaceutical components that are expensive or scarce naturally (Paddon and Keasling 2014), or even advance research on vaccines and drugs for diseases with limited to no vaccine, treatment, and/or cure (Barocchi and Rappuoli 2015; Bugaj and Schaffer 2012). Such benefits have worldwide implications for delivering treatments to areas around the world suffering from debilitating disease (Barocchi and Rappuoli 2015) and improving public health response times and treatment capabilities to various pandemics (Dormitzer et al. 2013).

However, synthetic biology may also yield potential novel health risks. While synthetic biology product development may generate conventional health risks that are relatively well understood, considerations of how the technology may generate problems for biosecurity and biosafety require a measured response by regulators and policymakers (Trump et al. 2019; Kelle 2009). From a biosafety perspective, this includes the concept of horizontal gene transfer (the transfer of genes between organisms in a manner other than traditional reproduction), where horizontal gene transfer is a particular problem of concern for synthetic biology as such gene transfer "is a common and somewhat uncontrolled trait through the microbial biosphere" (Schmidt 2008; Cardinale and Arkin 2012). A specific concern of horizontal gene transfer includes the notion that modified cells may transfer synthetic information to the natural environment and yield negative or unanticipated results (Schmidt 2008; Dröge et al. 1998).

Likewise for biosecurity, concerns by policymakers and regulators reflect concern that a nefarious agent or bioterrorist could utilize principles of synthetic biology to produce a biological weapon and with disastrous consequences (Kelle 2009; National Research Council 2004). The central issue here includes the notion of "dual use concerns" raised in the PCSBI (2010) report noted earlier, where such nefarious actors utilize synthetic biology research in a manner that deliberately yields harms to humans, animals, or the environment. These and other concerns of the risks that synthetic biology development may pose to human and environmental health require considerations of balancing the technology's potential risks and benefits as it continues to develop (Mandel and Marchant 2014).

The novel and uncertain health risks produced by synthetic biology research include the substantial genetic modification of cells that, under certain circumstances, could have deleterious effects upon humans and/or the natural environment (Mukunda et al. 2009; Moe-Behrens et al. 2013). Given such uncertainties, regulators and key stakeholders may or may not seek to consider whether or not traditional measures of governance (or the actions, processes, traditions, and institutions by which authority is exercised and decisions are taken and implemented) are sufficient to protect humans and the environment from significant health risk (Kuzma and Tanji 2010; Cummings and Kuzma 2017). The pathways of such risk may include, among others, the following:

(i) Exposure in a laboratory setting (Rabinow and Bennett 2012)
(ii) Accidental releases in an occupational/production setting (biosafety) (Schmidt 2008)

(iii) Intentional release of potentially harmful microorganisms (biosecurity) (Vogel 2014)

(iv) Acute risk concerns to individual human health in the workplace and upon commercialization (Howard et al. 2017; Fatehi and Hall 2015)

 (v) Improper disposal of such microorganisms upon their end-of-life disposal and their unintended proliferation in the environment (Myhr and Traavik 2002; Schmidt and de Lorenzo 2012; Trump et al. 2018a)

In this way, attempts to review the risk of synthetic biology products must consider collective biosafety and biosecurity concerns that could generate health concerns for humans, animals, and/or the environment (Trump et al. 2018b). Normatively, to protect against uncertain technological risks associated with synthetic biology's biosecurity and biosafety concerns, policymakers and key stakeholders within a given country must engage in active governance of the field based upon their perceptions of how serious such risks actually are.

The Growing Fields of Biosafety and Biosecurity

Two such early topics of risk-based discussion in the early 2000s related to synthetic biology research include the concepts of *biosafety* and *biosecurity* (Kelle 2009; Guan et al. 2013; Carter et al. 2014). Specific discussion on these topics centers on the potential for irreversible and/or hazardous outcomes from the process of synthetic biology product development, either from deliberate misuse (biosecurity) or unintended consequences (biosafety) (Kelle 2009; Guan et al. 2013; White and Vemulpad 2015).

Biosafety

Biosafety is a classic concern and governance challenge with any venture dealing with material or technology production, with specific safeguards for material handling, packaging, transportation, and safe use required by statutory law and/or regulatory practice. Research with biological organisms has specific biosafety requirements based upon the type of organism, its toxicity/pathogenicity/other concerns, and whether the organism is considered native or invasive to the local environment (Burnett et al. 2009). What makes synthetic biology different is the incomplete understanding of the types of phenotypic traits and hazard scenarios the engineered organism might possess, as well as how the engineered organism might persist within various unintended environments.

Biosafety considerations generally consider the unintentional release of genetically modified material that may subsequently alter or overwhelm its local environment and incur negative health consequences (Wright et al. 2013; Schmidt 2008;

Seyfried et al. 2014). Such concerns may occur across the life cycle of a given synthetic biology material, including at the research and development stage (i.e., biological material accidentally escapes lab containment and reaches unintended human or environmental hosts), the manufacturing stage (i.e., concerns of occupational health due to unintended exposure to modified cells), the commercial stage (i.e., unintended use among consumers), and the end-of-life stage (i.e., improper disposal or treatment of synthetic biology byproducts and waste) (Bates et al. 2015; Seager et al. 2017). Across all stages, an important consideration includes how such an unintended event could occur alongside the magnitude of health consequences that it may produce.

Unintentional release outcomes have been described as potentially problematic for several reasons, including (i) the potential for engineered organisms to act as invasive species and negatively impact biodiversity, (ii) concerns of exposure of engineered organisms to unintended human and animal targets, and (iii) the inability to control engineered organisms – particularly bacteria – once they are taken outside of a contained environment. Biosafety concerns are inherent within most laboratory research, yet synthetic biology adds another dimension of uncertainty and risk due to the potential release of substantially genetically modified organisms into the environment in an irreversible manner. Many scholars and researchers have identified various opportunities to limit or prevent such biosafety events, including (a) engineered control options like genetic kill switches or a reliance upon certain food or nutrient sources that are not easily found in the environment and (b) traditional safety checks such as oversight committees, proper biological containment protocol and safety gear, and approval requirements to edit or engineer certain genetic strains (Wright et al. 2013). The International Genetically Engineered Machine, or iGEM, is one example of a nonstate actor with rigorous biosafety requirements. IGEM is an international competition that encourages students to engage in synthetic biology research through mentored teams with strict safety and judging criteria (Guan et al. 2013).

One potential biosafety risk concern noted in the literature is the concept of horizontal gene transfer. Generally referring to the transfer of genes between organisms in a manner other than traditional reproduction, horizontal gene transfer is a particular problem of concern for synthetic biology as such gene transfer "is a common and somewhat uncontrolled trait through the microbial biosphere" (Wright et al. 2013; Dröge et al. 1998). Davison (1999) and Wright et al. (2013) state that horizontal gene transfer occurs by transduction, conjugation, and/or transformation of modified cells within the natural environment. For each of these three methods of transfer, transduction involves the active transfer through bacteriophages, conjugation through pili, and transformation via "sequence-independent uptake of free DNA from the environment" (Wright et al. 2013).

Synthetic biologists have begun to explore avenues to prevent horizontal gene transfer via one or more of these avenues, yet the process of fully resolving the "transformation" gene transfer avenue is challenging due to the potential for lingering cell DNA to persist in the environment well after cell death (Thomas and Nielsen 2005). Nielsen et al. (2007) note that even months after a cell is placed within certain environmental conditions, extracellular DNA can be detected. Further, such extracellular DNA may be actively assimilated by bacteria along with some unicellular and multicellular eukaryotes (Lorenz and Wackernagel 1994; Boschetti et al. 2012). While Khalil and Collins (2010) describe how engineering a "self-destruct" option can limit some vectors of horizontal gene transfer by programming the cells to die under certain conditions or time intervals, Wright et al. (2013) and Lorenz and Wackernagel (1994) discuss how even with cases of cell death, extracellular DNA may be scavenged and absorbed by other natural cells afterward. Callura et al. (2010) and Wright et al. (2013) discussed how self-destruct mechanisms serve as the best available tool to prevent synthetic material from escaping control and interacting with the environment, where engineered cells could be preprogrammed to self-destruct en masse if cell population density becomes too great.

Townsend et al. (2012) state that monitoring the rates of horizontal gene transfer is a complicated process due to the large swarms of cells required to monitor for gene transfer along with the extended timeframe needed to monitor whether or not a rare genetic mutation was able to grow into larger populations of cells. Nielsen and Townsend (2004) and Wright et al. (2013) further argue that horizontal gene transfer events are difficult to monitor due to their limited rate of occurrence, where the frequency of transformation of microbes in soil is less than 1×10^{-7} per bacterium exposed, with transformation generally capable only within a few hours to days after the release of novel cellular material into the environment. However, Pruden et al. (2012) note that despite the general rarity of horizontal gene transfer, certain DNA elements have been shown to proliferate through large and complex ecosystems. One of these includes antibiotic-resistant genes, which Mulvey and Simor (2009) describe as cases of horizontal gene transfer where antibiotic resistance spreads in environments such as hospitals and produces antibiotic-resistant superbugs. From the perspective of biosafety, Wright et al. (2013) argue that such antibiotic-resistant genes should not be utilized by synthetic biologists unless absolutely necessary, although such genes remain "commonly used as markers during plasmid construction."

Aside from cellular self-destruction, another approach to promoting biosafety involves making it easier to identify the origin of cells escaping containment in order to fix existing containment issues and prevent future breakout events. Wright et al. (2013) describe that synthetic operons within cellular DNA may be fashioned to contain a genetic "barcode" that may be indexed within a database in order to facilitate cellular recognition and communicate the cell's origin point to identifiers. Another approach described by Gibson et al. (2010) includes the introduction of a "DNA watermark" into several locations on the cell's genome, which acts as an identifier similar to that described earlier. Such a watermark or barcode may also have proprietary benefits, where such unique codes may be used for commercial purposes to "brand" a cell's DNA with unique identifying information in the event of theft.

Even should these approaches to reduce the opportunities for horizontal gene transfer fail, the chances for mutated genes that are harmful to humans to transfer and proliferate are minute (Arber 2014). In other words, it is rare that transferred traits are evolutionarily beneficial to targeted organisms that are also detrimental to human and ecosystem health in a natural setting (Rossi et al. 2014). However, White and Vemulpad (2015) note that synthetic biology may increase the potential for harmful gene transfer due to the use of artificial gene sequences. Even in such scenarios, however, Armstrong et al. (2012) argue that such concerns are more likely in deliberate biosecurity situations rather than through accidental release and random gene transfer.

Finally, one of the more recent developments in synthetic biology research includes gene editing, where engineers are able to insert, delete, substitute, or change a specific site of an organism's genome (historically, traditional approaches to genetic engineering inserted genetic material into the host genome with less precision or control). This advancement in gene editing is generally driven by engineered nucleases, or "molecular scissors," which can create double-strand breaks at specific sites within the host genome. These breaks are repaired via strategies of (a) nonhomologous end joining or (b) homologous recombination and ultimately foster a targeted change at that specific site of the host genome.

Currently, four gene editing approaches with engineered nucleases have been utilized, including (i) zinc finger nucleases (ZFNs), transcription activator-like effector-based nucleases (TALENs), clustered regularly interspaced short palindromic repeats (CRISPR/Cas9), and meganucleases (Joung and Sander 2013; Urnov et al. 2010; Cong et al. 2013). These approaches enable increasing levels of control over the organism engineering process and enable previously implausible breakthroughs such as improving or eliminating heritable disease to improving the resiliency of endangered or at-risk environmental biomes like the Great Barrier Reef from environmental stressors (Piaggio et al. 2017; van Oppen et al. 2015).

Despite the promise of gene editing research, many scholars and policymakers have voiced concern regarding the technology's risks. Such stakeholders typically focus on either the capacity of a careless or nefarious user to use gene editing approaches to create highly virulent diseases or dangerous engineered organisms or the potential for unintended secondary consequences associated with the release of gene-edited organisms into the environment.

A rising concern of gene editing in the biosafety landscape centers upon the use of gene drives or technologies that cause directed mutation to propagate a suite of genes throughout a specific target species far more efficiently than via natural, Mendelian inheritance. Accomplished through gene editing approaches such as CRISPR, a limited release of such organisms could trigger a widespread change consistent with the original genetic modification in wild-type populations (Esvelt et al. 2014). Such research has spawned many applications, such as the capacity to conduct more effective vector control of mosquito-borne diseases like malaria (Hammond et al. 2016; Finkel et al. 2018). However, limited insight exists regarding the risks and safety concerns associated with gene drives – or even what methods of analysis or tools of decision support may help close the gap on such uncertainty. For example, the UK-based Nuffield Council on Bioethics has argued that the low cost and relative ease of use of gene editing technologies could enable amateur or independent groups or individuals to conduct their own editing experiments outside of any necessary oversight or limitations (White and Vemulpad 2015; Frow 2017).

Scholars such as Kuzma et al. (2018), Oye et al. (2017), and Oye et al. (2014) have touched upon the core risk, governance, and ethical challenges associated with gene drive research within a variety of contexts. While risks associated with gene drives remain uncertain, such scholarship seeks to identify general concerns, risks, and potential strategies to address accidental or deliberate misuse of synthetic biology or new enabling technologies that facilitate gene drives. With exceptions, arguments in this space denote concerns that gene drives:

(a) Are, once released into the environment, nearly prohibitive to control
(b) Are, once percolated within a species, impossible to reverse
(c) Are currently not predictable in terms of long-term efficacy and sustainability

Concerns associated with environmental control and irreversibility are of particular concern to risk assessors and decision-makers concerned with the environmental release of gene-edited and modified material. In a March 2018 workshop, the US Army Engineer Research and Development Center hosted a workshop that sought to understand the hazards, exposure considerations, and other concerns associated with environmental release scenarios of engineered organisms. Figure 1 includes

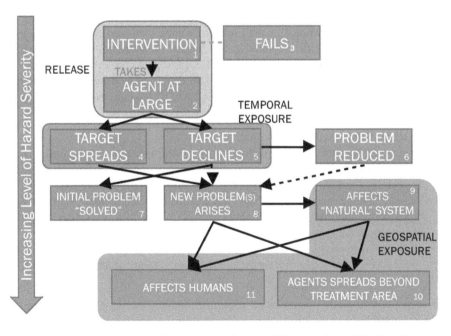

Fig. 1 Event tree for environmental release scenarios of modified organisms. (This figure is reproduced from: Trump et al. 2018b)

one of the workshop's key outputs that tracks concerns of risk as genetically modified constructs persist beyond their control area. Notably, gene drives theoretically cause engineered organisms to spread well beyond the initial treatment area, regardless of whether such species are intended to persist beyond a few generations.

Biosecurity

Biosecurity, or concerns of risk driven by the use of synthetic biology for nefarious or deliberately harmful means (i.e., bioterror), centers on the "dual use" concerns associated with emerging technology development (Perkins and Nordmann 2012; Marris 2015). Such concerns include fears that technological developments may also be utilized for deliberately harmful purposes. Dual use concerns have been discussed within synthetic biology research since at least 2004, when the World Health Organization outlined certain guidelines to promote lab safety while reducing the potential for malicious use of synthetic biology's concepts and tools for cellular manipulation (WHO 2004; Mandel and Marchant 2014).

Particularly within the United States, federal policymakers increasingly concerned with the potential for life sciences research to be misused in warfare or terrorism began to assize their own inquiries with respect to synthetic biology biosecurity, with the first such council including "The Committee on Research Standards and Practices to Prevent the Destructive Application of Biotechnology" of 2004 – colloquially known as the Fink Committee (National Research Council 2004; Kelle 2009). Specific to synthetic biology, the Fink Committee was asked to review those "practices that could improve US capacity to prevent the destructive application of biotechnology research while still enabling legitimate research to be conducted" (National Research Council 2004). Specific recommendations produced by the Fink Committee include the following:

1. To educate the scientific community
2. To review experiment proposals and plans related to genetic manipulation and experimentation
3. To review submitted manuscripts in this field prior to their publication
4. To foster the creation of a national science advisory board related to combatting bioterrorism and other threats arising from the misuse of life sciences research like synthetic biology
5. To "harmonize international oversight" (Kelle 2009)
6. To achieve a more active role for the life sciences in efforts to prevent biosecurity concerns

In response to the Fink Committee's recommendations, the US National Research Council established the Committee on Advances in Technology and the Prevention of Their Application to Next Generation Bioterrorism and Biological Warfare Threats, which later also became known as the Lemon-Relman Committee (Kelle 2009). To more specifically address potential biosecurity risks and threats of emerging life sciences research such as with early synthetic biology, the committee established a four-group classification methodology that included the following:

1. Technologies that seek to acquire novel biological or molecular diversity
2. Technologies that seek to generate novel but predetermined and specific biological or molecular entities through directed design
3. Technologies that seek to understand and manipulate biological systems in a more comprehensive and effective manner
4. Technologies that seek to enhance the production, delivery, and "packaging" of biologically active materials (National Research Council 2006; Kelle 2009)

Under the Lemon-Relman categorization, synthetic biology falls into categories 1 and 2, with Committee recommendations for the technology including increased awareness and oversight for biological capabilities to damage, for example, hosting homeostatic and defense systems or constructing synthetic organizations with limited control and/or the potential for deliberate negative health risk (National Research Council 2006). A primary outcome of the Lemon-Relman Committee and its subsequent categorization was an increased call for government oversight and monitoring related to the potential for dual use applications in life sciences research (inclusive of synthetic biology) (Choffnes et al. 2006).

Further, discussion from both committees, with particular discussion from the Lemon-Relman Committee, which called for increased consideration of the societal implications of and access to synthetic biology research, was discussed at the SB2.0. Specifically, the SB2.0 conference attendees produced a collective statement that discussed some of the biosecurity implications of DNA synthesis, including calling for an open working group to "improve existing software tools for screening DNA sequences" and promoting further discussion on options for national governments in Europe and the United States to govern DNA synthesis technology that may be produced in synthetic biology research (Conferees SB2.0 2006; Kelle 2009).

Aside from government-directed discussion on biosecurity issues, one of the early descriptive papers on the regulation of synthetic biology biosecurity concerns includes Church (2004). Specifically, Church called for a biosecurity paradigm where oversight agencies would screen any genetically modified material with various research projects based upon the product's oligonucleotide and DNA information to identify any similarities between the discussed organism and other traditional pathogenic organisms. To limit the proliferation of such research outside of institutions with clear external oversight and regulation, Church (2004) also called for the licensure of certain instruments and reagents involved in the production of genetically modified material deemed similar to harmful pathogens (Kelle 2009).

Despite these calls for expanded government oversight of synthetic biology research, Maurer and Zoloth (2007) and Bügl et al. (2007) argued instead for a governance paradigm driven by synthetic biologists instead of preemptive national regulation. Specifically, Maurer and Zoloth (2007) placed emphasis on the need for self-governance without external interventions or intrusive oversight. In a similar vein, Bügl et al. (2007) argued for a governance structure that, while incorporating external oversight from government, placed companies squarely within the governance-building process. In Bügl et al.'s (2007) proposal, biosecurity regulation would be driven by mutually agreed-upon guidelines between government and industry, with particular attention placed on the comparison of DNA sequences with a selection of those with known negative outcomes (i.e., pathogens).

Overall, Kelle (2009) identifies two distinct strands of discussion related to biosecurity regulation that emerged with the second and early third waves of synthetic biology. First, industry and DNA synthesis companies generally emphasize the "formation and implementation of best practices across the industry" where "oversight and enforcement of these standards [...] is not regarded as falling into the purview of industry itself, but rather as a governmental task" Kelle (2009). For the second element, Kelle (2009) acknowledges the growing discussion of self-governance within the synthetic biology community, which Maurer and Zoloth (2007) and Kelle (2009) describe as not easily reconciled with governmental wishes to strengthen external oversight over an industry they perceive as advancing potentially threatening technological capabilities if nefarious agents were able to gain access to them.

Governing Synthetic Biology: Mechanisms and Instruments of Governance

The sheer diversity of synthetic biology research presents regulators with a near-impossible problem of trying to assess risk in many differing technological processes as well as several potential product categories. In some areas, this impasse has spurred some (as with the European Union Court of Justice in a July 2018 ruling) to apply existing EU Directives from earlier generations of genetically modified organisms onto gene editing technologies like CRISPR, which may significantly slow progress on gene editing research in the European Union (Kupferschmidt 2018). Others such as the United States have applied existing regulatory authorities to capture the regulatory governance of synthetic biology like the Environmental Protection Agency (via TSCA), the US Department of Agriculture – the Animal and Plant Health Inspection Service (APHIS, via the Plant Pest Act) – or the Food and Drug Administration (FDA, via the Food, Drug, and Cosmetic Act) (Carter et al. 2014).

Despite these and other developments, the sheer uncertainty and scale of synthetic biology presents challenges that pressure many regulators to provide good guidance and safe operating procedures in a manner that is risk-informed and operates under principles of safety-by-design (Linkov et al. 2018). Though it is likely that these challenges can only be addressed with the luxury of time and adequate funding for safe testing and quantitative assessment for material hazard and exposure scenarios, it may be helpful to identify possible lessons from other emerging technologies where similar degrees of uncertainty were addressed through policy debate and various approaches of risk governance.

For synthetic biology, hard law approaches for technology coordination include international treaties like the Cartagena Protocol on Biosafety to the Convention on Biological Diversity, which supplemented the Convention on Biological Diversity (Schmidt 2012; Gupta and Falkner 2006). The Cartagena Protocol called for the use of the precautionary principle to evaluate new products from new technologies while allowing developing nations to balance public health risks against economic benefits of technological development. The broad vision and reach of this Protocol, as well as follow-on agreements like the Nagoya Protocol in 2010, are directly applicable to synthetic biology. This includes the recurring influence of the precautionary principle – the governing philosophy which allows policymakers to justify delaying or denying technological innovation in a specific product category or developmental process until its risks are better known and safe use guidelines have been established.

However, many scholars argue that international hard law such as the Cartagena Protocol are politically difficult to develop and implement and can discourage some from joining an agreement or maintaining membership in an agreement altogether. For nanotechnology, Marchant and Abbott (2012) argue that international harmonization of nanotechnology risk governance has and will continue to be through soft law rather than any legally binding treaty or hard law mechanism. It is difficult to tell now if synthetic biology will follow a similar route, yet international soft law agreements may help navigate broadly divergent regulatory traditions and differing cultures of risk acceptance within various pertinent product categories.

Past soft law efforts in the biotechnology realm have been quite influential, such as the Asilomar Conference on Recombinant DNA in February 1975 that still impacts the governance of biotechnology research today. The Conference, which brought about 140 practitioners and professionals to discuss biohazards and biotechnology regulation, established a series of voluntary principles and codes of conduct for genetic experimentation. This includes the prohibition of certain forms of research, such as the cloning of recombinant DNAs derived from highly pathogenic organisms, the cloning of DNA containing toxin genes, or large-scale recombination for products that are deliberately harmful to humans, animals, or the environment (Berg et al. 1975; Berg 2008). Alongside the development of such voluntary principles and operating procedures, the Asilomar Conference also served as a deliberate effort at public engagement and risk communication in order to improve public transparency – something that Pauwels (2013) notes as being a recurring concern of the public regarding synthetic biology today. Altogether, Asilomar serves as an example of how an effort at standards and norms harmonization can help further a field with tremendous uncertainty and might help advance synthetic biology for both (i) risk assessors that lack formal data to review synthetic biology product hazard, exposure, or dose-response effects and (ii) a concerned public that has limited scientific knowledge of the area yet remains concerned of its risks to humans and the environment.

The Growing Challenge of Synthetic Biology Development: The Need for Good Governance

While synthetic biology research exists along a general continuum of genetic and systems engineering scholarship, it does present some novel or complex risk concerns that decision-makers and risk assessors do not yet have reliable and tested methods to address. The earlier sections mentioned several of these ideas, ranging from horizontal gene transfer to extensive threats to biodiversity posed by the environmental impact of the engineered organism, to accidental release scenarios of dangerous engineered material, to various other concerns throughout the life cycle of organism development, manipulation, use, and disposal. Given these concerns, policymakers are faced with two critical questions: first, how should we make decisions with the limited risk data available, and second, how do we balance the need for safety alongside the potential benefits associated with future synthetic biology-derived products? More simply: how can we achieve good governance for synthetic biology?

For both biosafety and biosecurity, a critical challenge within any governing environment centers upon the dearth of data available for conducting hazard, exposure, and effects (i.e., dose response) assessments which are critical for many regulatory agencies with oversight responsibilities of the process of synthetic biology and/or its resulting products. Early synthetic biology research, such as the construction of synthetic circuits like synthetic toggle switches (Gardner et al. 2000) and transcriptional regulators (Elowitz and Leibler 2000), initiated new developments in biological engineering where engineered *E. coli* could "toggle between two stable expression states in response to external signals" (Cameron et al. 2014). These developments represented a greater control over cellular behavior and response to stimuli and helped trigger further advancements in a field dedicated to engineering and developing organisms with clear yet artificial design functions. Though these advancements and others that followed were foundational breakthroughs for synthetic biology, they also represented a departure from existing biotechnologies that could be evaluated via existing authorities like the Toxic Substances Control Act (TSCA) in the United States, the Biosafety Guidelines for GMOs in Singapore, the Gene Technology Act of 2000 and the Gene Technology Regulations of 2001 in Australia, and various Directives and Regulations within the European Union (i.e., Directive 2001/18/EC regarding ecosystem health and biodiversity or Directive 2009/41/EC on laboratory and workplace safety).

In other words, as synthetic biology research gains increasing levels of control over organism function and phenotypic expression, and organisms/genetic material becomes increasingly "synthetic," the capacity of existing regulatory authorities which capture the process or products of synthetic biology will become increasingly strained in their capacity to adequately review the technology's risks throughout all stages of its development. Scholarly literatures such as Oye (2012), Linkov et al. (2018), or Mandel and Marchant (2014) have argued that adaptive and proactive approaches may be needed to ensure good governance over an evolving synthetic biology landscape, yet a broader question remains on how such governing styles might be implemented and what veins of synthetic biology research present the greatest challenges for regulators going forward. Regulators, policymakers, and scholars have begun to identify general areas of concern such as horizontal gene transfer or threats to biodiversity, yet the ability to quantify synthetic biology risk as a measure of hazard, exposure, and consequence remains elusive. In the short term, much of the risk-based policymaking is likely to be semi-quantitative and expert-driven in nature. However, as specific areas of research begin to mature and risk data becomes more widely available (e.g., transgenic salmon, semi-synthetic pharmaceutical precursors, or options for vector control of tropical diseases), risk-based decision-making will become more quantified and in line with traditional practices of chemical and microbial risk assessment. The rate by which the adoption of quantified risk assessment options becomes accessible is dependent upon several factors, including the relative degree of controversy that the engineered product contains with the general public, the potential for unintended risk consequences, and the relative complexity of the engineered genome at use within a given product.

As an umbrella term for many avenues of biological research, synthetic biology comprises many activities such as the industrial production of semi-synthetic or fully synthetic compounds to environmental releases of engineered organisms for improving environmental health and sustainability. For the former, one prominent example includes the Swiss company Evolva, which edits yeast to synthesize synthetic vanilla (known as "vanillin") via its fermentation process (Cha 2014). Another prominent yet unrelated example includes the production of artemisinic acid via *E. coli* or *S. cerevisiae*, which serves as a sustainable precursor for antimalarial drugs (Paddon and Keasling 2014). Both cases have been met with significant yet diverging challenges. For the former, some groups such as Friends of the Earth have argued that the production of vanillin via synthetic biology has uncertain ecological and human health impacts that regulators must address prior to the product's entry into the market (Hayden 2014). For the latter, economic limitations have restricted the commercial success of semi-synthetic artemisinin, yet project developers still consider the approach a scientific success that passed through rigorous regulatory oversight (Peplow 2016). Regardless of outcome, these and similar cases raise differing levels of concern regarding the capacity of a regulator to review that which is "artificial" or "unnatural," with minimal data, and make an assessment regarding the safe use and good governance of such products.

Perhaps an even more controversial application of synthetic biology includes the environmental release of engineered organisms to incur some effect upon the environment. Recently, this has included several options to control the spread of several tropical viruses like dengue fever. One option includes vector control by engineering relevant mosquito species (specifically, *A. aegypti* produced by Oxford Insect Technologies) to produce offspring that are unable to reach maturity and further procreate (Finkel et al. 2018). A further approach includes the infection of mosquito-based viral vectors with gram-negative *Wolbachia* bacteria (specifically, *W. pipientis*), which can be engineered to foster immunity within the host mosquito (*A. aegypti*) (Murray et al. 2016). The latter case is currently being investigated by the Australian Commonwealth Scientific and Industrial Research Organisation (CSIRO), which is an independent Australian federal government agency that manages and oversees scientific research.

These and other applications of synthetic biology research are driven by a need for more efficacious, rapid, and cost-effective solutions to improve public health crises as with dengue fever, yet many have reflected upon the uncertainties and potential concerns associated with the sudden and broad-scale release of engineered organisms that are geared toward fostering crashes or changes in one or more species of a local ecology (Alphey and Beech 2012; Meghani and Kuzma 2018). Risk considerations in these cases are likely to be broad, including concerns related to secondary impacts of mosquito population crashes, unintentional mutation of engineered hosts, spread of engineered organisms beyond a release area and possibly across international borders, potential unintentional exposure to humans and other animals, and the highly unlikely yet concerning potential for horizontal gene transfer where genetic information from an engineered host is absorbed and adopted within a different and unintended species altogether (Oye et al. 2014; Schmidt and de Lorenzo 2016).

Synthetic biology is a field with explosive promise yet with broadly unresolved challenges of risk assessment and governance. Scholarship has begun to identify and characterize general concerns of risk, with research priorities forming across the life cycle of product development. The next step in synthetic biology governance is squarely focused on the use of more quantified assessment of various threats, such as horizontal gene transfer or disruption of local ecologies, in order to inform judgment regarding the efficacy, necessity, and safe use requirements of various products. Movement toward this quantified approach will not occur uniformly but instead will arise on a product-by-product basis as hazard and exposure data become more robust and reliable. Though no universal risk assessment process for all synthetic biology processes and products will likely to be implemented within any given country, it is likely that such assessment will be grounded in discussion of biosafety and biosecurity and review the potential novel risks or consequences that make such engineered organisms of particular concern to regulatory and policymaking audiences.

Disclaimer This chapter reflects the author's opinions alone, and may not be reflective of the positions held by his host institution.

References

Alphey, L., & Beech, C. (2012). Appropriate regulation of GM insects. *PLoS Neglected Tropical Diseases, 6*(1), e1496.

Arber, W. (2014). Horizontal gene transfer among bacteria and its role in biological evolution. *Life (Basel), 4*(2), 217–224.

Armstrong, R., Schmidt, M., & Bedau, M. (2012). Other developments in synthetic biology. In *Synthetic biology: Industrial and environmental applications* (pp. 145–156).

Barocchi, M. A., & Rappuoli, R. (2015). Delivering vaccines to the people who need them most. *Philosophical Transactions of the Royal Society of London B: Biological Sciences, 370*(1671), 20140150.

Bates, M. E., Grieger, K. D., Trump, B. D., Keisler, J. M., Plourde, K. J., & Linkov, I. (2015). Emerging technologies for environmental remediation: Integrating data and judgment. *Environmental Science & Technology, 50*(1), 349–358.

Berg, P. (2008). Meetings that changed the world: Asilomar 1975: DNA modification secured. *Nature, 455*(7211), 290.

Berg, P., Baltimore, D., Brenner, S., Roblin, R. O., & Singer, M. F. (1975). Summary statement of the Asilomar conference on recombinant DNA molecules. *Proceedings of the National Academy of Sciences, 72*(6), 1981–1984.

Boschetti, C., Carr, A., Crisp, A., Eyres, I., Wang-Koh, Y., Lubzens, E., et al. (2012). Biochemical diversification through foreign gene expression in bdelloid rotifers. *PLoS Genetics, 8*(11), e1003035.

Bugaj, L. J., & Schaffer, D. V. (2012). Bringing next-generation therapeutics to the clinic through synthetic biology. *Current Opinion in Chemical Biology, 16*(3), 355–361.

Bügl, H., Danner, J. P., Molinari, R. J., Mulligan, J. T., Park, H. O., Reichert, B., et al. (2007). DNA synthesis and biological security. *Nature Biotechnology, 25*(6), 627.

Burnett, L. C., Lunn, G., & Coico, R. (2009). Biosafety: Guidelines for working with pathogenic and infectious microorganisms. *Current Protocols in Microbiology, 13*(1), 1A. –1.

Callura, J. M., Dwyer, D. J., Isaacs, F. J., Cantor, C. R., & Collins, J. J. (2010). Tracking, tuning, and terminating microbial physiology using synthetic riboregulators. *Proceedings of the National Academy of Sciences, 107*(36), 15898–15903.

Cameron, D. E., Bashor, C. J., & Collins, J. J. (2014). A brief history of synthetic biology. *Nature Reviews Microbiology, 12*(5), 381.

Cardinale, S., & Arkin, A. P. (2012). Contextualizing context for synthetic biology–identifying causes of failure of synthetic biological systems. *Biotechnology Journal, 7*(7), 856–866.

Carter, S. R., Rodemeyer, M., Garfinkel, M. S., & Friedman, R. M. (2014). *Synthetic biology and the US biotechnology regulatory system: Challenges and options.* (No. DOE-JCVI-SC0004872). Rockville: J. Craig Venter Institute.

Cha, A. E. (2014). Companies rush to build 'biofactories' for medicines, flavorings and fuels. Washington Post. http://www. washingtonpost. com/national/healthscience/companies-rush-to-build-biofactoriesfor-medicines-flavorings-and-fuels/2013/10/24/f439dc3a-3032-11e3-8906-3daa2bcde110_story. html Accessed 31st of March.

Choffnes, E. R., Lemon, S. M., & Relman, D. A. (2006). A brave new world in the life sciences. *The Bulletin of the Atomic Scientists, 62*, 26–33.

Church, G. M. (2004). A synthetic biohazard non-proliferation proposal. http://arep.med.harvard.edu/SBP/Church_Biohazard04c.htm

Conferees SB2.0. (2006). Public draft of the declaration of the second international meeting on synthetic biology. http://hdl.handle.net/1721.1/32982

Cong, L., Ran, F. A., Cox, D., Lin, S., Barretto, R., Habib, N., et al. (2013). Multiplex genome engineering using CRISPR/Cas systems. *Science, 339*(6121), 819–823.

Cummings, C. L., & Kuzma, J. (2017). Societal risk evaluation scheme (SRES): Scenario-based multi-criteria evaluation of synthetic biology applications. *PLoS One, 12*(1), e0168564.

d'Espaux, L., Mendez-Perez, D., Li, R., & Keasling, J. D. (2015). Synthetic biology for microbial production of lipid-based biofuels. *Current Opinion in Chemical Biology, 29*, 58–65.

Dana, G. V., Kuiken, T., Rejeski, D., & Snow, A. A. (2012). Synthetic biology: Four steps to avoid a synthetic-biology disaster. *Nature, 483*(7387), 29–29.

Davison, J. (1999). Genetic exchange between bacteria in the environment. *Plasmid, 42*, 73–91.

Dormitzer, P. R., Suphaphiphat, P., Gibson, D. G., Wentworth, D. E., Stockwell, T. B., Algire, M. A., et al. (2013). Synthetic generation of influenza vaccine viruses for rapid response to pandemics. *Science Translational Medicine, 5*(185), 185ra68–185ra68.

Dröge, M., Pühler, A., & Selbitschka, W. (1998). Horizontal gene transfer as a biosafety issue: A natural phenomenon of public concern. *Journal of Biotechnology, 64*(1), 75–90.

Elowitz, M. B., & Leibler, S. (2000). A synthetic oscillatory network of transcriptional regulators. *Nature, 403*(6767), 335.

Esvelt, K. M., Smidler, A. L., Catteruccia, F., & Church, G. M. (2014). Emerging technology: Concerning RNA-guided gene drives for the alteration of wild populations. *eLife, 3*, e03401.

Fatehi, L., & Hall, R. F. (2015). Synthetic biology in the FDA realm: Toward productive oversight assessment. *Food and Drug Law Journal, 70*, 339.

Finkel, A. M., Trump, B. D., Bowman, D., & Maynard, A. (2018). A "solution-focused" comparative risk assessment of conventional and synthetic biology approaches to control mosquitoes carrying the dengue fever virus. *Environment Systems and Decisions, 38*(2), 177–197.

Frow, E. (2017). From "experiments of concern" to "groups of concern" constructing and containing citizens in synthetic biology. *Science, Technology, & Human Values*, 0162243917735382.

Gardner, T. S., Cantor, C. R., & Collins, J. J. (2000). Construction of a genetic toggle switch in Escherichia coli. *Nature, 403*(6767), 339.

Georgianna, D. R., & Mayfield, S. P. (2012). Exploiting diversity and synthetic biology for the production of algal biofuels. *Nature, 488*(7411), 329.

Gibson, D. G., Glass, J. I., Lartigue, C., Noskov, V. N., Chuang, R.-Y., Algire, M. A., Benders, G. A., Montague, M. G., Ma, L., et al. (2010). Creation of a bacterial cell controlled by a chemically synthesized genome. *Science, 329*, 52–56.

Guan, Z. J., Schmidt, M., Pei, L., Wei, W., & Ma, K. P. (2013). Biosafety considerations of synthetic biology in the international genetically engineered machine (iGEM) competition. *Bioscience, 63*(1), 25–34.

Gupta, A., & Falkner, R. (2006). The influence of the Cartagena protocol on biosafety: Comparing Mexico, China and South Africa. *Global Environmental Politics, 6*(4), 23–55.

Hammond, A., Galizi, R., Kyrou, K., Simoni, A., Siniscalchi, C., Katsanos, D., et al. (2016). A CRISPR-Cas9 gene drive system targeting female reproduction in the malaria mosquito vector Anopheles gambiae. *Nature Biotechnology, 34*(1), 78.

Harris, A. F., McKemey, A. R., Nimmo, D., Curtis, Z., Black, I., Morgan, S. A., et al. (2012). Successful suppression of a field mosquito population by sustained release of engineered male mosquitoes. *Nature Biotechnology, 30*(9), 828.

Hayden, E. (2014). Synthetic-biology firms shift focus. *Nature News, 505*(7485), 598.

Howard, J., Murashov, V., & Schulte, P. (2017). Synthetic biology and occupational risk. *Journal of Occupational and Environmental Hygiene, 14*(3), 224–236.

Joung, J. K., & Sander, J. D. (2013). TALENs: A widely applicable technology for targeted genome editing. *Nature Reviews Molecular Cell Biology, 14*(1), 49.

Kelle, A. (2009). Synthetic biology and biosecurity: From low levels of awareness to a comprehensive strategy. *EMBO Reports, 10*(1S), S23–S27.

Khalil, A. S., & Collins, J. J. (2010). Synthetic biology: Applications come of age. *Nature Reviews Genetics, 11*(5), 367.

Kupferschmidt, K. (2018). EU verdict on CRISPR crops dismays scientists.

Kuzma, J., & Tanji, T. (2010). Unpackaging synthetic biology: Identification of oversight policy problems and options. *Regulation & Governance, 4*(1), 92–112.

Kuzma, J., Gould, F., Brown, Z., Collins, J., Delborne, J., Frow, E., et al. (2018). A roadmap for gene drives: Using institutional analysis and development to frame research needs and governance in a systems context. *Journal of Responsible Innovation, 5*(Suppl 1), S13–S39.

Linkov, I., Trump, B. D., Wender, B. A., Seager, T. P., Kennedy, A. J., & Keisler, J. M. (2017). Integrate life-cycle assessment and risk analysis results, not methods. *Nature Nanotechnology, 12*(8), 740.

Linkov, I., Trump, B. D., Anklam, E., Berube, D., Boisseasu, P., Cummings, C., et al. (2018). Comparative, collaborative, and integrative risk governance for emerging technologies. *Environment Systems and Decisions, 38*(2), 170–176.

Lorenz, M. G., & Wackernagel, W. (1994). Bacterial gene transfer by natural genetic transformation in the environment. *Microbiological Reviews, 58*, 563–602.

Ludlow, K., Bowman, D. M., Gatof, J., & Bennett, M. G. (2015). Regulating emerging and future technologies in the present. *NanoEthics, 9*(2), 151–163.

Mandel, G. N., & Marchant, G. E. (2014). The living regulatory challenges of synthetic biology. *Iowa L. Rev., 100*, 155.

Marchant, G. E., & Abbott, K. W. (2012). International harmonization of nanotechnology governance through soft law approaches. *Nanotech. L. & Bus., 9*, 393.

Marris, C. (2015). The construction of imaginaries of the public as a threat to synthetic biology. *Science as Culture, 24*(1), 83–98.

Maurer, S. M., & Zoloth, L. (2007). Synthesizing biosecurity. *Bulletin of the Atomic Scientists, 63*(6), 16–18.

Maynard, A. D. (2007). Nanotechnology: The next big thing, or much ado about nothing? *Annals of Occupational Hygiene, 51*(1), 1–12.

Meghani, Z., & Kuzma, J. (2018). Regulating animals with gene drive systems: Lessons from the regulatory assessment of a genetically engineered mosquito. *Journal of Responsible Innovation, 5*(Suppl 1), S203–S222.

Moe-Behrens, G. H., Davis, R., & Haynes, K. A. (2013). Preparing synthetic biology for the world. *Frontiers in microbiology, 4*, 5.

Mukunda, G., Oye, K. A., & Mohr, S. C. (2009). What rough beast? Synthetic biology, uncertainty, and the future of biosecurity. *Politics and the Life Sciences, 28*(2), 2–26.

Mulvey, M. R., & Simor, A. E. (2009). Antimicrobial resistance in hospitals: How concerned should we be? *Canadian Medical Association Journal, 180*, 408–415.

Murray, J. V., Jansen, C. C., & De Barro, P. (2016). Risk associated with the release of Wolbachia-infected Aedes aegypti mosquitoes into the environment in an effort to control dengue. *Frontiers in Public Health, 4*, 43.

Myhr, A. I., & Traavik, T. (2002). The precautionary principle: Scientific uncertainty and omitted research in the context of GMO use and release. *Journal of Agricultural and Environmental Ethics, 15*(1), 73–86.

Nakano, T., Eckford, A. W., & Haraguchi, T. (2013). *Molecular communication*. Cambridge University Press.

National Research Council. (2004). *Biotechnology research in an age of terrorism*. Washington, D.C.: The National Academies Press.

National Research Council. (2006). *Globalization, biosecurity, and the future of the life sciences*. Washington, D.C.: The National Academies Press.

Neumann, H., & Neumann-Staubitz, P. (2010). Synthetic biology approaches in drug discovery and pharmaceutical biotechnology. *Applied Microbiology and Biotechnology, 87*(1), 75–86.

Nielsen, K. M., & Townsend, J. P. (2004). Monitoring and modeling horizontal gene transfer. *Nature biotechnology, 22*(9), 1110.

Nielsen, K. M., Johnsen, P. J., Bensasson, D., & Daffonchio, D. (2007). Release and persistence of extracellular DNA in the environment. *Environmental Biosafety Research, 6*, 37–53.

Oye, K. A. (2012). Proactive and adaptive governance of emerging risks: the case of DNA synthesis and synthetic biology. International Risk Governance Council, Geneva.

Oye, K. A., Esvelt, K., Appleton, E., Catteruccia, F., Church, G., Kuiken, T., et al. (2014). Regulating gene drives. *Science, 345*(6197), 626–628.

Oye, K. A., O'Leary, M., & Riley, M. F. (2017). Revisit NIH biosafety guidelines. *Science, 357*, 627.

Paddon, C. J., & Keasling, J. D. (2014). Semi-synthetic artemisinin: A model for the use of synthetic biology in pharmaceutical development. *Nature Reviews Microbiology, 12*(5), 355.

Pauwels, E. (2013). Public understanding of synthetic biology. *Bioscience, 63*(2), 79–89.

PCSBI (President's Commission on the Study of Bioethical Issues). (2010). *New directions: The ethics of synthetic biology and emerging technologies*. Washington, D.C.: President's Commission on the Study of Bioethical Issues.

Perkins, D., & Nordmann, B. (2012). Emerging technologies: Biosecurity and consequence management implications. In *Technological innovations in sensing and detection of chemical, biological, radiological, nuclear threats and ecological terrorism* (pp. 25–33). Dordrecht: Springer.

Peplow, M. (2016). Synthetic biology's first malaria drug meets market resistance. *Nature News, 530*(7591), 389.

Piaggio, A. J., Segelbacher, G., Seddon, P. J., Alphey, L., Bennett, E. L., Carlson, R. H., et al. (2017). Is it time for synthetic biodiversity conservation? *Trends in Ecology & Evolution, 32*(2), 97–107.

Pruden, A., Arabi, M., & Storteboom, H. N. (2012). Correlation between upstream human activities and riverine antibiotic resistance genes. *Environmental Science & Technology, 46*, 11541–11549.

Rabinow, P., & Bennett, G. (2012). *Designing human practices: An experiment with synthetic biology*. University of Chicago Press.

Rojahn, S. (2013). Synthetic biology could speed flu vaccine production. MIT Technology Review.

Rossi, F., Rizzotti, L., Felis, G. E., & Torriani, S. (2014). Horizontal gene transfer among microorganisms in food: Current knowledge and future perspectives. *Food Microbiology, 42*, 232–243.

Schmidt, M. (2008). Diffusion of synthetic biology: A challenge to biosafety. *Systems and Synthetic Biology, 2*(1–2), 1–6.

Schmidt, M. (Ed.). (2012). *Synthetic biology: Industrial and environmental applications*. Weinheim: Wiley.

Schmidt, M., & de Lorenzo, V. (2012). Synthetic constructs in/for the environment: Managing the interplay between natural and engineered biology. *FEBS Letters, 586*(15), 2199–2206.

Schmidt, M., & de Lorenzo, V. (2016). Synthetic bugs on the loose: containment options for deeply engineered (micro) organisms. *Current opinion in biotechnology, 38*, 90–96.

Schmidt, M., Torgersen, H., Ganguli-Mitra, A., Kelle, A., Deplazes, A., & Biller-Andorno, N. (2008). SYNBIOSAFE e-conference: Online community discussion on the societal aspects of synthetic biology. *Systems and Synthetic Biology, 2*(1–2), 7–17.

Seager, T. P., Trump, B. D., Poinsatte-Jones, K., & Linkov, I. (2017). Why life cycle assessment does not work for synthetic biology. *Environmental Science and Technology, 51*(11).

Seyfried, G., Pei, L., & Schmidt, M. (2014). European do it yourself (DIY) biology: Beyond the hope, hype and horror. *BioEssays, 36*(6), 548–551.

Tabor, J. J., Salis, H. M., Simpson, Z. B., Chevalier, A. A., Levskaya, A., Marcotte, E. M., et al. (2009). A synthetic genetic edge detection program. *Cell, 137*(7), 1272–1281.

Thomas, C. M., & Nielsen, K. M. (2005). Mechanisms of, and barriers to, horizontal gene transfer between bacteria. *Nature Reviews. Microbiology, 3*, 711–721.

Townsend, J. P., Bøhn, T., & Nielsen, K. M. (2012). Assessing the probability of detection of horizontal gene transfer events in bacterial populations. *Frontiers in Microbiology, 3*, 27.

Trump, B. D., Cegan, J. C., Wells, E., Keisler, J., & Linkov, I. (2018a). A critical juncture for synthetic biology: Lessons from nanotechnology could inform public discourse and further development of synthetic biology. *EMBO Reports, 19*(7), e46153.

Trump, B. D., Foran, C., Rycroft, T., Wood, M. D., Bandolin, N., Cains, M., et al. (2018b). Development of community of practice to support quantitative risk assessment for synthetic biology products: Contaminant bioremediation and invasive carp control as cases. *Environment Systems and Decisions, 38*(4), 517–527.

Trump, B., Cummings, C., Kuzma, J., & Linkov, I. (2018c). A decision analytic model to guide early-stage government regulatory action: Applications for synthetic biology. *Regulation & Governance, 12*(1), 88–100.

Trump, B. D., Cegan, J., Wells, E., Poinsatte-Jones, K., Rycroft, T., Warner, C., et al. (2019). Co-evolution of physical and social sciences in synthetic biology. *Critical Reviews in Biotechnology*, 1–15.

Tucker, J. B., & Zilinskas, R. A. (2006). The promise and perils of synthetic biology. *The New Atlantis, 12*, 25–45.

Urnov, F. D., Rebar, E. J., Holmes, M. C., Zhang, H. S., & Gregory, P. D. (2010). Genome editing with engineered zinc finger nucleases. *Nature Reviews Genetics, 11*(9), 636.

van Oppen, M. J., Oliver, J. K., Putnam, H. M., & Gates, R. D. (2015). Building coral reef resilience through assisted evolution. *Proceedings of the National Academy of Sciences, 112*(8), 2307–2313.

Vogel, K. M. (2014). Revolution versus evolution: Understanding scientific and technological diffusion in synthetic biology and their implications for biosecurity policies. *BioSocieties, 9*(4), 365–392.

Weber, W., & Fussenegger, M. (2012). Emerging biomedical applications of synthetic biology. *Nature Reviews Genetics, 13*(1), 21.

White, K., & Vemulpad, S. (2015). Synthetic biology and the responsible conduct of research. *Macquarie LJ, 15*, 59.

WHO. (2004). *Laboratory biosafety manual* (3rd ed.). Geneva: World Health Organization.

Wright, O., Stan, G. B., & Ellis, T. (2013). Building-in biosafety for synthetic biology. *Microbiology, 159*(Pt 7), 1221–1235.

Estimating and Predicting Exposure to Products from Emerging Technologies

Daniel A. Vallero

Introduction

Risk is a common metric for public health and environmental decision-making. Scientifically, credible risk assessments underpin decisions regarding the potential safety of emerging technologies (National Academy of Sciences National Research Council 1983). Furthermore, individuals who may be exposed to the products of these technologies must understand the exposure and decide whether the potential risks are acceptable. Most environmental exposure decisions have low probability of substantial risk, for example, wastewater treatment plant design, construction of barriers to prevent migration of pollutants to drinking water wells, or selection of air pollution control equipment for particulate matter (PM). The difference between these decisions and those involving emerging technologies is that the latter have much greater uncertainty and must often rely on comparisons to conventional technologies.

The more advanced the technology, the more uncertain the exposure and risk will be. In particular, emerging technological exposures involve the potential for low-probability, high consequence ("rare") events, which present special challenges to risk communications (Solomon and Vallero 2016). Scientific and engineering rigor are essential for rare events, as they are for any risk assessment scenario. Certainly, managing the risks presented by rare events requires many of the same fundamental communication elements of any credible risk-based decision analysis. Given the diversity of stakeholders and unconventional aspects of most rare events, a greater

D. A. Vallero (✉)
Center for Computational Toxicology and Exposure, U.S. Environmental Protection Agency, Research Triangle Park, NC, USA
e-mail: vallero.daniel@epa.gov

© Springer Nature Switzerland AG 2020   107
B. D. Trump et al. (eds.), *Synthetic Biology 2020: Frontiers in Risk Analysis and Governance*, Risk, Systems and Decisions, https://doi.org/10.1007/978-3-030-27264-7_6

understanding and application of numerous other factors are needed, especially psychosocial and ethical factors.

One means of determining whether a substance is handled properly is to determine if the actions lead to exceedances of a standard or limit, which is based on scientific evidence that the exposure and risk introduced by these actions are acceptable. These standards and limits are often based on the amount of product that is released, for example, the concentration in air or water that escapes during a process. However, exposure and risk information for emerging contaminants are often lacking or deficient, so other metrics are needed, for example, the expectation that an action is accompanied by measures to ensure that the actions' risk will be "as low as reasonably possible" (ALARP) for the potential to produce impurities, for example, genotoxic impurities (GTI) (Callis et al. 2010; Teasdale et al. 2013). In engineering and medicine, the size of safety factors increases indirectly with certainty (Arnaldi and Muratorio 2013; Falkner and Jaspers 2012; Finkel 1990; Kodell 2005; McNamara et al. 2014; Roca et al. 2017; Shepherd 2009). This measure of risk, then, is an expression of operational success or failure. Too much risk means the governance process has failed society. As mentioned, acceptable risk is defined by the standards and specifications, often developed by governmental or other certifying authorities. However, acceptable exposure and risk cannot be estimated solely from physical and biological information but must also factor in cultural, social, and communications information, for example, what is a person doing greatly affects the extent and intensity of exposure (Covello 2003; Cummings and Kuzma 2017; Kahan et al. 2009; Kuzma and Tanji 2010; Slovic 1987).

Within the environmental and public health communities, definitions of "acceptable risk" vary widely. The conventional metrics are incorporated into health codes and regulations, zoning and building codes and regulations, design principles, canons of professional engineering and medical practice, national standard-setting bodies, and standards promulgated by international agencies (e.g., ISO or the International Organization for Standardization). In the United States, for example, standards can come from a federal agency, such as hazardous waste landfill guidelines of the US Environmental Protection Agency, or material specifications for equipment, such as those of the National Institute of Standards and Technology (NIST). Unfortunately, these metrics are often absent or inappropriate for new technologies which vary from their conventional analogs. This begs the question as to whether existing disposal guidelines are sufficiently protective to reduce exposures in the myriad scenarios likely to arise when the new technology moves from research to application. Often, emerging technologies follow standards articulated by private groups and associations, but which may be so focused on the utility and other benefits of the technology that potential exposure and risk receive comparatively less rigorous and inadequate emphasis (Vallero 2010a).

Background

Risk is generally understood to be the likelihood that an unwelcome event will occur. Risk assessment is the scientific investigation into the factors that lead to a risk. An assessment may be retrospective, that is, to see what damage has occurred, or prospective, that is, to predict risk posed from reasonable present and future risk scenarios. Much of synthetic biological risk assessment is the latter. Risk management follows risk assessment. The dispassionate and objective scientific findings must underpin the decisions needed to reduce adverse outcomes. For example, an assessment may indicate that a synthetic organism poses a risk to human health if it were to escape and reach water supplies, whereas risk management would include the design and installation of building containment structures around a laboratory or test facility.

Articulating the hazard, that is, the physical, chemical, or biological agent of harm, is matched against the receptor's contact with that hazard, that is, exposure. The types of receptors range in scale and complexity, for example, the exposed receptor may be:

- An individual organism, for example, a human or other species
- A subpopulation, for example, asthmatic children or endangered plant species in a habitat
- An entire population, for example, all persons in a city or nation or the world
- A macro-system, for example, a forest ecosystem

Hazard is an inherent trait. Thus, the hazard may occur before a waste is generated, such as a component of a manufacturing process. For example, if 1,1,1-trichloroethane (TCE) is used as a solvent in a chemical processing plant, it may be hazardous to the workers because it is carcinogenic. It may also be hazardous if it finds its way to a landfill (in drums or in contaminated sawdust after a cleanup).

The second component of the risk assessment is the potential of exposure to the hazard. Most of the exposure science literature to date has addressed chemical hazards, but given the focus of this book, this chapter must also address exposure to biological agents, especially genetically modified organisms (GMOs) and products of synthetic biology. Organisms engineered from synthetic biology may contain genes/genomes that are derived de novo, possibly without naturally occurring homologs. Any biological agent, that is, natural, genetically modified, or synthetic, may have inherent properties that render it hazardous, for example, production of exotoxins or infection of higher organisms. The uncertainties are further increased when chemical and biological hazards are combined, for example, the use of solvents and biological materials in the synthesis phases.

In the previous TCE/biological agent example, people can come into contact with the solvent in occupational settings and the organism during environmental (e.g., escape) and use (e.g., drinking water) scenarios. Thus, the exposure to TCE varies by activities (high for workers who use it, less for workers who may not work with TCE, but are nearby and breathe the vapors, and even less for other workers). The exposure to the biological agent is zero if it is completely contained and potentially expansive if not. Also, worker exposure is commonly based on a 5-workday exposure (e.g., 8 or 10 hours), whereas environmental exposures, especially for chronic diseases like cancer, are based on lifetime, 24-hour per day exposures. Thus, environmental regulations are often more stringent than occupational regulations when aimed at reducing exposure to a substance.

Risk assessment requires a sound physical, chemical, and biological characterization of the hazard, a consideration of changes to the agent in time and space, and how they may act synergistically or antagonistically with abiotic and biotic components of the system to which they are introduced. To assess a given scenario, the severity of the effect and the likelihood that it will occur in that scenario are calculated. This combination of the hazard particular to that scenario constitutes the risk.

The relationship between the severity and probability of a risk follows a general equation (Doblhoff-Dier et al. 2000):

$$R = f(S, P) \tag{1}$$

where risk (R) is a function (f) of the severity (S) and the probability (P) of harm. The risk equation can be simplified to be a product of severity and probability:

$$R = S \times P \tag{2}$$

The traditional chemical risk assessment paradigm (see Fig. 1) is generally a step-wise process. It begins with the identification of a hazard, which summarizes an agent's physicochemical properties and routes and patterns of exposure and reviews toxic effects. The tools for hazard identification take into account the chemical structures that are associated with toxicity, metabolic and pharmacokinetic properties, short-term animal and cell tests, long-term animal (in vivo) testing, and human studies (e.g., epidemiology, such as longitudinal and case-control studies). These comprise the core components of hazard identification; however, additional hazard identification methods have emerged that provide improved reliability of characterization and prediction.

Characterizing the inherent properties of an individual constituent used in a process is the first step in risk assessment. A number of tools have emerged to assist in this characterization. Risk assessors now can apply biomarkers of genetic damage (i.e., toxicogenomics) for more immediate assessments, as well as improved structure-activity relationships (SAR), which have incrementally been quantified in terms of stereochemistry and other chemical descriptions, that is, using quantitative structure-activity relationships (QSARs) and computational chemistry. There are fewer tools available for biological agents, but incorporating quantitative microbial risk assessment into a life cycle analysis (LCA) is promising (Harder et al. 2015). Health-effects research has mainly focused on early indicators of outcomes, making it possible to shorten the time between exposure and observation of an adverse outcome (National Academy of Science National Research Council 2002).

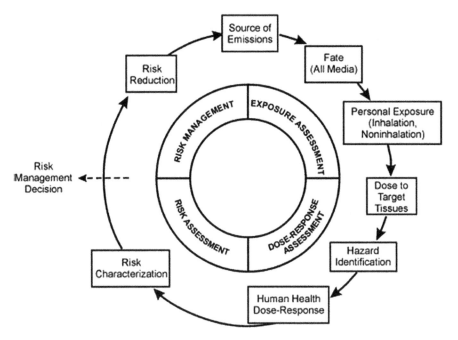

Fig. 1 Risk assessment and management paradigm as employed by environmental agencies in the United States. The inner circle includes the steps recommended by the National Research Council. The outer circle indicates the research and assessment activities that are currently used by regulatory agencies to meet these required steps. (Source: National Academy of Sciences National Research Council 1983. NRC (1983))

Microbes

Emerging technologies often generate nonchemical hazards. Notably, biological and infectious wastes present hazards from biological agents that differ from those posed by chemical-laden wastes. Of course, biological agents range from beneficial to extremely dangerous. The risks from microbes can be categorized. For genetically modified organisms, the categories are (Doblhoff-Dier et al. 2000):

1. Risk class 1. No adverse effect or very unlikely to produce an adverse effect. Organisms in this class are considered to be safe.
2. Risk class 2. Adverse effects are possible but are unlikely to represent a serious hazard with respect to the value to be protected. Local adverse effects are possible, which can either revert spontaneously (e.g., owing to environmental elasticity and resilience) or be controlled by available treatment or preventive measures. Spread beyond the application area is highly unlikely.
3. Risk class 3. Serious adverse local effects are likely with respect to the value to be protected, but spread beyond the area of application is unlikely. Treatment and/or preventive measures are available.

4. Risk class 4. Serious adverse effects are to be expected with respect to the value to be protected, both locally and outside the area of application. No treatment or preventive measures are available.

Future products of biotechnology also vary by novelty and complexity, that is, extent and method of genetic modification (e.g., transgenic or metagenome engineering), scale of impact, and comparators. As such, the National Academies of Sciences, Engineering, and Medicine has classified these products accordingly (National Academies of Sciences and Medicine 2017):

A. Organisms domesticated by transgenic/recombinant DNA, engineered along one or only a few gene pathways, and which have ample comparators
B. Undomesticated and domesticated organisms by transgenesis involving new genome engineering along multiple pathways and which have few or no comparators
C. Many candidate organisms generated by genome engineering and gene drives via genome refactoring, recoding, and cell-free synthesis and which have few or no comparators
D. Synthetic communities of microbes and individual synthetic, multicellular plants and animals generated by metagenome and microbiome engineering in a population or ecosystem and which have no or merely ambiguous comparators.

These classes indicate that even the safest microbes carry some risk, with uncertainty increasing with extent of synthesis. With more uncertainty about an organism, one cannot assume it to be safe, especially for synthetic protocells and larger organisms about which little is known, that is, Novelty Class D. The risks may be direct or indirect. An example of a direct risk would be the likelihood of a person contracting a pathogenic disease, whereas an indirect risk example is a change induced by the release of organism into an environment where there are no natural predators, allowing them to displace natural organisms. Thus, risk scenarios include not only the effects resulting from the intended purpose of the environmental application but also downstream and side effects that are not part of the desired purpose. For example, the European Union (EU) requires that a synthetic biology risk assessment define the "exposure chain," that is, the events leading to the adverse health or environmental outcome (Scientific Committee on Health and Environmental Risks 2015).

As mentioned, the large uncertainties associated with emerging technologies and synthetic biology call for conservative science and treating the potential hazards and exposure as risk class 4. An impact could be widespread and irreversible. The nature of emerging entities is that we cannot know with even a modicum of confidence the extent and effectiveness of any existing treatment or preventive measure. Risk can be extrapolated from available knowledge of chemical or biological agents with similar characteristics or to yet untested but similar environmental conditions (e.g., a field study's results in one type of field extrapolated to a different agricultural or environmental remediation setting). In chemical hazard identification, this is accomplished by structural activity relationships.

Complex Mixtures

Organisms are seldom exposed to a single hazard but are rather continuously exposed to complex mixtures. Until recently, toxicologists have considered a complex mixture to be a combination of two or more chemicals (Carpenter et al. 2002). From an exposure perspective, a mixture is actually a co-exposure. Humans and ecosystems are exposed to an array of compounds simultaneously (Kortenkamp et al. 2009). A key question is how do individual constituents' physical and chemical properties affect those of other chemical and biological constituents used during biological synthesis? The additive, synergistic, and antagonistic effects must be considered. Until relatively recently, toxicologists studied mixtures in a stepwise manner, adding substances one at a time to ascertain the response of an organism with each iteration (Feron and Groten 2002). Thus, toxicologists and exposure scientists have begun to look at multicomponent mixtures from a systems perspective.

Synthetic biology further complicates the concept of "mixtures," that is, the complex series of steps in synthetic biology, as mentioned, can result in exposures to mixtures that may contain both biological and chemical agents.

Exposure Probability

Following the hazard identification process for a chemical or a natural or synthetic microbe according to its inherent properties, the environmental conditions are examined to characterize different responses to doses in different populations. Both the hazard identification and dose-response information are based on research that is used in the risk analysis. For microbes, the highest score for any one effect determines the overall risk class for environmental application. In addition, the exposure estimate is the sum of all the exposures, that is, the evaluation of the likelihood of the occurrence of each potentially adverse outcome (Scientific Committee on Health and Environmental Risks 2015).

The factors leading to the exposure probability include the release, replication, dispersion, and ultimate contact with the microbe and other contaminants produced during and after the synthesis. The release may be intentional, for example, use of the product during medical, veterinary, agricultural, or consumer activities, and unintentional, for example, during laboratory studies and manufacturing.

Managing exposures to biological wastes (and any waste for that matter) must consider protecting the most vulnerable members of society, especially pregnant women and their yet-to-be-born infants, neonates, and immunocompromised subpopulations. Also, the exposure protections vary by threat. For example, adolescents may be particularly vulnerable to hormonally active agents, including many pesticides.

In the United States, ecological exposure and risk assessment paradigms have differed from those applied to human health risk. The ecological risk assessment framework (see Fig. 2) is based mainly on characterizing exposure and ecological effects. Both exposure and effects are considered during problem formulation (US Environmental Protection Agency 1992).

Interestingly, the ecological risk framework is driving current thinking in human risk assessment. The process shown in the inner circle of Fig. 1 does not target the technical analysis of risk so much as it provides coherence and connections between risk assessment and risk management. When scientific assessment and management are carried out simultaneously, decision-making could be influenced by the need for immediacy, convenience, or other political and financial motivations. The advantage of an arms-length, bifurcated approach is that decisions and management of risks are based on a rational and scientifically credible assessment (Loehr et al. 1992; Ruckelshaus 1983).

In both human health and ecological assessments, the final step is "characterization," that is, integrating the "quantitative and qualitative elements of risk analysis and of the scientific uncertainties in it" (National Academy of Sciences National Research Council 2009). The problem formulation step in the ecological framework has the advantage of providing an analytic-deliberative process early on, since it combines sound science with input from various stakeholders inside and outside of the scientific community.

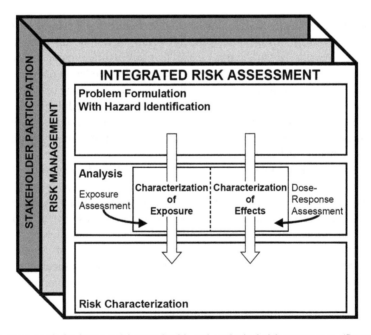

Fig. 2 Framework for integrated human health and ecological risk assessment. (Sources: US Environmental Protection Agency 1992; World Health Organization 2000)

The ecological risk framework calls for the characterization of ecological effects instead of hazard identification used in human health risk assessments. This is because the term "hazard" has been used in chemical risk assessments to connote either intrinsic effects of a stressor or a margin of safety by comparing a health effect with an estimate of exposure concentration. Thus, the term becomes ambiguous when applied to nonchemical hazards, such as those encountered in biological systems. Specific scientific investigations will often be needed to augment existing assessment methods, especially when adverse outcomes may be substantial and small changes may lead to very different functions and behaviors from unknown and insufficiently known chemicals or microbes. For example, a genetically modified microbe (GMM) may have only been used in highly controlled experiments with little or no information about how it would behave inside another organism. Often, the proponents of a product will conduct substantial research on the benefits and operational aspects of the chemical constituents, but the regulatory agencies and the public may call for more and better information about unintended and yet-to-be-understood consequences and side effects (Doblhoff-Dier et al. 2000).

Even when a GMM is well studied, there often remain large knowledge gaps when trying to estimate environmental impacts. The bacterium *Bacillus thuringiensis,* for instance, has been applied for several decades as a biological alternative to some chemical pesticides. It has been quite effective when sprayed onto cornfields to eliminate the European corn borer. The current state of knowledge indicates that this bacterium is not specific in the organisms that it targets. What if in the process, *B. thuringiensis* (Bt) also kills honeybees? Obviously, this would be a side effect that would not be tolerable from either an ecological or agricultural perspective (the same corn crop being protected from the borer needs the pollinators). Furthermore, physical, chemical, and biological factors can influence these effects, for example, type of application of Bt can influence the amount of drift toward nontarget species. Downstream effects can be even more difficult to predict than side effects, since they not only occur within variable space but also in variable time regimes. For example, exposure potential can arise from both the application method and from the buildup of toxic materials and gene flow following the use of a GMM.

Dosimetry and Exposure Calculation

The typical routes of exposure are by inhalation or ingestion or through the skin (Dionisio et al. 2015; Jones-Otazo et al. 2005; Weschler and Nazaroff 2012). Humans and other organisms can also come into contact with synthetic organisms or other substances generated during the life cycle of an emerging technology, for example, nanomaterials or chemicals through various exposure routes simultaneously. Inhalation is the most likely route for human exposure when a substance reaches the air, which can occur during manufacturing processes, consumer use, and other scenarios involving synthetically derived substances. Airborne exposures do not always involve the respiratory system, such as nasal exposures where the substance passes from the nose to the brain or when airborne contaminants penetrate the skin via the dermal route (Genter et al. 2015; Maheshwari et al. 2019; Schiffman et al. 1995). Likewise, waterborne substances may include routes other than ingestion, for example, inhalation of volatile substances during showering and cooking (Northcross et al. 2015; Zhang et al. 2018).

Emerge technologies may also generate aerosols, which may be living, for example, a modified cell or GMM, or nonliving, for example, an engineered nanoparticle (NP). Numerous synthetic biology processes can produce aerosols (Scientific Committee on Health and Environmental Risks 2015). In addition to atmospheric concentrations, exposure calculations must also account for the scale and extent of the activity, the concentrations of the substance of concern in reactors and other vessels, the production volume (cultures, supernatants, etc.), the industrial use or other types of setting, and the kind of biological processes used during synthesis (e.g., in vivo or in vitro).

Identifying potential hazards is the first step in risk assessment. Sometimes the physicochemical structures of a substance can provide clues of potential hazards. If an unknown compound is similar to better known substances, statistical and mathematical modeling based structural activity relationships can be used as a first step in screening for hazard and exposure (Lagunin et al. 2011; Liu and Gramatica 2007; Roy and Mitra 2012; U.S. Environmental Protection Agency 2015; Vilar et al. 2008). For example, partitioning coefficients, such as the octanol-water coefficient (K_{ow}) of known compounds, can be used to estimate and model the chemical and biological behavior of lesser known substances (Kimura et al. 1996). However, the greater the divergence from the known to the unknown, the less reliable such chemometric methods, for example, QSARs, become. For synthetic biology, there are little or no reliable data and information available for even crude QSARs. This is also true for other emerging technologies, for example, genetic engineering and nanomaterials, but the databases are much larger and more reliable (Tropsha et al. 2017; Winkler et al. 2015). Often, preliminary or screening toxicity data may be available for a substance, but potential uses and exposures are almost completely uncertain. Regulatory programs are beginning to identify and categorize substances according to potential toxicity *and* potential exposure. Notable examples include REACH in Europe (Kortelainen 2015), exposure-based prioritization in the United States (Egeghy et al. 2011), and rapid exposure and dosimetry in North America (Barber et al. 2017; Dionisio et al. 2015; Egeghy et al. 2016; Wambaugh et al. 2014). Unfortunately, these almost exclusively address chemicals.

Most of the exposure knowledgebase consists of chemical compounds, aerosols, and pathogenic microbes. To demonstrate the steps involved in human exposure, this chapter will focus on chemical exposure routes and pathways generally and aerosols specifically. However, it is important to keep in mind that synbio and other emerging technologies may produce substances and organisms that do not follow every concept discussed here. We will also focus on the inhalation route and air pathway.

Much can happen internally after substances are absorbed. The mass at the interface between the organism and the environment, for example, breathing zone, is merely the potential dose. Applied does occurs once the chemical crosses the interface. The dose experienced by different organs and tissues within the body is the focus of toxicokinetics (TK) studies. TK models have been developed to predict the chemical's internal fate, which begins with absorption, followed by distribution, metabolism and elimination (ADME). Therefore, the uptake into an organism begins with the absorbed dose. Exposure is completed at biologically effective or target dose, that is, when the aerosol or its metabolic products reach the organ/tissue that is the site of effect/outcome, for example, the liver for a hepatotoxin and brain for a neurotoxin. The amount of the parent compound and its metabolites remaining in the organs and tissues is known as the chemical body burden. Any damage that results from this exposure falls in the realm of effects. For example, an exposure biomarker would show that the xenobiotic has hit the target (e.g., release of a liver enzyme), whereas an effects biomarker would show liver damage (perhaps a different liver enzyme, or the same enzyme, but at higher concentrations to indicate hepatotoxicity).

Aerosol size and shape determine the rate and extent of exposure. The differences between the dosimetry of nanoscale and bulk materials are not well understood. Measuring the hazard of a chemical substance is difficult in part because the applied dose will not be the same as the absorbed and biologically effective dose, given the losses to container wall, dissolution, aggregation and other mechanisms that may be much more important for nanoscale materials, but also much more difficult to quantify at the nanoscale (Ivask et al. 2018; Lead et al. 2018; Sekine et al. 2015).

The human respiratory tract can be divided into three regions, that is, the extrathoracic, tracheobronchial, and alveolar (see Fig. 3). The extrathoracic region consists of airways within the head, that is, nasal and oral passages, through the larynx and represents the areas through which inhaled air first passes. From there, the air enters the tracheobronchial region at the trachea. From the level of the trachea, the conducting airways then undergo dichotomous branching for a number of generations. The terminal bronchiole is the most peripheral of the distal conducting airways and leads to alveolar region where gas exchange occurs in a complex of respiratory bronchioles, alveolar ducts, alveolar sacs, and alveoli. Except for the trachea and parts of the mainstem bronchi, the airways surrounded by parenchymal tissue are composed mainly of the alveolated structures and blood and lymphatic vessels. The respiratory tract regions are made up of various cell types, and the distribution of cells that line the airway surfaces has different anatomical qualities in the three regions (EPA 2004).

The first exposure characterization of a particle is its size and shape, because the way that a particle of any size behaves in the lung depends on the aerodynamic characteristics of particles in flow streams. In contrast, the major factor for gases is the solubility of the gaseous molecules in the linings of the different regions of the respiratory system (see Fig. 3). However, given the size of nanoparticles, they may behave at times as an aerosol and other times as a gas.

The deposition of particles in different regions of the respiratory system depends on their size. The nasal openings permit very large dust particles to enter the nasal region, along with much finer airborne particulate matter. Air pollution scientists and engineers consider particulate matter (PM) the same size as engineered nanoparticles. PMs with aerodynamic diameters of less than 100 nm, that is, the upper size

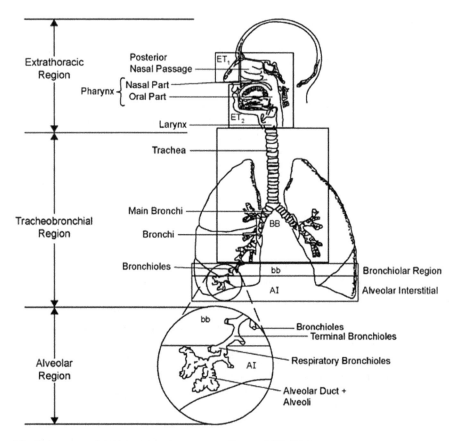

Fig. 3 Anatomy of the human respiratory tract. (Source: EPA 2004)

range of nanoparticles, are known as ultrafine PM. For example, drug delivery research applying synthetic biology may involve the engineering of nanoparticles, as well as the unintentional release of variously sized PMs, ranging from ultrafine aerosols to coarse particles, for example, those with aerodynamic diameters larger than 2.5 microns, that is, $PM_{2.5}$.

Coarse particles deposit in the nasal region by impaction on the hairs of the nose or at the bends of the nasal passages (Fig. 4). Smaller particles pass through the nasal region and are deposited in the tracheobronchial and pulmonary regions. Particles are removed from the airflow by impacts with the walls of the bronchi when they are unable to follow the gaseous streamline flow through subsequent bifurcations of the bronchial tree. As the airflow decreases near the terminal bronchi, the smallest particles are removed by Brownian motion, which pushes them to the alveolar membrane (Vallero 2014).

The aerodynamic properties of particles are determined not only by size but also by their shape and density. The behavior of a chain type or fiber may also be dependent on its orientation to the direction of flow. Thus, another variable introduced by

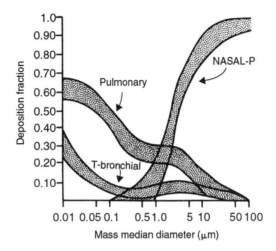

Fig. 4 Particle deposition as a function of particle diameter in various regions of the lung, from nanoparticles (10–100 nm) to coarse particles (>10 μm). The nasopharyngeal region consists of the nose and throat; the tracheobronchial (T-bronchial) region consists of the windpipe and large airways; and the pulmonary region consists of the small bronchi and the alveolar sacs. (Source: International Commission on Radiological Protection Task Force on Lung Dynamics and Task Group on Lung Dynamics 1966)

synthetic biology is uniquely shaped PM. Morphology and size are important factors in aerosol exposure, including nanoparticles, but others will be more critical for products generated by other emerging technologies like synthetic biology, in which novel biological functions are likely to lead to toxicity, exposure, and risk (Pauwels et al. 2013). For example, a synthetic organism may have traits that allow it to be undetected by immune cells or have unprecedented toxicokinetics and dynamics after taken up by an organism (SCHER).

The highly complex mechanisms controlling the inhaled particle behavior also control the extent to which an aerosol is eliminated from the body. The walls of the nasal and tracheobronchial regions are coated with a mucous fluid. The tracheobronchial walls have fiber cilia which sweep the mucous fluid upward, transporting particles to the top of the trachea, where they are swallowed. This mechanism is often referred to as the mucociliary escalator. In the pulmonary region of the respiratory system, foreign particles can move across the epithelial lining of the alveolar sac to the lymph or blood systems, or they may be engulfed by scavenger cells called alveolar macrophages. The macrophages can move to the mucociliary escalator for removal. For gases, solubility controls removal from the airstream. Highly soluble gases such as SO_2 are absorbed in the upper airways, whereas less soluble gases such as NO_2 and ozone (O_3) may penetrate to the pulmonary region. Irritant gases are thought to stimulate neuroreceptors in the respiratory walls and cause a variety of responses, including sneezing, coughing, bronchoconstriction, and rapid, shallow breathing. The dissolved gas may be eliminated by biochemical processes or may diffuse to the circulatory system (Vallero 2008).

Since the location of particle deposition in the lungs is a function of aerodynamic diameter and density, then changing the characteristics of aerosols can greatly affect their likelihood to elicit an effect. Larger particles (>5 μm) tend to deposit before reaching the lungs, especially being captured by ciliated cells that line the upper airway. Moderately sized particles (1–5 μm) are more likely to deposit in the central and peripheral airways and in the alveoli but are often scavenged by macrophages. Particles with an aerodynamic diameter less than 1 μm remain suspended in air and will be exhaled if they do not adhere to lung tissue. Thus, smaller aerosols that deposit will do so deeply in the lung.

Inhaled NPs may alter the lung tissue, changing the respiratory system either directly (e.g., airway inflammation) or indirectly (e.g., by altering its immune response). Susceptibility to air pollutants differs among individuals, as exemplified by several diseases and conditions (e.g., asthma), but the fluid dynamics are the same, that is, disruption of the movement of air into the lungs to provide oxygen.

The motion of air and gases in the respiratory system follows the fundamental fluid dynamics theory (Isaacs et al. 2012; European Union 2015). The motion of these fluids is governed by the conservation of mass (continuity) equation and conservation of momentum (Navier-Stokes) equation. Under most conditions, the flow of air in the respiratory airways is assumed to be incompressible. For incompressible flow, the continuity equation is expressed as (Grotberg 2011):

$$\nabla \cdot V = 0 \tag{3}$$

And, the continuity equation is:

$$\rho \left[\frac{\partial V}{\partial t} + (V \cdot \nabla)V \right] = \rho f - \nabla p + \mu \nabla^2 V \tag{4}$$

where ∇ is a gradient operator; ∇^2 is a Laplacian operator; V is velocity; ρ is fluid density; μ is absolute fluid viscosity; p is the hydrodynamic density; and f is a volumetric force that is applied externally, for example, gravity.

For cylindrical profiles like bronchi, the gradient operator ∇ can be expressed in cylindrical coordinates:

$$\frac{\partial}{\partial r} + \frac{1}{r} \frac{\partial}{\partial \theta_\theta} + \frac{\partial}{\partial z} \tag{5}$$

Thus, the continuity equation can also be expressed cylindrically:

$$\frac{1}{r} \frac{\partial}{\partial r}(rV_r) + \frac{1}{r} \frac{\partial}{\partial \theta}V_\theta + \frac{\partial}{\partial z}V_z = 0 \tag{6}$$

where V_r, V_θ and V_z are the components of the fluid velocity, which are depicted in Fig. 5, that is, radial (r), circumferential (θ), and axial (z) directions, respectively. Thus, the momentum equations in these directions can be expressed as:

Fig. 5 Coordinate system for an ideal cylindrical airway, depicting velocity component at an arbitrary point. (Source: Vallero 2014 Adapted from: Isaacs et al. 2012)

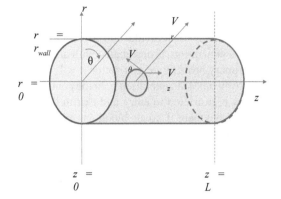

$$\frac{\partial V_r}{\partial t}+\left(\boldsymbol{V}\cdot\nabla\right)V_r-\frac{1}{r}V_\theta^2=-\frac{1}{\rho}\frac{\partial p}{\partial r}+f_r+\frac{\mu}{\rho}\left(\nabla^2 V_r-\frac{V_r}{r^2}-\frac{2}{r^2}\frac{\partial V_\theta}{\partial\theta}\right) \qquad (7)$$

$$\frac{\partial V_\theta}{\partial t}+\left(\boldsymbol{V}\cdot\nabla\right)V_\theta+\frac{V_r V_\theta}{r}=-\frac{1}{\rho r}\frac{\partial p}{\partial\theta}+f_\theta+\frac{\mu}{\rho}\left(\nabla^2 V_\theta-\frac{V_\theta}{r^2}+\frac{2}{r^2}\frac{\partial V_r}{\partial\theta}\right) \qquad (8)$$

$$\frac{\partial V_z}{\partial t}+\left(\boldsymbol{V}\cdot\nabla\right)V_z=-\frac{1}{\rho}\frac{\partial p}{\partial z}+f_z+\frac{\mu}{\rho}\nabla^2 V_z \qquad (9)$$

where:

$$\boldsymbol{V}\cdot\nabla=V_r\frac{\partial}{\partial r}+\frac{1}{r}V_\theta\frac{\partial}{\partial\theta}+V_z\frac{\partial}{\partial z} \qquad (10)$$

The first terms (i.e., time derivatives) in these three equations can be ignored under steady-state conditions. The Laplacian operator can be defined in cylindrical airways as:

$$\nabla^2=\frac{1}{r}\frac{\partial}{\partial r}\left(r\frac{\partial}{\partial r}\right)+\frac{1}{r^2}\frac{\partial^2}{\partial\theta^2}+\frac{\partial^2}{\partial z^2} \qquad (11)$$

Airway velocities are complicated by numerous factors, including lung and other tissue morphologies and the airway generations, that is, the levels of branching through which the air is flowing. Equations can be tailored to these conditions, or idealized velocity profiles can be assumed for the cascade of generations. These include parabolic flow (laminar fully developed), plug flow (laminar undeveloped), and turbulent flow (Isaacs et al. 2012). For example, the upper tracheobronchial airways may be assumed to be turbulent, but in the pulmonary region, plug and parabolic profiles may be assumed.

The right lung and left lung are connected via their primary bronchi to the trachea and upper airway of the nose and mouth (see Fig. 3). From there, the bronchi, that is, airways, subdivide into a branching network of many levels. Each level, called a generation, is designated with an integer. The tracheas are generation $n = 0$, the primary bronchi are generation $n = 1$, and so forth. Thus, theoretically there are $2n$ airway tubes at generation n. In the conducting zone (i.e., generations $0 \leq n \leq 16$), airflow is restricted to entry and exit in the airway (Grotberg 2011). That is, air is moving, but there is no air-blood gas exchange of O_2 and CO_2.

Air exchange occurs in generation $n > 16$, known as the respiratory zone. Generations $17 \leq n \leq 19$ are the locations of the airway walls' air sacs (alveoli), which range from 75 to 300 μm in diameter (Grotberg 2011). Alveoli are thin-walled and, owing to the rich capillary blood supply in them, are designed for gas exchange. The respiratory bronchioles are the vessels by which air passes to alveoli. The walls of the tubes or ducts in generations $20 \leq n \leq 22$ consist entirely of alveoli. At generation $n = 23$, terminal alveolar sacs are made up of clusters of alveoli (Isaacs et al. 2012). Thus, Fig. 4 shows that this is the pulmonary region where aerosols can deposit (International Commission on Radiological Protection Task Force on Lung Dynamics & Task Group on Lung Dynamics, 1966).

Two principal factors that are relevant to gas exchange are the airway volume (V_{aw}) and airway surface area (A_{aw}), which are proportional to the size of the person. Air exchange increases in proportion to A_{aw}. The V_{aw} (mL) for children is proportional to height and is approximated as (Kerr 1976):

$$V_{aw} = 1.018 \times \text{Height}\,(\text{cm}) - 76.2 \qquad (12)$$

V_{aw} (mL) can be estimated for adults by adding the ideal body weight (pounds) plus age in years (Bouhuys 1964). For example, a 40-year-old adult whose ideal body weight is 160 pounds has an estimated V_{aw} of 200 mL (George and Hlastala 2011).

The average human lung has from 300 to 500 million of these air sacs. In an average adult lung, the total alveolar surface area is 70 m². This large A_{aw} allows for efficient gas exchange to supply O_2 for normal respiration, but also large increases in gas exchange are needed when a person is stressed (e.g., during exercise, injury, or illness). The Reynolds number varies according to the branching level through which the air is flowing, that is, the generation (very high in the trachea, but low in the alveoli) (Grotberg 2011). Airways have liquid lining, with two layers in the first generations (up to about $n = 15$). A watery, serous layer is next to the airway wall, behaving as a Newtonian fluid. This layer has cilia that pulsate toward the mouth. Atop the serous layer is a mucus layer that possesses several non-Newtonian fluid properties, for example, viscoelasticity, shear thinning, and a yield stress.

Alveolar cells produce surfactants that orient at the air-liquid interface and reduce the surface tension significantly. Air pollutants can adversely affect the surfactant chemistry, which can make the lungs overly rigid, thus hindering inflation (Grotberg 2011) A pulmonary surfactant is a surface-active lipoprotein complex (phospholipoprotein) produced by type II pneumocytes, which are also known as alveolar type II cells. These pneumocytes are granular and comprise 60% of the alveolar lining cells. Their morphology allows them to cover smaller surface areas than type I pneumocytes. Type I cells are highly attenuated, very thin (25 nm) cells that line the alveolar surfaces and cover 97% of the alveolar surface. Surfactant molecules have both a hydrophilic head and a lipophilic tail. Surfactants adsorb to the air-water interface of the alveoli with the hydrophilic head that collects the water, while the hydrophobic tail is directed toward the air. The principal lipid component is dipalmitoylphosphatidylcholine, which is a surfactant that decreases surface tension. The actual surface tension decrease depends on the surfactant's concentration on the interface. This concentration's saturation limit depends on temperature and the presence of other compounds in the interface. Surface area of the lung varies during compliance (i.e., lung and thorax expansion and contraction) during ventilation. Thus, the surfactant's interface concentration is seldom at the level of saturation. When the lung expands, the surface increases, opening space for new surfactant molecules to join the interface mixture. During expiration, lung surface area decreases, compressing the surfactant and increasing the density of surfactant molecules, thus further decreasing the surface tension. Therefore, surface tension varies with air volume in the lungs, which protects the lungs from collapsing at low air volume and from tissue damage at high air volume (Schurch et al. 1992; George & Hlastala 2011).

Transport by concentration gradient at the molecular scale, that is, Fickian diffusion, is important only for very small particles (≤ 0.1 μm diameter) because the Brownian motion allows them to move in a "random walk" away from the airstream. Interception works mainly for particles with diameters between 0.1 and 1 μm. During interception, the particle does not leave the airstream but comes into contact with the filter medium (e.g., a strand of fiberglass). Inertial impaction collects particles that are sufficiently large to leave the airstream by inertia (diameters ≥ 1 μm). Electrostatics consist of electrical interactions between the atoms in the filter and those in the particle at the point of contact (Van der Waal's forces), as well as electrostatic attraction (charge differences between particle and filter medium). Other important factors affecting filtration efficiencies include the thickness and pore diameter or the filter, the uniformity of particle diameters and pore sizes, the solid volume fraction, the rate of particle loading onto the filter (e.g., affecting particle "bounce"), the particle phase (liquid or solid), capillarity and surface tension (if either the particle or the filter media are coated with a liquid), and characteristics of air or other carrier gases, such as velocity, temperature, pressure, and viscosity.

Basically, lung filtration consists of four mechanical processes: (1) diffusion, (2) interception, (3) inertial impaction, and (4) electrostatics (see Fig. 6). Diffusion is important only for very small particles (≤ 0.1 μm diameter) because the Brownian motion allows them to move away in a "random walk" away from the airstream. This can be an important process for NPs.

All of these filtration processes apply to capture and escape of synbio and nanoparticles. Notably, interception occurs when a particle stays in the airstream but comes into contact with matter (e.g., lung tissue), mainly for particles in the upper nanoscale size range, that is, diameters near 100 nm and up to 1 μm. Impaction

Fig. 6 Mechanical processes leading to the deposition of particulate matter. Diffusion can be an important filtration mechanism for nanoparticles. (Source: Vallero 2013, 2014; adapted from: Rubow et al. 2004)

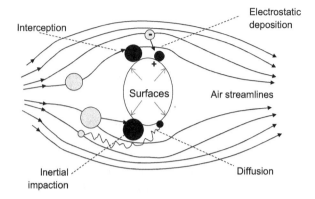

collects sufficiently large particles to leave the airstream by inertia (diameters ≥ 1 μm); hence this is commonly referred to as "inertial impaction." Given their size, nanoscale and ultrafine aerosols are strongly affected by electrostatics given the electrical interactions between the atoms in a surface and those in the particle at the point of contact (Van der Waal's force), as well as electrostatic attraction (charge differences between particle and surface). Other important factors affecting lung filtration are surface stickiness, uniformity of particle diameters, the solid volume fraction, the rate of particle loading onto tissue surfaces, the particle phase (whether liquid or solid), capillarity and surface tension, and characteristics of air in the airway, such as humidity, velocity, temperature, pressure, and viscosity.

In addition to aerosol size, the chemical composition also determines the fate of symbiotic products in the respiratory system. Endogenously, varying amounts of the parent substance (e.g., zero-valent metal), any salts and ions formed, and other chemical species (e.g., organometallic compounds) are absorbed and distributed within the body. For metal NPs, the principal difference between the way that nanomaterials and other forms of metal will partition among zero-valence, ions, and metallic compounds is determined by its relative volume and mass. The greater amount of surface area in NPs means that compared to even fine particulate matter, the NP has a greater number of potentially active sites for sorption and solution. The low mass also means that the NP can remain suspended for a very long time. Such nano-suspensions in surface waters mean that the metal tends to remain in the water column, rather than settle onto the surface, so it is more likely to be exposed to free oxygen than to the more reduced and anoxic conditions of the sediment. In the air, these features mean that the NP will be more likely to remain airborne for longer time periods and to undergo atmospheric transformation.

These differences in mass and volume from bulk materials can also translate into endogenous differences, meaning that absorption, distribution, metabolism, excretion, and toxicity could also be different for a NP. The fraction of the metal species or its transformation products that accumulates in lipids and other tissue substrates could be higher, and the amount excreted decreased, so that the difference results in bioaccumulation and increased body burden (see Fig. 7).

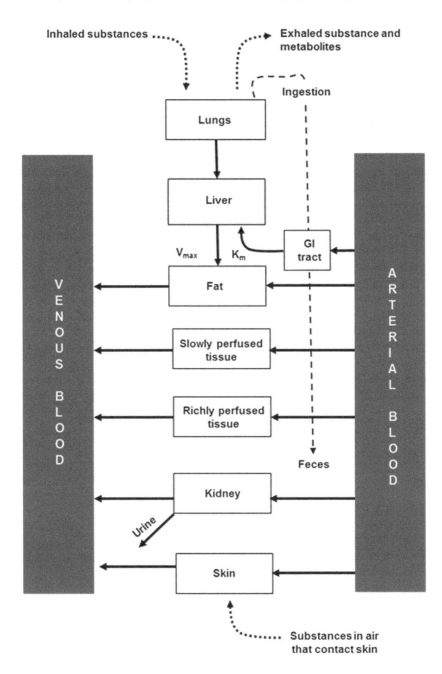

Fig. 7 Toxicokinetics for a hypothetical nanomaterial that has been inhaled, ingested, or contacted dermally. (Based on: Agency for Toxic Substances and Disease Registry 2002; adapted from: Vallero 2014)

The metal NP, cations, and its metabolites thereafter induce toxicity in various ways. For toxicity (e.g., metal-induced neuropathologies) to occur, a metal must reach a target (e.g., a neuron) at a concentration sufficient to alter mechanistically the normal functioning of the tissue. Metal toxicity can involve the types of membrane receptor-ligand disruptions. However, it may also involve intracellular receptors and ion channels. Metals tend to react with nucleophilic macromolecules, for example, proteins, amino acids, and nucleic acids. A nucleophile donates an electron pair to an electrophile, an electron pair acceptor, to form a chemical bond. Mercury, for example, reacts with sulfur (S) in thiols, cysteinyl protein residues, and glutathione and S in thiols and thiolates. However, other metals, for example, lithium, calcium, and barium, preferentially react with harder nucleophiles, for example, the oxygen in purines. Lead (Pb) tends to fall between these two extremes, that is, exhibits universal reactivity with all nucleophiles (Shanker 2008).

Again, these effects have been observed in metals and metalloids in various forms, with nanomaterials playing a role of either degrading or improving environmental conditions. How metal NPs differ is a subject of current research. In addition, metals in various forms and sizes are influenced by the presence of NPs. For example, Pb mobility and bioavailability can be adjusted by inserting iron (Fe) NPs (e.g., $Fe_3(PO_4)_2 \cdot 8H_2O$) into Pb-contaminated soil, that is, converting highly aqueous soluble and exchangeable forms to less soluble and less exchangeable forms (Liu and Zhao 2007). Such findings can greatly improve environmental remediation efforts.

Much of the toxicology resulting from inhalation exposure (E) can be expressed as (Derelanko 2014; Vallero 2014):

$$E = \frac{(C) \cdot (PC) \cdot (IR) \cdot (RF) \cdot (EL) \cdot (AF) \cdot (ED) \cdot (10^{-6})}{(BW) \cdot (TL)} \tag{13}$$

where:

C = concentration of the contaminant on the aerosol/particle (mg kg^{-1})
PC = particle concentration in air (gm m^{-3})
IR = inhalation rate (m^3 h^{-1})
RF = respirable fraction of total particulates (dimensionless, usually determined by aerodynamic diameters, e.g., 2.5 μm)
EL = exposure length (h d^{-1})
ED = duration of exposure (d)
AF = absorption factor (dimensionless)
BW = body weight (kg)
TL = typical lifetime (d)
10^{-6} = a conversion factor (kg to mg)

The human body and other biological systems have a capacity for the uptake of myriad types of substances and utilize them to support some bodily function or eliminate them. In work or exercise scenarios, for example, the exposure to NPs is greatly increased because of the elevated IR and PC values.

The quality and amount of data from which to base nanomaterial exposures vary. As analytical capabilities have improved, increasingly lower concentrations of chemicals have been observed in various parts of the body. Some of these chemicals enter the body by inhalation, whereas the dominant pathway for others could be in drinking water, food, and skin contact. Equations for each of these pathways are analogous to Eq. 1.

Engineers and scientists document and try to quantify uncertainty by working within the known domain and using tools to extend knowledge to the lesser known domains, that is, extrapolating information and knowledge from the data-rich to data-poor domains. If something has failed under specified conditions and did not fail under different, specified conditions, this may inform decisions within unknown domains. However, if this is all the information available, the gap between the two domains is the region of uncertainty. From both an engineering and biomedical perspective, uncertainties are addressed by conservative safety, including protective factors of safety. For example, regulatory agencies may have reference doses (RfD) and concentrations (RFC) for chemical compounds that have been based largely on in vitro and in vivo studies of pure compounds. However, when these compounds are constituents of synbio products and nanoparticles, additional levels of protection are required, given the additional uncertainties about exposure and toxicity.

Exposure Models

Risk management depends on models to estimate exposures. Such models range from "screening-level" to "high-tiered." Screening models generally overpredict exposures because they are based on conservative default values and assumptions. They provide a first approximation that screens out exposures not likely to be of concern (Chemical Computing Group 2013; Guy et al. 2008; Hilton et al. 2010; Judson et al. 2010; U.S. Environmental Protection Agency 2017; Zhang et al. 2014). Conversely, higher-tiered models typically include algorithms that provide specific site characteristics and time activity patterns and are based on relatively realistic values and assumptions. Such models require data of higher resolution and quality than the screening models and, in return, provide more refined exposure estimates (U.S. Environmental Protection Agency 2017).

Environmental stressors can be modeled in a unidirectional and one-dimensional fashion. A conceptual framework can link exposure to environmental outcomes across levels of biological organization (Fig. 8). Thus, environmental exposure assessment considers coupled networks that span multiple levels of biological organization and can describe the interrelationships within the biological system. Mechanisms can be derived by characterizing and perturbing these networks, for example, behavioral and environmental factors (Hubal et al. 2010). This can apply

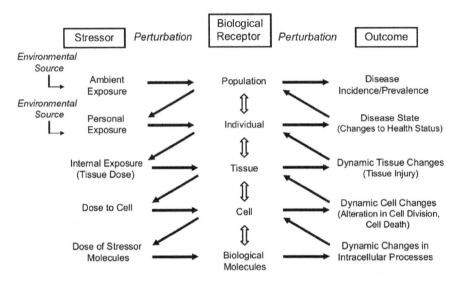

Fig. 8 Systems cascade of exposure-response processes. In this instance, scale and levels of biological organization are used to integrate exposure information with biological outcomes. The stressor (chemical or biological agent) moves both within and among levels of biological organization, reaching various receptors, thereby influencing and inducing outcomes. The outcome can be explained by physical, chemical, and biological processes (e.g., toxicogenomic mode-of-action information). (Source: Hubal et al. 2010)

to a food chain or food web model (Fig. 9) or a kinetic model (Fig. 10) or numerous other modeling platforms.

Exposure Estimation

Exposure results from sequential and parallel processes in the environment, from release to environmental partitioning, movement through pathways to uptake, and fate in the organism (see Fig. 11). The substances often change to other chemical species as a result of the body's metabolic and detoxification processes. From a precautionary perspective, it may be necessary to assume that synthesis and genetic modifications will affect such processes. New substances, known as degradation products or metabolites, are produced as cells use the parent compounds as food and energy sources. These metabolic processes, such as hydrolysis and oxidation, are the mechanisms by which chemicals are broken down.

The exposure pathway also includes the ways that humans and other organisms can come into contact with a hazard. The pathway has five parts:

1. The source of contamination (e.g., fugitive dust or leachate from a landfill)
2. An environmental medium and transport mechanism (e.g., soil with water moving through it)

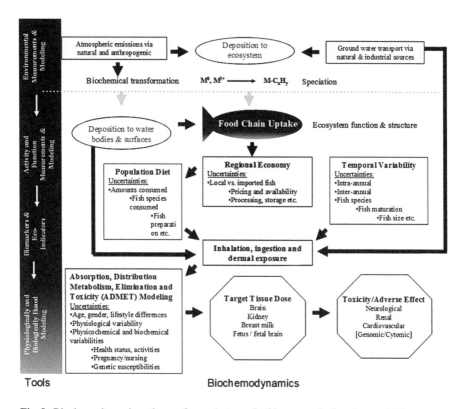

Fig. 9 Biochemodynamic pathways for a substance (in this case, a single substance). The receptor is mammalian tissue. Various modeling tools are available to characterize the movement, transformation, uptake, and fate of the compound. Similar biochemodynamic paradigms can be constructed for multiple chemicals (e.g., mixtures) and microorganisms. (Source: Vallero 2010b)

3. A point of exposure (such as a well used for drinking water)
4. A route of exposure (e.g., inhalation, dietary ingestion, nondietary ingestion, dermal contact, and nasal)
5. A receptor population (those who are actually exposed or who are where there is a potential for exposure)

If all the five parts are present, the exposure pathway is known as a completed exposure pathway. In addition, the exposure may be short term, intermediate, or long term. Short-term contact is known as an acute exposure, that is, occurring as a single event or for only a short period of time (up to 14 days). An intermediate exposure is one that lasts from 14 days to less than 1 year. Long-term or chronic exposures are greater than 1 year in duration.

Determining the exposure for a neighborhood can be complicated. For example, even if we do a good job identifying all of the contaminants of concern and possible sources (no small task), we may have little idea of the extent to which the receptor population has come into contact with these contaminants (steps 2 through 4). Thus,

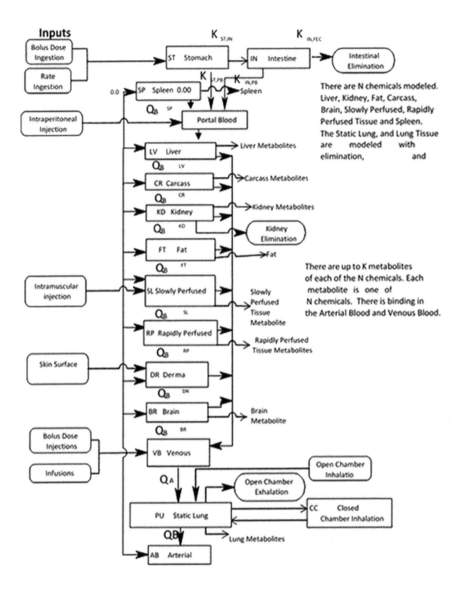

Fig. 10 Toxicokinetic model used to estimate dose as part of an environmental exposure. This diagram represents the static lung, with each of the compartments (brain, carcass, fat, kidney, liver, lung tissue, rapidly and slowly perfused tissues, spleen, and the static lung) having two forms of elimination, an equilibrium binding process, and numerous metabolites. Notes: K refers to kinetic rate; Q to mass flow; and Q_B to blood flow. A breathing lung model would consist of alveoli, lower dead space, lung tissue, pulmonary capillaries, and upper dead space compartments. Gastrointestinal (GI) models allow for multiple circulating compounds with multiple metabolites entering and leaving each compartment, that is, the GI model consists of the wall and lumen for the stomach, duodenum, lower small intestine, and colon, with lymph pool and portal blood compartments included.

assessing exposure involves not only the physical sciences but the social sciences, for example, psychology and behavioral sciences. People's activities greatly affect the amount and type of exposures. That is why exposure scientists use a number of techniques to establish activity patterns, such as asking potentially exposed individuals to keep diaries, videotaping, and using telemetry to monitor vital information, for example, heart and ventilation rates.

General ambient measurements, such as air pollution monitoring equipment located throughout cities, are often not good indicators of actual population exposures. For example, metals and their compounds comprise the greatest mass of toxic substances *released* into the environment. This is largely due to the large volume and surface areas involved in metal extraction and refining operations. However, this does not necessarily mean that more people will be exposed at higher concentrations or more frequently to these compounds than to others. A substance that is released or even that if it resides in the ambient environment is not tantamount to its coming in contact with a *receptor*. Conversely, even a small amount of a substance under the right circumstances can lead to very high levels of exposure (e.g., handling raw materials and residues at a waste site).

The simplest quantitative expression of exposure is:

$$E = D / t \tag{14}$$

where E is the human exposure during the time period t (units of concentration $(mg\ kg^{-1}d^{-1})$; D is the mass of pollutant per body mass $(mg\ kg^{-1})$; and t is time (day).

D, the chemical concentration of a pollutant, is usually measured near the interface of the person and the environment, during a specified time period. This measurement is sometimes referred to as the potential dose (i.e., the chemical has not yet crossed the boundary into the body but is present where it may enter the person, such as on the skin, at the mouth, or at the nose).

Expressed quantitatively, exposure is a function of the concentration of the agent and time. It is an expression of the magnitude and duration of the contact. That is, exposure to a contaminant is the concentration of that contact in a medium integrated over the time of contact:

$$E = \int_{t=t_1}^{t=t_2} C(t)\,dt \tag{15}$$

where E is the exposure during the time period from t_1 to t_2 and $C(t)$ is the concentration at the interface between the organism and the environment, at time t.

The concentration at the interface is the potential dose (i.e., the agent has not yet crossed the boundary into the body but is present where it may enter the receptor).

Fig. 10 (continued) Bile flow is treated as an output from the liver to the duodenum lumen. All uptaken substances are treated as circulating. Nonspecific ligand binding, for example, plasma protein binding, is represented in arterial blood, pulmonary capillaries, portal blood, and venous blood. Source: (C. C. Dary, 2007); adapted from: (Blancato, Power, Brown, & Dary, 2006)

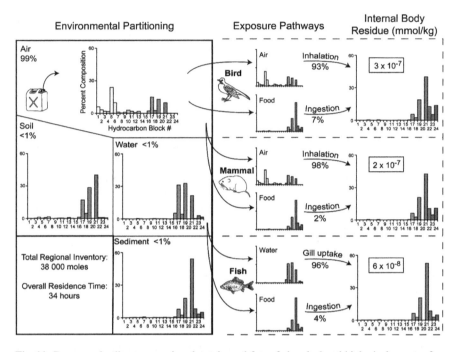

Fig. 11 Processes leading to organismal uptake and fate of chemical and biological agents after release into the environment. In this instance, the predominant sources are air emissions, and the predominant pathway of exposure is inhalation. However, due to deposition to surface waters and the agent's affinity for sediment, the ingestion pathways are also important. Dermal pathways, in this case, do not constitute a large fraction of potential exposure. (Source: McKone et al. 2006)

Since the amount of a chemical agent that penetrates from the ambient atmosphere into a control volume affects the concentration term of the exposure equation, a complete mass balance of the contaminant must be understood and accounted for; otherwise, exposure estimates will be incorrect. Recall that the mass balance consists of all inputs and outputs, as well as chemical changes to the contaminant:

$$\begin{aligned}\text{Accumulation or loss of contaminant } A &= \text{Mass of } A \text{ transported in}\\ &\quad -\text{Mass of } A \text{ transported out} \pm \text{Reactions}\end{aligned} \tag{16}$$

The reactions may be either those that generate substance A (i.e., *sources*) or those that destroy substance A (i.e., *sinks*). Thus, the amount of mass transported in is the inflow to the system that includes pollutant discharges, transfer from other control volumes and other media (e.g., if the control volume is soil, the water and air may contribute mass of chemical A), and formation of chemical A by abiotic chemistry and biological transformation. Conversely, the outflow is the mass transported out of the control volume, which includes uptake, by biota, transfer to other compartments (e.g., volatilization to the atmosphere), and abiotic and biological degradation of chemical A. This means the rate of change of mass in a control volume is equal to the rate of chemical A transported in less the rate of chemical A transported out, plus the rate of production from sources, and minus the rate of elimination by sinks. Stated as a differential equation, the rate of change contaminant A is:

$$\frac{d[A]}{dt} = -v \cdot \frac{d[A]}{dx} + \frac{d}{dx}\left(\Gamma \cdot \frac{d[A]}{dx}\right) + r \tag{17}$$

where:

v is the fluid velocity.

Γ is a rate constant specific to the environmental medium.

$\dfrac{d[A]}{dx}$ is the concentration gradient of chemical A.

r is the internal sinks and sources within the control volume.

Reactive compounds can be particularly difficult to measure. For example, many volatile organic compounds in the air can be measured by collection in stainless steel canisters and followed by chromatography analysis in the lab. However, some of these compounds, like the carbonyls (notably aldehydes like formaldehyde and acetaldehyde), are prone to react inside the canister, meaning that by the time the sample is analyzed, a portion of the carbonyls are degraded (underreported). Therefore, other methods may need to be applied, such as trapping the compounds with dinitrophenylhydrazine (DNPH)-treated silica gel tubes that are frozen until being extracted for chromatographic analysis. The purpose of the measurement is to see what is in the air, water, soil, sediment, or biota at the time of sampling, so any reactions before the analysis give measurement error.

The general exposure in Eq. 13 is rewritten to address each route of exposure, accounting for chemical concentration and the activities that affect the time of contact. The exposure calculated from these equations is actually the chemical intake (I) in units of concentration (mass per volume or mass per mass) per time, such as mg kg^{-1} d^{-1}:

$$I = \frac{C \cdot CR \cdot EF \cdot ED \cdot AF}{BW \cdot AT} \tag{18}$$

where:

C is the chemical concentration of contaminant (mass per volume).

CR is the contact rate (mass per time).

EF is the exposure frequency (number of events, dimensionless).

ED is the exposure duration (time).

These factors are further specified for each route of exposure, such as the lifetime average daily dose (LADD) as shown in Table 1. The LADD is obviously based on a chronic, long-term exposure.

Table 1 Equations for calculating lifetime average daily dose (LADD) for various routes of exposure

Route of exposure	Equation LADD (in mg kg^{-1}d^{-1})=	Definitions
Inhaling aerosols (particulate matter)	$$\dfrac{(C) \cdot (PC) \cdot (IR) \cdot (RF) \cdot (EL) \cdot (AF) \cdot (ED) \cdot (10^{-6})}{(BW) \cdot (TL)}$$	C = concentration of the contaminant on the aerosol/particle (mg kg^{-1}) PC = particle concentration in air (gm m^{-3}) IR = inhalation rate (m^{-3} h^{-1}) RF = respirable fraction of total particulates (dimensionless, usually determined by aerodynamic diameters, e.g., 2.5 μm) EL = exposure length (h d^{-1}) ED = duration of exposure (d) AF = absorption factor (dimensionless) BW = body weight (kg) TL = typical lifetime (d) 10^{-6} is a conversion factor (kg to mg)
Inhaling vapor phase contaminants	$$\dfrac{(C) \cdot (IR) \cdot (EL) \cdot (AF) \cdot (ED)}{(BW) \cdot (TL)}$$	C = concentration of the contaminant in the gas phase (mg m^{-3}) Other variables the same as above
Drinking water	$$\dfrac{(C) \cdot (CR) \cdot (ED) \cdot (AF)}{(BW) \cdot (TL)}$$	C = concentration of the contaminant in the drinking water (mg L^{-1}) CR = rate of water consumption (L d^{-1}) ED = duration of exposure (d) AF = portion (fraction) of the ingested contaminant that is physiologically absorbed (dimensionless) Other variables are the same as above
Contact with soil-borne contaminants	$$\dfrac{(C) \cdot (SA) \cdot (BF) \cdot (FC) \cdot (SDF) \cdot (ED) \cdot (10^{-6})}{(BW) \cdot (TL)}$$	C = concentration of the contaminant in the soil (mg kg^{-1}) SA = skin surface area exposed (cm^{-2}) BF = bioavailability (percent of contaminant absorbed per day) FC = fraction of total soil from contaminated source (dimensionless) SDF = soil deposition, the mass of soil deposited per unit area of skin surface (mg cm^{-1} d^{-1}) Other variables are the same as above

Source: M. Derelanko, 1999, *CRC Handbook of Toxicology*, "Risk Assessment," M. J. Derelanko and M.A. Hollinger, editors, CRC Press, Boca Raton, FL

Acute and subchronic exposures require different equations, since the exposure duration (ED) is much shorter. For example, instead of LADD, acute exposures to noncarcinogens may use maximum daily dose (MDD) to calculate exposure (see discussion box). However, even these exposures follow the general model given in Eq. 15.

Hypothetical Example of an Exposure Calculation

Over an 18-year period, VICHLOSYN has successfully applied synthetic biology to detoxify soil contaminated with vinyl chloride. Contaminated soil has been trucked to their facility. However, storing the soil and treatment have contaminated the soil on its property. Complaints and audits led to VICHLOSYN closing the facility 2 years ago but vinyl chloride vapors continue to reach the neighborhood surrounding the plant at an average concentration of 1 mg m^{-3}. Assume that people are breathing at a ventilation rate of 0.5 m^3 h^{-1} (about the average of adult males and females over 18 years of age) (Moya et al. 2011). The legal settlement allows neighboring residents to evacuate and sell their homes to the company. However, they may also stay. The neighbors have asked for advice on whether to stay or leave, since they have already been exposed for 20 years.

Vinyl chloride is highly volatile, so its phase distribution will be mainly in the gas phase rather than the aerosol phase. Although some of the vinyl chloride may be sorbed to particles, we will use only vapor phase LADD equation, since the particle phase is likely to be relatively small. Also, we will assume that outdoor concentrations are the exposure concentrations. This is unlikely, however, since people spend very little time outdoors compared to indoors, so this may provide an additional factor of safety. To determine how much vinyl chloride penetrates living quarters, indoor air studies would have to be conducted. For a scientist to compare exposures, indoor air measurements should be taken.

Find the appropriate equation in Table 1 and insert values for each variable. Absorption rates are published by the EPA and the Oak Ridge National Laboratory's Risk Assessment Information System (http://risk.lsd.ornl.gov/cgi-bin/tox/TOX_select?select=nrad). Vinyl chloride is well absorbed, so for a worst case we can assume that AF = 1. We will also assume that the person staying in the neighborhood is exposed at the average concentration 24 hours a day (EL = 24) and that a person lives the remainder of entire typical lifetime exposed at the measured concentration.

Although the ambient concentrations of vinyl chloride may have been higher when the plant was operating, the only measurements we have are those taken recently. Thus, this is an area of uncertainty that must be discussed with the clients. The common default value for a lifetime is 70 years, so we can assume the longest exposure would be is 70 years (25,550 days). Table 2 gives some of the commonly used default values in exposure assessments. If the person is now 20 years of age and has already been exposed for that time and lives the remaining 50 years exposed at 1 mg m^{-3}, then:

$$LADD = \frac{(C)\cdot(IR)\cdot(EL)\cdot(AF)\cdot(ED)}{(BW)\cdot(TL)}$$

$$= \frac{(1)\cdot(0.5)\cdot(24)\cdot(1)\cdot(25550)}{(70)\cdot(25550)}$$

$$= 0.2\,mg\,kg^{-1}day^{-1}$$

If the 20-year-old leaves today, the exposure duration would be for the 20 years that the person lived in the neighborhood. Thus, only the ED term would change, that is, from 25,550 days to 7300 days (i.e., 20 years).

Thus, the LADD falls to 2/7 of its value:

$$LADD = 0.05\,mg\,kg^{-1}day^{-1}.$$

Note that this is a straightforward, chemical exposure estimate in the gas phase. Often, a chemical will exist as a vapor, an aerosol, or sorbed to an aerosol. In this case, the inhalation exposure would have to be calculated for the gas and the PM, that is, the concentration of PM and the concentration of the chemical in the PM. Furthermore, if this were an exposure involving an emerging technology, it would be much more complex and uncertain, since the routes and pathway information may be more difficult to ascertain, for example, GMMs do not behave like chemical compounds. The risk assessment may be even more uncertain, since it is likely that at least some of the products may lack data on toxicity and hazard, including genetically modified organisms, so even if the exposure probability is reliable, the risk assessment will be weakened.

Once the hazard and exposure calculations are complete, risks can be characterized quantitatively. There are two general ways that such risk characterizations are used in environmental problem-solving, that is, direct risk assessments and risk-based cleanup standards.

Conclusion

The benefits of emerging technologies must be weighed against the amount of risk that they introduce. The risks to health and the environment must be reduced or avoided by proper management. Risk management decisions must be underpinned by scientifically credible and reliable assessments of both the hazards and the likeli-

Table 2 Commonly used human exposure factors

Exposure factor	Adult male	Adult female	Child (3–12 years of age)
Body weight (kg)	70	60	15–40
Total fluids ingested (L d⁻¹)	2	1.4	1.0
Surface area of the skin, without clothing (m²)	1.8	1.6	0.9
Surface area of the skin, wearing clothes (m²)	0.1–0.3	0.1–0.3	0.05–0.15
Respiration/ventilation rate (L min⁻¹) – resting	7.5	6.0	5.0
Respiration/ventilation rate (L min⁻¹) – light activity	20	19	13
Volume of air breathed (m³ d⁻¹)	23	21	15
Typical lifetime (years)	70	70	NA
National upper-bound time (90th percentile) at one residence (years)	30	30	NA
National median time (50th percentile) at one residence (years)	9	9	NA

Sources: Centers for Disease Control and Prevention 2005; Moya et al. 2011

hood and extent of exposure to those hazards. Thus, reliable exposure estimates are required for decisions involving products of synthetic biology and other emerging technologies.

This chapter introduced exposure assessment approaches, identifying where conventional methods may fail, along with possible ways to augment them to address the large uncertainties in assessing and managing the risks posed by these technologies.

Among the challenges of substances generated in synthetic biology processes is that they are not limited to chemical contaminants but will include mixtures and biological agents generated during various life stages of synthesis and use. The agents may include products during various stages of synthesis, beginning with chassis bacteria. They may also include genetically modified biological agents, as well as pathogens and other natural organisms which induce harm when released into a human population or ecosystem. Methods for estimating and predicting exposures to these agents are much more uncertain than those employed in traditional chemical risk assessment. Assessment methodologies must be adapted to address the various routes of exposure and adverse outcomes introduced from new technologies that generate unprecedented biological entities, such as (Epstein and Vermeire 2016):

1. Integration of protocells into living organism
2. Xenobiology
3. DNA synthesis and direct genome editing of zygotes that can lead to multiplexed genetic modifications
4. Increased modifications introduced in parallel by large-scale DNA synthesis and highly parallel genome editing

These and other synthetic process will result in increased genetic distance between the synthetic organism and any natural organism or any previously modified organism (Epstein and Vermeire 2016). Thus, existing exposure and risk science provide a pathway to exposure assessment for synthetic biology but are wholly insufficient given these differences. Research is needed to compare and contrast synthetic biology-generated contaminants and agents with chemicals.

Disclaimer Drs. Jay Reichman and Caroline Stevens of EPA's Office of Research and Development provided substantive reviews and technical recommendations that enhanced this chapter. Mention of trade names commercial products does not constitute endorsement nor recommendation for use. The views expressed in this chapter are those of the author and do not necessarily reflect the views or policies of the U.S. EPA.

Bibliography

Agency for Toxic Substances and Disease Registry. (2002). *Toxicological profile for DDT, DDE, and DDD*. Atlanta. U.S. Government, Center for Disease Control and Prevention.

Arnaldi, S., & Muratorio, A. (2013). *Nanotechnology, uncertainty and regulation. A guest editorial*. Springer.

Barber, M. C., Isaacs, K. K., & Tebes-Stevens, C. (2017). Developing and applying metamodels of high resolution process-based simulations for high throughput exposure assessment of organic chemicals in riverine ecosystems. *Science of the Total Environment, 605*, 471–481.

Blancato, J., Power, F. W., Brown, R. N., & Dary, C. C. (2006). Exposure Related Dose Estimating Model (ERDEM) a Physiologically-Based Pharmacokinetic and Pharmacodynamic (PBPK/PD) model for assessing human exposure and risk. U.S. Environmental Protection Agency. Las Vegas, Nevada, EPA/600/R-06/061.

Bouhuys, A. (1964). Respiratory dead space. *Handbook of Physiology. Section III, 1*, 699–714.

Callis, C. M., Bercu, J. P., DeVries, K. M., Dow, L. K., Robbins, D. K., & Varie, D. L. (2010). Risk assessment of genotoxic impurities in marketed compounds administered over a short-term duration: Applications to oncology products and implications for impurity control limits. *Organic Process Research & Development, 14*(4), 986–992.

Carpenter, D. O., Arcaro, K., & Spink, D. C. (2002). Understanding the human health effects of chemical mixtures. *Environmental Health Perspectives, 110*(Suppl 1), 25–42.

Centers for Disease Control and Prevention, A. f. T. S. a. D. R. (2005). *Public health assessment guidance manual*. Atlanta. Retrieved from https://www.atsdr.cdc.gov/hac/PHAManual/toc.html

Chemical Computing Group. (2013). Molecular operating environment: Chemoinformatics and structure based tools for high throughput screening. Montreal, Canada.

Covello, V. T. (2003). Best practices in public health risk and crisis communication. *Journal of Health Communication, 8*(S1), 5–8.

Cummings, C. L., & Kuzma, J. (2017). Societal risk evaluation scheme (SRES): Scenario-based multi-criteria evaluation of synthetic biology applications. *PLoS One, 12*(1), e0168564.

Dary, C. C., P. J. G., Vallero, D.A., Tornero-Velez, R., Morgan, M., Okino, M., Dellarco, M., Power, F. W., & Blancato, J. N. (2007). Characterizing chemical exposure from biomonitoring data using the exposure related dose estimating model (ERDEM). 17th Annual Conference of the International Society of Exposure Analysis, Durham, North Carolina.

Derelanko, M. J. (2014). Risk assessment. In M. J. Derelanko & C. S. Auletta (Eds.), *Handbook of toxicology*. CRC Press. Boca Raton, Florida.

Dionisio, K. L., Frame, A. M., Goldsmith, M.-R., Wambaugh, J. F., Liddell, A., Cathey, T., et al. (2015). Exploring consumer exposure pathways and patterns of use for chemicals in the environment. *Toxicology Reports, 2,* 228–237.

Doblhoff-Dier, P., Bachmayer, H., Bennett, A., Brunius, G., Býrki, K., Cantley, M., et al. (2000). Safe biotechnology 10: DNA content of biotechnological process waste. The Safety in Biotechnology Working Party on the European Federation of Biotechnology. *Trends in Biotechnology, 18*(4), 141–146.

Egeghy, P. P., Vallero, D. A., & Hubal, E. A. C. (2011). Exposure-based prioritization of chemicals for risk assessment. *Environmental Science & Policy, 14*(8), 950–964.

Egeghy, P. P., Sheldon, L. S., Isaacs, K. K., Özkaynak, H., Goldsmith, M.-R., Wambaugh, J. F., et al. (2016). Computational exposure science: An emerging discipline to support 21st-century risk assessment. *Environmental Health Perspectives (Online), 124*(6), 697.

EPA, U. (2004). *Air quality criteria for particulate matter.* (EPA/600/P-99/002bF).

Epstein, M. M., & Vermeire, T. (2016). Scientific opinion on risk assessment of synthetic biology. *Trends in Biotechnology, 34*(8), 601–603.

European Union (2015). Opinion on synthetic biology II-risk assessment methodologies and safety aspects. Scientific Committee on Health and Environmental Risks, and Scientific Committee on Consumer Safety, European Commission. Brussels, Belgium.

Falkner, R., & Jaspers, N. (2012). Regulating nanotechnologies: Risk, uncertainty and the global governance gap. *Global Environmental Politics, 12*(1), 30–55.

Feron, V., & Groten, J. (2002). Toxicological evaluation of chemical mixtures. *Food and Chemical Toxicology, 40*(6), 825–839.

Finkel, A. M. (1990). *Confronting uncertainty in risk management; a guide for decision makers a report* (No. GTZ 828). Resources for the Future, Washington, DC (EUA).

Genter, M. B., Krishan, M., & Prediger, R. D. (2015). The olfactory system as a route of delivery for agents to the brain and circulation. In *Handbook of olfaction and gustation* (pp. 453–484). Wiley-Blackwell. Hoboken, New Jersey.

George, S. C., & Hlastala, M. P. (2011). Airway gas exchange and exhaled biomarkers. *Comprehensive Physiology, 1*(4), 1837–1859.

Grotberg, J. B. (2011). Respiratory fluid mechanics. *Physics of Fluids (1994-present), 23*(2), 021301.

Guy, A., Gauthier, C., & Griffin, G. (2008). *Adopting alternative methods for regulatory testing in Canada.* Paper presented at the Proceedings of the 6th World Congress on Alternatives & Animal Use in the Life Sciences. AATEX.

Harder, R., Holmquist, H., Molander, S., Svanström, M., & Peters, G. M. (2015). Review of environmental assessment case studies blending elements of risk assessment and life cycle assessment. *Environmental Science & Technology, 49*(22), 13083–13093.

Hilton, D. C., Jones, R. S., & Sjödin, A. (2010). A method for rapid, non-targeted screening for environmental contaminants in household dust. *Journal of Chromatography A, 1217*(44), 6851–6856.

Hubal, E. A. C., Richard, A. M., Shah, I., Gallagher, J., Kavlock, R., Blancato, J., & Edwards, S. W. (2010). Exposure science and the US EPA National Center for computational toxicology. *Journal of Exposure Science and Environmental Epidemiology, 20*(3), 231–236.

International Commission on Radiological Protection Task Force on Lung Dynamics, & Task Group on Lung Dynamics. (1966). Deposition and retention models for internal dosimetry of the human respiratory tract. *Health Physics, 12*(2), 173.

Isaacs, K., Rosati, J. A., & Martonen, T. B. (2012). Modeling deposition of inhaled particles. In *Aerosols handbook: Measurement, dosimetry, and health effects.* Boca Raton: CRC press.

Ivask, A., Mitchell, A. J., Malysheva, A., Voelcker, N. H., & Lombi, E. (2018). Methodologies and approaches for the analysis of cell–nanoparticle interactions. *Wiley Interdisciplinary Reviews: Nanomedicine and Nanobiotechnology, 10*(3), e1486.

Jones-Otazo, H. A., Clarke, J. P., Diamond, M. L., Archbold, J. A., Ferguson, G., Harner, T., et al. (2005). Is house dust the missing exposure pathway for PBDEs? An analysis of the urban fate and human exposure to PBDEs. *Environmental Science & Technology, 39*(14), 5121–5130.

Judson, R. S., Houck, K. A., Kavlock, R. J., Knudsen, T. B., Martin, M. T., Mortensen, H. M., et al. (2010). In vitro screening of environmental chemicals for targeted testing prioritization: The ToxCast project. *Environmental Health Perspectives (Online), 118*(4), 485.

Kahan, D., Braman, D., & Mandel, G. (2009). Risk and culture: Is synthetic biology different?

Kerr, A. (1976). Dead space ventilation in normal children and children with obstructive airways diease. *Thorax, 31*(1), 63–69.

Kimura, T., Miyashita, Y., Funatsu, K., & Sasaki, S.-i. (1996). Quantitative structure– activity relationships of the synthetic substrates for elastase enzyme using nonlinear partial least squares regression. *Journal of Chemical Information and Computer Sciences, 36*(2), 185–189.

Kodell, R. L. (2005). Managing uncertainty in health risk assessment. *International Journal of Risk Assessment and Management, 5*(2), 193–205.

Kortelainen, M. (2015). The REACH authorisation procedure–Follow-up and prediction as a downstream user.

Kortenkamp, A., Backhaus, T., & Faust, M. (2009). State of the art report on mixture toxicity. *Contract, 70307*(2007485103), 94–103.

Kuzma, J., & Tanji, T. (2010). Unpackaging synthetic biology: Identification of oversight policy problems and options. *Regulation & Governance, 4*(1), 92–112.

Lagunin, A., Zakharov, A., Filimonov, D., & Poroikov, V. (2011). QSAR modelling of rat acute toxicity on the basis of PASS prediction. *Molecular Informatics, 30*(2–3), 241–250.

Lead, J. R., Batley, G. E., Alvarez, P. J., Croteau, M. N., Handy, R. D., McLaughlin, M. J., et al. (2018). Nanomaterials in the environment: Behavior, fate, bioavailability, and effects—An updated review. *Environmental Toxicology and Chemistry, 37*(8), 2029–2063.

Liu, H., & Gramatica, P. (2007). QSAR study of selective ligands for the thyroid hormone receptor β. *Bioorganic & Medicinal Chemistry, 15*(15), 5251–5261.

Liu, R., & Zhao, D. (2007). Reducing leachability and bioaccessibility of lead in soils using a new class of stabilized iron phosphate nanoparticles. *Water Research, 41*(12), 2491–2502.

Loehr, R., Goldstein, B., Nerode, A., & Risser, P. (1992). Safeguarding the future: Credible science, credible decisions. *The report of the expert panel on the role of science at EPA*. Washington, DC: Environmental Protection Agency.

Maheshwari, R., Joshi, G., Mishra, D. K., & Tekade, R. K. (2019). Bionanotechnology in pharmaceutical research. In *Basic fundamentals of drug delivery* (pp. 449–471). London: Elsevier.

McKone, T., Riley, W., Maddalena, R., Rosenbaum, R., & Vallero, D. (2006). Common issues in human and ecosystem exposure assessment: The significance of partitioning, kinetics, and uptake at biological exchange surfaces. *Epidemiology, 17*(6), S134.

McNamara, J., Lightfoot, S. B.-Y., Drinkwater, K., Appleton, E., & Oye, K. (2014). *Designing safety policies to meet evolving needs: iGEM as a testbed for proactive and adaptive risk management*. ACS Publications. Washington, DC.

Moya, J., Phillips, L., Schuda, L., Wood, P., Diaz, A., Lee, R., et al. (2011). *Exposure factors handbook: 2011 edition*. Washington: US Environmental Protection Agency.

National Academies of Sciences, E., & Medicine. (2017). *Preparing for future products of biotechnology*. Washington, D.C.: National Academies Press.

National Research Council, N. A. o. S. (1983). *Risk assessment in the federal government: Managing the process*. Washington, D.C.: National Academy Press.

National Research Council, N. A. o. S. (2002). *Biosolids applied to land: Advancing standards and practices*. Washington, D.C.: National Academies Press.

National Research Council, N. A. o. S. (2009). *Science and decisions: Advancing risk assessment*. Washington, D.C.: National Academies Press.

Northcross, A. L., Hwang, N., Balakrishnan, K., & Mehta, S. (2015). Exposure to smoke from the use of solid fuels and inefficient stoves for cooking and heating is responsible for approximately 4 million premature deaths yearly. As increasing investments are made to tackle this important public health issue, there is a need for identifying and providing guidance on best practices for exposure and stove performance monitoring, particularly for public health research. *EcoHealth, 12*(1), 196–199.

Pauwels, K., Mampuys, R., Golstein, C., Breyer, D., Herman, P., Kaspari, M., et al. (2013). Event report: SynBio Workshop (Paris 2012)–Risk assessment challenges of Synthetic Biology. *Journal für Verbraucherschutz und Lebensmittelsicherheit, 8*(3), 215–226.

Roca, J. B., Vaishnav, P., Morgan, M. G., Mendonça, J., & Fuchs, E. (2017). When risks cannot be seen: Regulating uncertainty in emerging technologies. *Research Policy, 46*(7), 1215–1233.

Roy, K., & Mitra, I. (2012). Electrotopological state atom (E-state) index in drug design, QSAR, property prediction and toxicity assessment. *Current Computer-Aided Drug Design, 8*(2), 135–158.

Rubow, K. L., Stange, L. L., & Huang, B. (2004). *Advances in filtration technology using sintered metal filters*. Paper presented at the 3rd China Int. Filtration Exhibition and Conf.

Ruckelshaus, W. D. (1983). Science, risk, and public policy. *Science, 221*(4615), 1026–1028.

Schiffman, S. S., Miller, E. A. S., Suggs, M. S., & Graham, B. G. (1995). The effect of environmental odors emanating from commercial swine operations on the mood of nearby residents. *Brain Research Bulletin, 37*(4), 369–375.

Schurch, S., Lee, M., & Gehr, P. (1992). Pulmonary surfactant: Surface properties and function of alveolar and airway surfactant. *Pure and Applied Chemistry, 64*(11), 1745–1750.

Scientific Committee on Health and Environmental Risks, S. C. o. E. a. N. I. H. R., and Scientific Committe on Consumer Safety. (2015). *Opinion on synthetic biology II-risk assessment methodologies and safety aspects*. Luxembourg: European Commission.

Sekine, R., Khurana, K., Vasilev, K., Lombi, E., & Donner, E. (2015). Quantifying the adsorption of ionic silver and functionalized nanoparticles during ecotoxicity testing: Test container effects and recommendations. *Nanotoxicology, 9*(8), 1005–1012.

Shanker, A. K. (2008). 21 mode of action and toxicity of trace elements.

Shepherd, J. (2009). *Geoengineering the climate: Science, governance and uncertainty*. London: The Royal Society.

Slovic, P. (1987). Perception of risk. *Science, 236*(4799), 280–285.

Solomon, J. D., & Vallero, D. A. (2016). From our partners – Communicating risk and resiliency: Special considerations for rare events. *The CIP Report*. Retrieved from https://cip.gmu.edu/2016/06/01/partners-communicating-risk-resiliency-special-considerations-rare-events/

Teasdale, A., Elder, D., Chang, S.-J., Wang, S., Thompson, R., Benz, N., & Sanchez Flores, I. H. (2013). Risk assessment of genotoxic impurities in new chemical entities: Strategies to demonstrate control. *Organic Process Research & Development, 17*(2), 221–230.

Tropsha, A., Mills, K. C., & Hickey, A. J. (2017). Reproducibility, sharing and progress in nanomaterial databases. *Nature Nanotechnology, 12*(12), 1111.

U.S. Environmental Protection Agency. (2015). *Quantitative structure activity relationship*. Retrieved from http://www.epa.gov/nrmrl/std/qsar/qsar.html

U.S. Environmental Protection Agency. (2017). *Guidelines for human exposure assessment: Peer review draft*. Washington, D.C.: U.S. EPA. Retrieved from https://www.epa.gov/sites/production/files/2016-02/documents/guidelines_for_human_exposure_assessment_peer_review_draftv2.pdf.

US Environmental Protection Agency. (1992). *Framework for ecological risk assessment*. Washington, D.C.: USEPA Risk Assessment Forum. Retrieved from https://www.epa.gov/sites/production/files/2014-11/documents/framework_eco_assessment.pdf.

Vallero, D. A. (2008). *Fundamentals of air pollution* (4th ed.). Amsterdam; Boston: Elsevier.

Vallero, D. (2010a). *Environmental biotechnology: A biosystems approach*. Academic Press. Burlington, Massachusetts.

Vallero, D. (2010b). *Environmental contaminants: Assessment and control*. Academic Press. Burlington, Massachusetts.

Vallero, D. A. (2013). Measurements in environmental engineering. In M. Kutz (Ed.), *Handbook of measurement in science and engineering*. Wiley, Hoboken, New Jersey.

Vallero, D. A. (2014). *Fundamentals of air pollution* (5th ed.). Waltham: Elsevier Academic Press.

Vilar, S., Cozza, G., & Moro, S. (2008). Medicinal chemistry and the molecular operating environment (MOE): Application of QSAR and molecular docking to drug discovery. *Current Topics in Medicinal Chemistry, 8*(18), 1555–1572.

Wambaugh, J. F., Wang, A., Dionisio, K. L., Frame, A., Egeghy, P., Judson, R., & Setzer, R. W. (2014). High throughput heuristics for prioritizing human exposure to environmental chemicals. *Environmental Science & Technology, 48*(21), 12760–12767.

Weschler, C. J., & Nazaroff, W. (2012). SVOC exposure indoors: Fresh look at dermal pathways. *Indoor Air, 22*(5), 356–377.

Winkler, J. D., Halweg-Edwards, A. L., & Gill, R. T. (2015). The LASER database: Formalizing design rules for metabolic engineering. *Metabolic Engineering Communications, 2,* 30–38.

World Health Organization. (2000). *Environmental health criteria 214: Human exposure assessment.* Geneva: WHO.

Zhang, X., Arnot, J. A., & Wania, F. (2014). Model for screening-level assessment of near-field human exposure to neutral organic chemicals released indoors. *Environmental Science & Technology, 48*(20), 12312–12319.

Zhang, Y., Zhang, N., & Niu, Z. (2018). Health risk assessment of trihalomethanes mixtures from daily water-related activities via multi-pathway exposure based on PBPK model. *Ecotoxicology and Environmental Safety, 163,* 427–435.

Mosquitoes Bite: A Zika Story of Vector Management and Gene Drives

David M. Berube

Most of the articles in this volume involve biosafety and biosecurity and strict biosafety and laboratory biosecurity protocols and whether current standards sufficiently assure the safety of both the public and the overall ecosystem. This chapter suggests that developments occurring in mosquito vector control using synthetic biology will introduce new genetic bugaboos into the debate over releasing genetically altered species, both accidentally and purposefully.

Mosquitoes are incredibly dangerous insects. There are around 3600 species of mosquitoes and about 100 spread human disease. Two species will be the subject of this chapter: *Aedes aegypti* and *Aedes albopictus*. This focus is derived from the research I have been doing for 2 years on the Zika virus and what we should have learned about the recent outbreak.

Ae. aegypti, the main mosquito species known to transmit ZIKV (Zika virus), can be found in the Middle East, Africa, and Asia. Half of the planet's population lives in areas where *Ae. aegypti* is present. They have been documented in 258 counties in the United States. The most concentrated populations are still in Southern California, Arizona, Texas, Louisiana, and Florida, but they have been found as far north as New Hampshire.

A close relative, *Ae. albopictus*, is also thought to be a potential carrier and could bring ZIKV to new areas because it can survive in cooler temperatures than its cousin. In 2016, *Ae. albopictus* has been reported in 1368 counties, including 127 counties that have no previously known populations. Additionally, 177 counties are home to both *Ae. aegypti* and *Ae. albopictus* species (Rose 2016).

According to UC Riverside entomologist Omar Akbari "*Aegypti* is literally probably the most dangerous animal in the world" (McCay 2016). Bill Gates has blogged that they are even more dangerous to humans than humans themselves (Gates 2014). Mosquitoes kill nearly 15 times more people than snakes and 72,000 times more

D. M. Berube (✉)
NCSU, Raleigh, NC, USA

© Springer Nature Switzerland AG 2020
B. D. Trump et al. (eds.), *Synthetic Biology 2020: Frontiers in Risk Analysis and Governance*, Risk, Systems and Decisions, https://doi.org/10.1007/978-3-030-27264-7_7

than sharks (Beck 2016). According to the World Health Organization's (WHO) report, mosquitoes kill an estimated 700,000 people a year (Guardian 2017).

Female mosquitoes suck human blood. In the process, some transfer dangerous bacteria and viruses. Efforts to reduce viral infections from mosquitoes have been frustrated by a plethora of problems, one of which is that viruses evolve and mutate. There is a powerful tradition in modern science to turn to vaccines as a solution to infectious diseases. Regrettably, vaccines are not effective against many viruses as they mutate. This frustration makes vaccine research not only ineffective but also discouraging. Pharmaceutical companies profit less from vaccine research than many of their other ventures which helps account for the few companies still in the field. In addition, vaccines are subject to vaccine hesitancy (MacDonald 2015) whereby the public is hesitant to vaccinate themselves and their children fearing the vaccine will cause an infection and/or fearing the vaccine will have undesirable side effects such as autism in the case of the MMR (measles, mumps, and rubella) and promiscuity with the HPV (human papillomavirus) vaccine.

Alternative approaches have included making the victims less attractive to female mosquitoes through repellants in the form of ultrasonic devices, fire activated coils, bracelets, citronella candles, barrier yard sprays, aerosols, oil of lemon eucalyptus, picaridin, etc. (Consumer Reports 2018). The most highly recommended is DEET (N,N-Diethyl-meta-toluamide). DEET is hardly ideal (ATSDR 2015). The problem with most of these approaches is that they fail. Those that work require physical application and reapplication.

Other approaches involve vector management and are the subject of this chapter. Traditionally, vector management (reducing populations of biting mosquitoes) has included larvicides and adulticides (technical term of insecticides). Larvicides include *Bacillus thuringiensis israelensis* (Bti), *Bacillus sphaericus* (Bs), *metho-prene*, *temephos*, and *spinosad* or oil dispersants such as *Kontrol* or *CocoBear*.

Pyriproxyfen is an insect juvenile-hormone analogue used in Brazil. It is active against many arthropods and has been in use for agricultural pest control for about 15 years. It is effective at inhibiting adult *Ae. aegypti* emergence at concentrations of less than or equal to one part per billion, can be applied in various formulations (e.g., sticks, granules), and is cost-competitive. It remains effective up to 5 months, longer than *Bacillus thuringiensis israelensis*, *methoprene*, or *temephos*, and is less toxic. Adult mosquitoes exposed to *pyriproxyfen* have decreased fecundity. Importantly, contaminated adults can disseminate lethal doses from treated to untreated sites (Morrison et al. 2008). Though a group of Argentinian doctors and another from Brazil known as PCST (Physicians in Crop-Sprayed Towns) claimed *pyriproxyfen* caused Zika, their claims have been summarily debunked.

Light traps, biologicals (such as fish and other insects), and irradiation are inef-fective or minimally effective and are used as components within a regimen of other approaches.

Adulticides or insecticides including *malathion*, *chlorpyrifos*, *dibrom*, *naled*, *dichlorvos*, *permethrin*, and *sumithrin* are generally ineffective. The mosquitoes responsible for Zika are resistant. *Ae. aegypti* live in homes. The US Centers for Disease Control and Prevention (CDC) has written, "Adulticiding, application of chemicals to kill adult mosquitoes by ground or aerial applications, is usually the least efficient mosquito control technique" (NALED 2016). Furthermore, insecticides have dangerous ecological signatures. Simply put, pesticides may be far more dangerous to beneficial creatures in the environment than they are harmful to the *Ae. aegypti* mosquito populations.

Some creative approaches that surfaced recently with the rise of the Zika pandemic of 2015–2016 involved using bacteria known as *Wolbachia* and a genetically engineered (GE) mosquito by a company called Oxitec. These approaches have been highly controversial.

Two companies have developed a bacterial approach using *Wolbachia*: Eliminate Dengue and MosquitoMate. MosquitoMate seems to be the top player, having garnered Gates Foundation support to use this approach to target malaria. MosquitoMate Inc. working with researches at the University of Kentucky claims if female *Aedes* mate with a male that has *Wolbachia*, her eggs will not hatch. The EPA has given commercial approval for MosquitoMates's *albopictus* ZAP mosquitoes for 5 years.

Neither the MosquitoMate nor the Eliminate Dengue group has met much resistance to its technology, presumably because their approaches involve naturally occurring bacteria, not engineered genes. Their experiments have met with some mixed results but are mostly positive. Many of the problems, such as separating male from female mosquitoes after application and before release and heat stress negatively affecting overall effectiveness, are being resolved.

Oxitec's approach differs from the *Wolbachia* approach, and the company has received support by the Gates Foundation as well. The Oxitec mosquito control program involves the repeated controlled release of (GE) male *Ae. aegypti* mosquitoes (strain *OX513A*), expressing a conditional lethality trait and a fluorescent marker.

When *OX513A* is reared in the presence of tetracycline as a dietary supplement, expression of tTAV (a cellular protein) is repressed, allowing normal cell function and survival. tTAV quantities can cause cell malfunction and >95% mortality before adulthood (Gorman et al. 2015). Developmental failure occurs when the cells cannot make the proteins they require to function normally, which then causes cell death. This is known as transcriptional squelching (Oxitec 2016).

The strain was first constructed in 2002, and a publication about it appeared in a peer-reviewed scientific journal in 2007 (Atkinson et al. 2007). It has been characterized for over 10 years. Oxitec uses a self-limiting strategy, meaning the modification tends to disappear from the target population unless replenished by periodic release of additional modified insects.

While it has been tested abroad, efforts to release the *OX513A* mosquito in Florida were resisted. Oxitec generally cites successes in tests conducted in Brazil, Panama, and the Cayman Islands. Between 2011 and 2013, the team released Oxitec mosquitoes in three neighborhoods in the northeastern state of Bahia. It reported at least 90% population reductions in all three…. In Brazil, it has been called the "Friendly *Aedes aegypti* Project." Oxitec reported roughly a 96% reduction in the mosquito population in the tiny 0.16 square kilometer release area in the Cayman Islands (Servick 2016). The results in the Cayman Island have been contradicted. GeneWatch UK has reported that new information shows that the releases have been ineffective and large numbers of biting female GM mosquitoes have been released. Dr. Helen Wallace, director of GeneWatch UK, said: "Oxitec's GM technology is failing in the field and poses unnecessary risks. Islanders' money should not be thrown away on an approach which has not been successful" (Caribbean News Now 2018).

However, field tests in Brazil have been generally successful. A field test in Itaberaba, a suburb of Juazeiro, was 95% successful based on adult trap data (Fang 2015). There, mosquito larvae were reduced by 82% last year, which led to a massive 91% drop in dengue fever cases. Now, Oxitec has demonstrated that its approach can sustain mosquito reduction in the long term. After releases in a neighborhood of Piracicaba, the company achieved an 81% reduction in mosquito larvae in the second year. In addition, a new trial in a second, central neighborhood of Piracicaba has shown a 78% mosquito reduction in only 6 months (Fernandez 2017).

Oxitec announced in May 2018 it was adding to its Friendly™ *Aedes aegypti* mosquito line. *OX5034* began open field trials on May 23, 2018. OX5034 is the next generation of Oxitec's non-biting Friendly™ Aedes mosquitoes, designed to reduce populations of the disease-spreading *Ae. aegypti* mosquito (PRNewswire 2018).

Absent a vaccine, Oxitec's technology could be our most effective tool in fighting Zika. GM mosquitoes could go a long way toward fighting some of the world's deadliest viruses including yellow fever and dengue. The bigger question is whether we will let them (Brown 2016a). This is not a "gene drive" approach, discussed below, but trying to convince the opponents on the difference might be one of the most challenging scientific arguments of the decade.

Gene Drives and Mosquitoes

The next generation of engineered mosquitoes may involve gene drives. The term gene drive refers to the ability of a gene to be inherited more frequently than Mendelian genetics would dictate, thus, increasing in frequency, perhaps even to fixation (Adelman and Tu 2016).

Detailed descriptions of gene drives, such as CRISPR (clustered regularly interspaced short palindromic repeats)/Cas9 (CRISPR-associated protein 9 enabled), will be left to the other chapters in this volume. However, it must be noted mosquitoes modified with gene drive systems are being proposed as new tools that will complement current practices aimed at reducing or preventing transmission of vector-borne diseases, especially malaria (James et al. 2018).

Even if Oxitec is chased away, the idea of undermining *Ae. aegypti* at a genetic level will persist. More sophisticated gene-editing techniques have been developed, and new businesses will emerge to take advantage of them. Gene drive systems have the potential to spread beneficial traits through interbreeding populations of mosquitoes carrying infectious diseases.

However, the characteristics of this technology have raised concerns that necessitate careful consideration of the product development pathway (James et al. 2018). The permanent presence of a mosquito with novel traits is an inherently difficult topic with which to deal, mainly due to the unforeseen future risks (Sikka et al. 2016).

Synthetic Biology

Synthetic biology is in its infancy and the first commercial applications are likely to appear incrementally. Depending on the pace of development, we might expect to see commercialized outputs from synthetic biologists in full production within the next 10 years (Lloyd's 2009).

In general, synthetic biology is variously described or treated as the application of engineering principles to genetic modification; or a generic set of tools, technologies, and approaches (essentially services) for achieving biotechnology objectives; or as simply a synonym for biotechnology, with no meaningful difference between the two (Trump et al. 2019).

Industry research estimated that equity funding to private synthetic biology companies topped $1bn in 2016, which is helping to drive market forecasts to an estimate of almost $40bn by 2020(Polizzi et al. 2018). One of the first examples of synthetic biology's application was the production of the antimalarial therapy artemisinin in yeast by introducing additional genes encoding the biosynthesis of artemisinic acid from natural fatty acid precursors (Polizzi et al. 2018).

Gene drive systems for population modification of vector mosquitoes have been proposed for nearly half a century (see Curtis 1968). "Gene drives have enormous potential for the control of populations of insect vectors and pests," mosquito researchers Tony Nolan and Andrea Crisanti wrote in *The Scientist*. "They are species-specific, self-sustaining, and have the potential to be long-term and cost-effective" (Brown 2017).

Because gene drives could rapidly propagate novel DNA through an entire population in the wild, they could be used, proponents say, to eradicate marauders. They might make mosquitoes resistant to the microbes that cause malaria or dengue fever or even block the gene that makes locusts swarm, saving millions of tons of crops every year (Begley 2015).

Risks associated with synthetic biology research include accidental release of biological organisms (bioerror), construction of biological weapons (bioterror), and the unintended consequences of biological research. The likelihood of bioerror and bioterror are low relative to unintended consequences. At present, there is a high degree of uncertainty about the types of things that can go wrong, let alone the probability of these risks occurring. These risks are examined elsewhere in this volume.

Three Current Approaches Using Gene Drives to Engineer Mosquitoes

This field is very young and new research findings and approaches are inevitable. Before getting to this non-exclusive list of approaches, there is the prerequisite entomology.

Mosquitoes are highly sexually dimorphic. For the major vectors of malaria, dengue, and Zika, there is also a clear separation between harmless, nectar-feeding males and deadly blood-feeding females. The identification of molecular switches and genetic programs that control the decision made in the early developing mosquito embryo to proceed as male or female may be used to improve existing sterile insect strategies for controlling mosquito-borne disease agents (Adelman and Tu 2016):

> Scientists could alter mosquito genetics to spread a fatal flaw through the entire population, reducing overall numbers; they could modify mosquitoes to produce more male offspring than female offspring, reducing the number of mosquito bites; or they could equip mosquitoes with genes to help them fend off malaria, reducing transmission of the disease within mosquito populations and thus to humans, too. (Brown 2017)

Sexual Biasing

Gene drives for mosquitoes have recently been designed to cause little or no disadvantage to offspring receiving a copy of the gene drive from only one parent, but to cause sterility in females which receive the gene drive from both parents. This design under development by Zach Adelman and Zhijian Tu from Virginia Tech (2016) and others is intended to allow the gene drive to spread rapidly through populations until it accumulates to a high level, at which point the population numbers will crash, with little opportunity for mosquitoes to escape this fate through natural selection. The aim is to deploy this system in the environment to rapidly reduce the mosquito population to below the threshold level that supports the spread of diseases like malaria and dengue fever and therefore to massively reduce or even eliminate these diseases (Polizzi et al. 2018).

Expression of this so-called M factor, a sex-determination gene called Nix, in female embryos triggers the development of external and internal male genitalia. Nix is both necessary and sufficient to initiate male development in *Ae. aegypti*, a major carrier for dengue, yellow fever, Zika, and chikungunya viruses. "This discovery sets the stage for future efforts to leverage the CRISPRCas9 system to drive maleness genes such as Nix into mosquito populations, thereby converting females into males or simply killing females," Tu says (Adelman and Tu 2016; Cell Press 2016).

Since females have one kind of sex chromosome (XX) and males have both kinds (XY), scientists aimed to prevent the passing on of an X chromosome to the next generation. This leads to mosquito populations with over 95% male offspring. Andrew Hammond and colleagues from Imperial College in London developed their gene drive system to ensure low proportions of female mosquitoes. Their drive is designed such that female mosquitoes carrying only one copy would be fertile whereas females carrying two copies would be sterile. However, inheritance of even one gene drive copy reduced female fertility by 90–95%. This seemed to result from the element copying itself in somatic as well as germline cells, so that enough somatic cells carried two copies to render the females infertile. As a result, the ability of the element to spread was greatly reduced (Tome 2017).

Using an enzyme that targets some 200 sites on the X chromosomes in mosquito eggs, the team has managed to shred X chromosomes so thoroughly "that it's too much for the cell to repair," Burt said. The result is that eggs carry only the Y chromosome, which makes sons, and no X, which makes daughters. So far, they have gotten 95% sons using an editing tool other than CRISPR (Alphey 2016).

In Hammond's research, the researchers were able to spot three genes (AGAP005958, AGAP011377, and AGAP007280) which grant a recessive female-sterility characteristic when interfered. CRISPR-Cas9 constructs engineered to target and edit each gene were inserted into each targeted locus. Results showed that for each targeted locus, a strong gene drive was exhibited at the molecular level, with transmission rates ranging from 91.4% to 99.6% (Begley 2015).

Omar Akbari, cited earlier, from UC Riverside is also using CRISPR to inactivate a fertility gene in female *Ae. aegypti* to sterilize future generations of females (Tome 2017).

Egg Shells

Jun Isoe and Roger Miesfield, a team from the University of Arizona, believe they have isolated the gene responsible for hardening (tanning) the egg shells laid by disease-carrying mosquitoes. Shortly after the eggs are deposited in a moist area such as a flower pot or the edge of a pond, the eggshell hardens through a process called "tanning" in which the eggshell turns from white to brown during the maturation process (McCay 2016). Because the drought-resistant larvae will hatch from the hardened mosquito eggs in the presence of water during the monsoon season in areas such as Tucson, the complete formation of the eggshell within a few hours of egg laying is essential to mosquito reproduction (Brown 2017):

> To discover mosquito genes that are uniquely required for mosquito eggshell synthesis in blood-fed female *aegypti* mosquitoes, we used a computer-based approach to identify genes that have evolved to be unique to mosquitoes and are not found in closely-related insects such as fruit flies and honeybees, nor in animals such as ourselves. Among the roughly 100 mosquito-specific genes we disrupted in blood-fed *aegypti*, we found one we call Eggshell Organizing Factor 1 (EOF1) that was absolutely required for completion of eggshell synthesis. We discovered that 100 percent of the eggs laid by mosquitoes lacking the EOF1 protein were missing a tanned eggshell, and none of the larvae survived. The lethal effect of an EOF1 deficiency was in part because the eggs did not complete the tanning process required for eggshell maturation (Miesfeld and Isoe 2017).

Miesford and Isoe claim: This does not mean that mosquitoes should be eliminated from our ecosystem, as that could have unknown consequences. Instead, we want to selectively reduce mosquito populations by decreasing reproductive rates at specific times of the year, such as the rainy season (Miesfeld and Isoe 2017).

Disrupting Blood Meal Activity

Sharon Begley (2015) argues that gene drive might be harnessed to target insect-borne disease. If CRISPR replaced the gene in mosquitoes that lets them detect the odor of people, and substituted a dud, and if gene drive ensured the dud was carried by both chromosomes, then every offspring would have a double dose of the dud. Eventual result: mosquitoes that can't smell humans, reducing their odds of biting (Begley 2015).

Implication: Species-Cide

Species-cide is the engineering of extinction of an entire species. The three gene drive approaches mentioned above could result in the collapse of the entire population of a species of mosquito.

There is nothing sinister about extinction. Species go extinct all the time. The disappearance of a few species, while a pity, does not bring a whole ecosystem crashing down: we're not left with a wasteland every time a species vanishes. Removing one species sometimes causes shifts in the populations of other species, but different need not mean worse (Judson 2003).

The scientific community is largely unperturbed by the idea of removing the *Ae. aegypti* mosquito from the face of the Earth. Over 2 years of research have failed to uncover in the technical literature any significant drawback from *Ae. aegypti* or *Ae. albopictus* extinction. There are over 3000 species of mosquito, and there seems to be no species dependent on the specific *Ae. aegypti* or *Ae. albopictus* subspecies as food sources. Some insects, birds, and even bats include mosquitoes in their diet, but mosquitoes do not compose the exclusive diet for any of them.

"The disappearance of a few species, while a pity, does not bring a whole eco-system crashing down," evolutionary biologist Olivia Judson has written (Kolker 2016).

Gene Drives, Controversy, and Science Communication

Why would a science and technology communication scholar be interested in vector management of mosquitoes and the Zika virus? As the pioneering synthetic biologist Jack Newman put it to Kristen Brown, "What stands between us and addressing one of the biggest public health issues (mosquitoes and infectious diseases) in the world is not science. It's how we talk about science" (Brown 2016a).

Over a decade ago, Xi et al. wrote that "a gene-drive vehicle is an important component of vector population replacement strategies, providing a mechanism for the autonomous spread of desired transgenes into the targeted population. Compared with strategies that rely on inundated releases and Mendelian inheritance, gene drive strategies would require relatively small 'seedings' of transgenic individuals into a field population. Perhaps more important than increased cost efficacy, gene drive strategies can facilitate population replacement with transgenic individuals that have a lower fitness relative to the natural population" (Xi et al. 2005).

The anxiety here is clearly that of the unknown. About three-quarters of adults said they thought the technology would be used before the health effects are fully understood. Unsurprisingly, among those who were already aware of gene editing technology, a higher percentage (57%) said that they were inclined to give it a whirl (Brown 2016b).

Key Haven, Florida: Attitudes About Genetic Engineering

It is important to note that the Oxitec approach does not involve a gene drive. Most importantly, the status of the environment is restored when releases are stopped (i.e., the released mosquitoes all die, and the environment reverts to the pre-trial status) (Oxitec 2016). Nonetheless, the public is reticent to expose themselves to this technology. In comparison, insecticides influence "insect life right across the spectrum" and destroy the food chain, while the Oxitec method targets just one species, which contributes little to the environment, Oxitec CEO Haydn Parry maintained (Caldera 2017).

The fact that MosquitoMate's killer mosquito is not genetically modified works in its favor. While Oxitec's genetically modified mosquito was rejected in Florida, MosquitoMate has already successfully conducted several tests there as there was no public resistance to it (Yeoh 2017).

Back in 2013, 59% of residents in Key Haven, the neighborhood where the trial is planned, supported Oxitec's project, with only 9% opposed. Now, 58% are in opposition (Brown 2016b). When Oxitec attempted field testing in Key Haven, Florida, in 2016, it met with significant resistance. Eventually, Oxitec cancelled its original efforts in 2016 after release was rejected in a referendum.

The Monroe County Mosquito Board approved trials elsewhere in the Keys at a location still to be determined. Parry said, "While we did not win over every community in the Keys, Oxitec appreciates the support received from the community and is prepared to take the next steps with the Florida Keys Mosquito Control Board" (Bluth and Kopp 2016).

Oxitec will get their release in one of six places under consideration by the district in Monroe County though not in Key Haven. This solidified a long-anticipated agreement. Oxitec will need to return to the US Environmental Protection Agency for approval of a new site like it did for the Key Haven site. If approved, the approval would last a maximum of 2 years (Atkins 2016).

Relevant Variable Explaining US Attitudes

The *Ae. aegypti* is a tiny insect responsible for infecting millions with debilitating and sometimes deadly viruses. Surely this can be viewed as a rare case when genetic modification is a positive undertaking (Moses 2016). However, when it comes to how the public responds toward scientific, especially genetics, efforts to control an epidemic, nothing is guaranteed. They are leery about what seems unnatural. "Anything GMO freaks me out because it's like you're putting your hand in something that God has set up," said Kayla Efrece, a Labelle resident (Polansky 2016).

Opposition to the use of GE mosquitoes may merely reflect public concerns of genetic engineering in general rather than the case instance. In psychology, this is called transference.

In Oxitec's case, we have high levels of uncertainty and unknowns as well as the two-headed specter of a private business interest, Oxitec and genetic engineering as a proposed remedy (Finkel et al. 2018). For this set of stakeholders, the phrase "genetically engineered" alone inspires an immediate resistance to Oxitec's mosquitoes (Brown 2016a). The public tends to worry about the effect of adding a mosquito that contains DNA cobbled together from *E. coli*, coral, a vinegar fly, and a cabbage looper moth (Brown 2016a).

As well, we note NIMBY ("Not in my backyard") effects if we are to believe the results of the referendum in Key Haven, Florida (Burgess et al. 2018). Given the dengue outbreak in Key West, Florida, in 2009–2010, we might expect a more generous attitude toward field tests in Key Haven. However, Key Haven is a section of Key West and does not represent the population at large; hence, Key Haven is not Key West's "backyard."

This takes on more meaning when we consider the costs to tourism. The Keys depend almost solely on visitors to the warm teal waters for income, with the hotel occupancy rates near 100% in February and March, and tens of thousands of visitors arriving from around the world in winter months. The CDC previously warned people travelling in Brazil and the Caribbean because of the Zika virus outbreaks. Zika virus-prone areas lost 30% of their regular profits, while Zika-free zones like Hawaii earned twice as they used to.

For some, tourism was more threatened by a Zika outbreak than a release of genetically engineered mosquitoes. For example, in a letter to Secretary Burwell who headed the Department of Health and Human Services, Pinellas County, a partisan coalition pled with her to use "emergency use" authorization to allow the county to deploy Oxitec mosquitoes (see below). Their rationale: As you are aware, the ZIKV poses serious risk to our tourism industry and we must do everything in our power to stop the spread of the ZIKV to protect our economy. Last year, Florida saw over one million visitors to the state creating over $89 billion in economic impact and over one million jobs. We cannot afford to have visitors cancel their vacation plans due to the ZIKV and urge you to provide Pinellas County authorization to combat the ZIKV immediately (Gartner 2016).

On the other hand, there is a contrary opinion. Jeffrey Smith, the executive director of the Institute for Responsible Technology in Iowa, testified the release of modified mosquitoes could backfire, resulting in unintended consequences that could harm the state's economy and environment. "They are using this emergency (in Florida) to rush into production this very risky technology." "And even if the modified mosquitoes worked as planned, foes of GE organisms could decide to stay away from Florida in droves, hurting the Florida tourism industry (Gartner 2016).

Some Surveys

In a 2016 survey by the University of Pittsburgh Medical Center (UPMC) of Fort Haven, for those residents who did not support GM mosquito use, the survey provided seven possible reasons for opposition of this method. Of these reasons, respondents chose the following reasons most often: "I am concerned about the overall safety of GM mosquitoes" ($n = 37$); "Introduction of GM mosquitoes could upset the local ecosystem by eliminating mosquitoes from the food chain" ($n = 20$); and "The use of GM mosquitoes could lead to the use of other GMO products in [community]" ($n = 19$). Other reasons included concerns that the mosquito could make people sick and/or pass on modified genes to people and animals but were chosen less and thus were less important to the respondents (Carroll 2016). These survey findings are reported here because they underlie the entire crisis. Public support may have had little to do whatsoever with this case. The concern may simply have been once GM organisms are introduced they develop a foothold and become the basis for the deployment or marketing of other GMOs.

A 2017 Pew Research Center survey asked 4726 people how they feel about gene editing and other human-enhancing technologies. More than 60% said that they were "very" or "somewhat" worried about such technologies (Adaja et al. 2016). According to another Pew survey last year, 88% of scientists believe it's safe to eat genetically modified foods, but only 37% of the public does. That gulf is even wider than the one between the public and scientists about climate change (Brown 2016a).

A 2018 Pew survey of 2537 adults found about 7 in 10 Americans (72%) say that changing an unborn baby's genetic characteristics to treat a serious disease or condition that the baby would have at birth is an appropriate use of medical technology, while 27% say this would be taking technology too far. A somewhat smaller share of Americans says gene editing to reduce a baby's risk of developing a serious disease or condition over their lifetime is appropriate (60% say this, while 38% say it would be taking medical technology too far). But just 19% of Americans say it would be appropriate to use gene editing to make a baby more intelligent; eight in ten (80%) say this would be taking medical technology too far (Pew Research Center 2018).

As more data is generated, we may be able to design a communication strategy that can rehabilitate the public's view of things genetic, like gene drive mosquitoes. What we learn from these surveys and others will need to play through what we already know about inferential shortcuts used by the general public to make sense out of things.

Conflation Effects

For the public, there seems to be little if any difference from what Oxitec has done and what an engineering using CRISPR-Cas9 technology may do in designing a gene drive. The meta-debate is over the release of gene drive-modified organisms not the finer delineations of experimental procedures and protocols.

For the public, the debate over GE mosquitoes is not much different from the debate over GE foods. The primary challenge in infectious disease communication is helping the public to understand that GE mosquitoes as a vector management strategy is not just another genetic intrusion; there is a contextual imperative at work in the case of Zika.

There are two powerful heuristics at play with the public understanding of all risks, especially those that are difficult for a member of the general public to understand. They are similar, and both lead to a bias whereby the public overgeneralizes about the risk of something based on what they already know or feel is right and can recall and remember. If they know little, then they find something they understand in what they have learned before regardless of its actual pertinence. They are the confirmation and availability biases (Funk and Hefferon 2018; Nickerson 1998). These, respectively, denote judgments based on confirming what a member of the public may already believe is true and based on what a member of the public may have heard before. Examples abound of sensational and hyperbolic press and digital media reports hyperbolically conflating all biotechnology together regardless of approach or context. This has encouraged stereotypes and generalizations about all biotechnology, including the approach used by Oxitec. A negative sign associated with one instance of biotechnology transfers itself to a different instance regardless of the incommensurability.

The fear mongering is significant from both sides of the debate. Proponents of GE mosquitoes appeal to human tragedy, mostly pregnancy-associated birth defects, while opponents generate often unsupportable hypotheses about the impact of releasing Oxitec's mosquitoes and gene drive mosquitoes. These counterclaims were recently debunked by a team of journalists working out of Key West and Kristie Wilcox, a staff writer, who blogs for *Discovery Magazine* (Wilcox 2016).

Why are so many members of the public highly opposed to genetically modified mosquitoes given the risks of the ZIKV and its danger to fetuses, young children, and the public at large? Why does it make more sense to employ intensively pesticide spraying which is both dangerous and, in the case of the vector for the ZIKV, mostly ineffective?

As cognitive psychologist Susan Fiske has put it, we are cognitive misers (Fiske and Taylor 2013). We do not want to spend the time and energy it takes to make sense out of much we know little about. Instead as Sherry Chaiken and Daniel Kahneman have argued, we use mental shortcuts (Chaiken 1980; Kahneman 2011). Sometimes we get things right but more than not we can get them very wrong. The dominant affirmation and availability biases blend all this GE as a single phenomenon. Teasing out the differences requires a new strategy toward public engagement which is the subject of the much longer work (see Acknowledgment).

Enter CRISPR-Cas9

Several mechanisms are being examined to achieve gene drive. There are two approaches: population suppression and population replacement.

Population suppression strategies are intended to reduce the size of the vector population to such an extent that it will not be able to sustain malaria transmission. This is an extension of the goal of all current vector-control products and does not require driving a population to extinction. Population suppression strategies are based on inactivation, or knockout, of genes involved in the target mosquito's survival or reproduction (e.g., reducing fertility or production of female progeny) and/or bias of the sex ratio toward males. These may be termed "loss of function" techniques (James et al. 2018).

Population replacement strategies are intended to reduce the inherent ability of individual mosquitoes to transmit the malaria pathogen. These strategies may be built around inactivation of a gene or genes that facilitate parasite survival in the mosquito vector or that are required for the mosquito to transmit malaria, such as a tendency to feed on humans (James et al. 2018).

Until recently, the attempted methods either did not work in mosquitoes or were difficult to engineer; however, discovery of the CRISPR-Cas9 system for gene editing has provided a widely accessible and versatile molecular tool for creating driving transgenes (James et al. 2018).

Computational modeling based on other gene drive systems suggests that the type of drive that can be achieved with the CRISPR-Cas9 system can be so effective that release of low numbers of modified mosquitoes into the environment could result in establishment of the genetic modification in the natural interbreeding population. Computer simulations and population genetic analyses suggest that gene drive strategies for reducing or modifying the population of vector mosquitoes both have the potential to provide a transformative new tool for conquering malaria and to make a valuable contribution toward the elimination, and ultimate eradication, of this disease (James et al. 2018).

The history of the development of CRISPR-Cas9 will be left to another author in this book. Noteworthy is the report that CRISPR-Cas9 has produced promising results in the *Ae. aegypti* species. In a study published in 2015, researchers were able to generate specific mutations and insertions to better understand how genetic manipulation of a vector might affect the transmission of diseases such as dengue and chikungunya (Kistler et al. 2015).

As mentioned earlier, Texas Tech's Adelman and Tu from Virginia Tech are using CRISPR-Cas9 to build a gene drive to explode the male population and produce a local population crash. The Hammond team from Imperial commits to reducing the virus-carrying mosquitoes to levels that there are too few left to transmit pathogens from one person to another. Using CRISPR-Cas9, they plan to disrupt genes involved in producing eggs in females and then build a gene drive that passes that trait along to as many as 99.6% of their offspring (McCay 2016).

Finally, many others have demonstrated that CRISPR-based editing is also highly effective in *Ae. aegypti*, where mutant phenotypes can be detected in injection survivors and characterized as somatic mosaics or used to generate heritable mutations (Hall et al. 2015; Basu et al. 2015; Kistler et al. 2015; Dong et al. 2015).

But gene drives offer a significant advantage. While Oxitec will have to release tens of thousands of mosquitoes over many years to significantly impact the wild population, in theory, the gene drive automates the spread of any lab-made alteration by relying on "selfish genes" to force desired traits into offspring. Non-gene drive approaches represent a long-term financial and administrative commitment that must be maintained even in the absence of continued transmission.

Alternatively, the extinction gene itself might prove unstable and jump into a different species entirely. Though such jumping is not unknown for wild selfish genetic elements, it is rare, and the chance of this being a problem seems remote. (The risk to humans from this technology is negligible. Even supposing an extinction gene appeared in humans by accident or by malice, it would take thousands of years for extinction to be affected. During this time, it is inconceivable the gene's spread would go unnoticed; once noticed, it could easily be stopped)(Hall et al. 2015).

These observations have led some scientists including Andrea Crisanti at Imperial College to conclude the vector may be the Achilles' heel (Judson 2003). Entomologist Zach Adelman believes it is our moral duty to eliminate this mosquito (Adler 2016). Gregory Kaebnick at the Hastings Center admits wiping a species off the planet is "an unfortunate thing to do" and "we ought to try not to do it," but a serious public health threat could be an ethical justification (McCay 2016).

However, gene drives as part of the genetic engineering remain highly suspicious to the general public, which expresses fears from the same old panoply of boogeymen of earlier GM debates over food (Trump et al. 2018). Meanwhile, experts have their own reservations about gene drives such as CRISPR.

Public Concerns About CRISPR

The public decodes gene drives in terms of their applications. The fear is that germline engineering is a path toward a dystopia of superpeople and designer babies for those who can afford it. Want a child with blue eyes and blond hair? Why not design a highly intelligent group of people who could be tomorrow's leaders and scientists? Just 3 years after its initial development, CRISPR technology is already widely used by biologists as a kind of search-and-replace tool to alter DNA, even down to the level of a single letter. It's so precise that it's expected to turn into a promising new approach for gene therapy in people with devastating illnesses (McCay 2016).

The ease with which CRISPR gene editing can be carried out has raised worries that humans could be next. Those fears were stoked in April when a Chinese team based at the Sun-Yat-Sen University reported altering human embryos in the laboratory to correct a genetic defect that causes beta-thalassemia.

For some, when they hear GE, they think Jurassic Park; or they believe that just because something is natural, it is somehow better. "The public fears genetic engineering. Nearly all politicians don't understand it," said Arthur Caplan, the founding director of the Division of Medical Ethics at NYU School of Medicine. "I don't think the issue is economic. It is ignorance, distrust, fear of the unknown, fear of prior efforts to use biology to combat pests which went sour" (Regalado 2015).

Expert Concerns About CRISPR

The expert community has other observations and reservations about the technology itself, only a few of which are mentioned below. For example, the University of Hawaii biologist Floyd Reed who works in avian malaria stresses caution when it comes to population modification gene technology since changes to a single small release could theoretically spread to the entire species (Lafrance 2016). Members of the IUCN voted for caution on gene drive technology at the ongoing World Conservation Congress in early September 2016. They passed a non-binding motion calling on their members to refrain "from supporting or endorsing research, including field trials, into the use of gene drives for conservation or other purposes" until a rapid assessment was completed by 2020 (Agence France Presse 2016).

CRISPR-specific reservations are beginning to surface. A Chinese team report showed the method is not yet very accurate, confirming scientific doubts around whether gene editing could be practical in human embryos and whether GE people are going to be born anytime soon (Liang et al. 2015). The tool can cause large DNA deletions and rearrangements near its target site on the genome, according to a paper published on July 16, 2018, in *Nature Biotechnology*. Such alterations can muddle the interpretation of experimental results and could complicate efforts to design therapies based on CRISPR (Kosicki et al. 2018).

Kosicki et al. (2018) write: exploration of Cas9-induced genetic alterations has been limited to the immediate vicinity of the target site and distal off-target sequences, leading to the conclusion that CRISPR-Cas9 was reasonably specific. They report significant on-target mutagenesis, such as large deletions and more complex genomic rearrangements at the targeted sites in mouse embryonic stem cells, mouse hematopoietic progenitors, and a human differentiated cell line. Researchers often use CRISPR to generate small deletions in the hope of knocking out a gene's function. But when examining CRISPR edits, Bradley and his colleagues found large deletions — often several thousand DNA letters long — and complicated rearrangements of DNA sequences in which previously distant DNA sequences were stitched together. The phenomenon was prevalent in all three of the cell types they tested, including a kind of human cell grown in the laboratory (Ledford 2018).

Gene Drives Are Here

Scientists in China say they are the first to use gene editing to produce customized dogs. They created a beagle with double the amount of muscle mass by deleting a gene called myostatin: "The goal of the research is to explore an approach to the generation of new disease dog models for biomedical research," says Liangxue Lai, a researcher at the Key Laboratory of Regenerative Biology at the Guangzhou Institutes of Biomedicine and Health. "Dogs are very close to humans in terms of metabolic, physiological, and anatomical characteristics" (Regalado 2015).

Recent press reports have not helped assure the public that scientists have it right. Enter micropigs. Micropigs have already proved useful in studies of stem cells and of gut microbiota, because the animals' smaller size makes it easier to replace the bacteria in their guts. They will also aid studies of Laron syndrome, a type of dwarf-ism caused by a mutation in the human GHR gene. Known as Bama pigs, they weigh 35–50 kg (by contrast, many farm pigs weigh more than 100 kg) and have been used in research (Cyranoski 2015). However, markets are what they are, and in no time this development extended itself into the pet industry with the decision from a leading Chinese biotech company to sell their micropigs as pets. A Chinese insti-tute, BGI, said in September 2015 that it had begun selling miniature pigs, created via gene editing, for $1600 each as novelty pets (Regalado 2015). The decision to sell the pigs as pets surprised Lars Bolund, a medical geneticist at Aarhus University in Denmark who helped BGI in Shenzhen to develop its pig gene-editing program, but he admits that they stole the show at the Shenzhen summit. "We had a bigger crowd than anyone," he says. "People were attached to them. Everyone wanted to hold them" (Cyranoski 2015).

In September 2015, Duanqing Pei, a representative of the Chinese Academy of Sciences, highlighted Lai's work as part of what he called a large Chinese effort to modify animals using CRISPR. The list of animals already engineered using gene editing in China includes goats, rabbits, rats, and monkeys (Regalado 2015).

Scientists and ethicists agree that gene-edited pets are not very different from conventional breeding – the result is just achieved more efficiently. But that doesn't make the practice a good idea, says Jeantine Lunshof, a bioethicist at the Harvard Medical School in Boston, Massachusetts, who describes both as "stretching physi-ological limits for the sole purpose of satisfying idiosyncratic aesthetic preferences of humans" (Cyranoski 2015).

With gene editing taking biology by storm, the field's pioneers say that the appli-cation to pets was no big surprise. Some also caution against it. "It's questionable whether we should impact the life, health and well-being of other animal species on this planet lightheartedly," says geneticist Jens Boch at the Martin Luther University of Halle-Wittenberg in Germany (Cyranoski 2015).

Josiah Zayner, a biochemist who once worked for NASA, appears to be the first person known to have edited his own genes with CRISPR. Zayner's experiment was intended to boost his strength by removing the gene for myostatin, which regulates muscle growth. A similar experiment in 2015 showed that this works in beagles whose genomes were edited at the embryo stage. He injected himself with the CRISPR system to remove the gene. Robin Lovell-Badge, a leading CRISPR researcher at the Francis Crick Institute in London, says Zayner's experiment was "foolish" and could have unintended consequences, including tissue damage, cell death, or an immune response that attacks his own muscles (Pearlman 2017).

Concluding Remarks

We were lucky with ZIKV. We either reached herd immunity or the ZIKV that surfaced in 2016–2017 mutated into a less dangerous form. Most likely, it was a combination of both these events. However, ZIKV has not disappeared and may return. If not ZIKA, then there will be another zoonotic disease that crosses over and affects human populations. We learned a lot about ZIKV over the last 3 years. There have been hundreds of technical articles and thousands of popular and digital articles written about the virus and the South American pandemic.

Vector management is an important strategy to reduce the transference of infectious diseases when mosquitoes depend upon blood from humans. Most theories of integrative pest management include vector control as a feature. The roles played by GE and especially gene drives on vector control will increase as the technology matures.

How we engage human populations in fields test, scheduling releases, and assessing the consequences of these activities will determine whether we can effectively use gene drives in contexts involving direct contact with humans. Gene drives may be here, but a communication strategy to engage multiple stakeholding publics is not! To overcome the ignorance, uncertainty, fear, and reservations, we need to develop proactive and ongoing infectious diseases policies, advance anticipatory surveillance, internal bio-surveillance, and an integrated pest and vector management strategy. These activities must involve science communication specialists as well as entomologists, infectious diseases specialists, caregivers, and policy makers.

Taking what we have learned and applying it to a comprehensive approach to technological pest management is becoming increasingly urgent with infectious outbreaks driven by ecological changes involving agriculture and economic development, human demographic changes and attendant behavior, increasing global travel and commerce, microbial and viral adaptation and change, anomalies in climate, developments in industry and technology, and a serious breakdown in national and global public health measures.

Acknowledgments This project has been directly supported by a fellowship from the Genetic Engineering and Society Center on the Centennial Campus of North Carolina State University. This project has been indirectly supported by the National Science Foundation, an organization that has continued to encourage my critical forays into science communication. The chapter was distilled in part from two book manuscripts under simultaneous preparation on Zika: one technical and the other popular. The popular text includes a substantial focus on how science and technology communication and engagement in infectious disease and epidemic may be optimized in a digital environment. All comments are my own and do not represent those of the GES and its membership, the RTNN (Research Triangle Nanotechnology Network), and North Carolina State University.

References

Adaja, A., et al. (2016, May 25). Genetically Modified (GM) mosquito use to reduce mosquito transmitted disease in the US: A community opinion survey. *PLOS*. http://currents.plos.org/outbreaks/article/genetically-modified-mosquito-use-to-reduce-mosquito-transmitted-disease-in-the-us-opinion-survey/. Accessed 29 Sept 2016.

Adelman, Z., & Tu, Z. (2016, March). Control of mosquito-borne infectious diseases: Sex and gene drive. *Trends in Parasitology, 32*, 3. https://doi.org/10.1016/j.pt.2015.12.003. Accessed 7 Apr 2017.

Adler, J. (2016, June). A world without mosquitoes. *The Smithsonian, 38*. http://www.smithsonianmag.com/innovation/kill-all-mosquitos-180959069/?no-ist. Accessed 2 Oct 2016.

Agence France Presse. (2016, September 2). Life-altering science is moving fast, sparking debate . Physics.org. http://phys.org/news/2016-09-controversial-dna-ethical-debate.html. Accessed 18 Sept 2016.

Alphey, L. (2016, February). Can CRISPR-Cas9 gene drives curb malaria? *Nature Biotechnology, 34*, 2. http://www.nature.com/nbt/journal/v34/n2/full/nbt.3473.html. Accessed 3 Apr 2017.

Atkins, K. (2016, December 14). Bug board to talk Lower Keys building project Friday. *Florida Keys News*. http://www.flkeysnews.com/news/local/article120794933.html. Accessed 10 Jan 2017.

Atkinson, M. P., et al. (2007). Analyzing the control of mosquito-borne diseases by a dominant lethal genetic system. *Proceedings of the National Academy of Sciences, 104*. http://www.pnas.org/content/104/22/9540.full.pdf. Accessed 1 Oct 2018.

ATSDR. (2015). *Public health statement for DEET* (N,N-Diethyl-meta-toluamide). https://www.atsdr.cdc.gov/phs/phs.asp?id=1447&tid=201. Accessed 23 Sept 2018.

Basu, S., et al. (2015, March 31). Silencing of end-joining repair for efficient site-specific gene insertion after TALEN/CRISPR mutagenesis in Aedes aegypti. *Proceedings of the National Academies of Sciences, 112*, 13. http://www.pnas.org/content/112/13/4038. Accessed 1 Oct 2018.

Beck, J. (2016, September 15). Tiny vampires. *The Atlantic*. https://www.theatlantic.com/health/archive/2016/09/tiny-vampires/500069/. Accessed 7 Feb 2017.

Begley, S. (2015, November 15). Gene drive gives scientists power to hijack evolution. *STAT*. https://www.statnews.com/2015/11/17/gene-drive-hijack-evolution/. Accessed 98 Mar 2017.

Bluth, R., & Kopp, E. (2016, November 10). Genetically engineered mosquitoes split A straw poll vote in the Florida keys. *Kaiser Health News*. http://khn.org/news/genetically-engineered-mosquitoes-split-a-straw-poll-vote-in-the-florida-keys/. Accessed 7 Feb 2017.

Brown, K. (2016a, September 19). Genetically modified mosquitoes could wipe out the world's most deadly viruses. If we let them. *Fusion.net*. http://fusion.net/story/347298/oxitec-genetically-modified-mosquitoes/. Accessed 25 Sept 2016.

Brown, K. (2016b, October 31). Americans are terrified genetic enhancements will turn the rich into sci-fi supermen. *Fusion.net*. http://fusion.net/americans-are-terrified-genetic-enhancements-will-turn-1793861544. Accessed 4 May 2017.

Brown, K. (2017, January 14). This controversial genetic engineering technology could eliminate malaria. *GIZMODO*. http://gizmodo.com/controversial-genetic-engineering-technology-could-elim-1790727517. Accessed 14 Mar 2017.

Burgess, M., Mumford, J., & Lavery, J. (2018). Public engagement pathways for emerging GM insect technologies. *BMC Proceedings, 12*, Supp. 8. https://bmcproc.biomedcentral.com/articles/10.1186/s12919-018-0109-x. Accessed 30 Sept 2018.

Caldera, C. (2017, July 14). Genetically engineered mosquitoes could wipe out Zika, but some in Dallas County oppose local trials. *Dallas News*. https://www.dallasnews.com/news/zika-virus/2017/07/14/genetically-engineered-mosquitoes-wipe-zika-dallas-county-oppose-local-trials. Accessed 21 Jul 2017.

Caribbean News Now. (2018, September 13). New documents show genetically modified mosquitoes released in the Cayman Islands are ineffective and risky. *Caribbean News Now*. https://wp.caribbeannewsnow.com/2017/09/13/new-documents-show-genetically-modified-mosquitoes-released-cayman-islands-ineffective-risky/. Accessed 23 Jul 2018.

Carroll, M. (2016, September 13). Florida officials want to enlist frankenskeeters' in Zika fight. *AMI Newswire*. https://aminewswire.com/stories/511009888-florida-officials-want-to-enlist-frankenskeeters-in-zika-fight. Accessed 24 Sept 2016.

Cell Press. (2016, February 17). Can CRISPR help edit out female mosquitoes? *Science News*. https://www.sciencedaily.com/releases/2016/02/160217125535.htm. Accessed 2 Jun 2017.

Chaiken, S. (1980). Heuristic versus systematic information processing and the use of source versus message cues in persuasion. *Journal of Personality and Social Psychology, 39*, 752–766.

Consumer Reports. (2018). *Insect repellents*. https://www.consumerreports.org/cro/insect-repellent/buying-guide/index.htm. Accessed 23 Sept 2018.

Curtis, C. (1968, April 27). Possible use of translocations to fix desirable genes in insect pest populations. *Nature, 218*(5139). https://www.nature.com/nature/journal/v218/n5139/abs/218368a0.html. Accessed 14 May 2017.

Cyranoski, D. (2015, October 1). Gene-edited 'micropigs' to be sold as pets at Chinese institute. *Nature, 526*, 18. https://www.nature.com/news/gene-edited-micropigs-to-be-sold-as-pets-at-chinese-institute-1.18448. Accessed 7 Aug 2018.

Dong, S., et al. (2015, March 27). Heritable CRISPR/Cas9-mediated genome editing in the yellow fever mosquito, Aedes aegypti. *PLoS ONE, 10*. https://www.ncbi.nlm.nih.gov/pubmed/25815482. Accessed 1 Oct 2018.

Fang, J. (2015, July 7). Genetically modified mosquitoes released in Brazil. *IFLSCIENCE!*. http://www.iflscience.com/plants-and-animals/dengue-fighting-mosquitoes-are-suppressing-wild-populations-brazil/. Accessed 4 Sept 2016.

Fernandez, C. (2017, March 4). New Results Show GM Mosquitoes Keep Dengue and Zika at Bay in Brazil. *LABIOTECH*. https://labiotech.eu/oxitec-dengue-zika-brazil/. Accessed 24 Jul 2018.

Finkel, A. M., Trump, B. D., Bowman, D., & Maynard, A. (2018). A "solution-focused" comparative risk assessment of conventional and synthetic biology approaches to control mosquitoes carrying the dengue fever virus. *Environment Systems and Decisions, 38*(2), 177–197.

Fiske, S., & Taylor, S. (2013). *Social cognition: From brains to culture* (2nd ed.). Thousand Oaks, CA: SAGE.

Funk, C., & Hefferon, M. (2018, July 26). Public views of gene editing for babies depend on how it would be used. *Press Release-Pew Research Center*. http://www.pewinternet.org/2018/07/26/public-views-of-gene-editing-for-babies-depend-on-how-it-would-be-used/. Accessed 10 Aug 2018.

Gartner, L. (2016, August 30). Fearing Zika, local businesses join the call for genetically modified mosquitoes. *Tampa Bay Times*. http://www.tampabay.com/blogs/baybuzz/fearing-zika-local-businesses-join-the-call-for-genetically-modified/2291478. Accessed 25 Sept 2016.

Gates, B. (2014, April 25). *Gatenotes: The blog of Bill Gates*. https://www.gatesnotes.com/Health/Most-Lethal-Animal-Mosquito-Week. Accessed 7 Feb 2017.

Gorman, K., et al. (2015, October 16). Short-term suppression of Aedes aegypti using genetic control does not facilitate Aedes albopictus. *Pest Management Science*. https://www.ncbi.nlm.nih.gov/pubmed/26374668. Accessed 18 Apr 2017.

Guardian. (2017, January 4). Can bacteria-infected mosquitoes stop spread of dengue, Zika? *The Guardian*. https://guardian.ng/features/health/can-bacteria-infected-mosquitoes-stop-spread-of-dengue-zika/. Accessed 14 May 2017.

Hall, A. B., et al. (2015, May 12). A male-determining factor in the mosquito, Aedes aegypti. *Science, 348*. https://www.ncbi.nlm.nih.gov/pmc/articles/PMC5026532/. Accessed 1 Oct 2018.

James, S., et al. (2018). Pathway to deployment of gene drive mosquitoes as a potential biocontrol tool for elimination of malaria in Sub-Saharan Africa: Recommendations of a Scientific Working Group. *American Journal of Tropical Medicine and Hygiene, 98*, Supp 6. https://www.ncbi.nlm.nih.gov/pubmed/29882508. Accessed 10 Sept 2018.

Judson, O. (2003, September 25). A Bug's Death. *The New York Times*. http://www.nytimes.com/2003/09/25/opinion/a-bug-s-death.html. Accessed 5 Jun 2017.

Kahneman, D. (2011). *Thinking fast and slow*. New York, NY: Farrar, Straus and Giroux.

Kistler, K., Vosshall, L., & Matthews, B. (2015). Genome engineering with CRISPR-Cas9 in the mosquito Aedes aegypti. *Cell Reports, 11*, 51–60. https://www.ncbi.nlm.nih.gov/pubmed/25818303. Accessed 1 Oct 2018.

Kolker, R. (2016, October 6). Florida's feud over Zika-fighting GMO mosquitoes. *Bloomberg Businessweek*. http://www.bloomberg.com/features/2016-zika-gmo-mosquitos/. Accessed 25 Oct 2016.

Kosicki, M., Tomberg, K. & Bradley, A. (2018, July 31). Repair of double-strand breaks induced by CRISPR–Cas9 leads to large deletions and complex rearrangements. *Nature Biotechnology*. https://www.nature.com/articles/nbt.4192. Accessed 28 Aug 2018.

Lafrance, A. (2016, April 26). Genetically modified mosquitoes: What could possibly go wrong? *The Atlantic*. https://www.theatlantic.com/technology/archive/2016/04/genetically-modified-mosquitoes-zika/479793/. Accessed 18 May 2017.

Ledford, H. (2018, July 16). CRISPR gene editing produces unwanted DNA deletions. *Nature*. https://www.nature.com/articles/d41586-018-05736-3. Accessed 28 Aug 2018.

Liang, P., et al. (2015, May). CRISPR/Cas9-mediated gene editing in human tripronuclear zygotes. *Protein & Cell, 6*, 5. https://link.springer.com/article/10.1007/s13238-015-0153-5. Accessed 11 Aug 2018.

Lloyd's. (2009, July). *Lloyd's emerging risks team report: Synthetic biology influencing development*. https://www.lloyds.com/news-and-risk-insight/risk-reports/library/technology/synthetic-biology. Accessed 4 Sept 2018.

MacDonald, N. E. (2015, August). Vaccine hesitancy: Definition, scope and determinants. *Vaccine, 33*, 34. https://www.ncbi.nlm.nih.gov/pubmed/25896383. Accessed 1 Oct 2018.

McCay, B. (2016, September 2). Mosquitoes are deadly, so why not kill them all? *Wall Street Journal*. http://www.wsj.com/articles/mosquitoesaredeadlysowhynotkillthemall1472827158. Accessed 15 Sept 2016.

Miesfeld, R., & Isoe, J. (2017, January 31). Mosquito eggs without eggshells disrupt the ability to reproduce. *Tucson.com*. http://tucson.com/news/local/mosquito-eggs-without-eggshells-disrupt-the-ability-to-reproduce/article_7fb6d4fa-f853-5f1a-ab88-47701e9a9403.html. Accessed 23 May 2017.

Morrison, A., et al. (2008, March). Defining challenges and proposing solutions for control of the virus vector Aedes aegypti. *PLOS Medicine, 5*, 3. http://journals.plos.org/plosmedicine/article?id=10.1371/journal.pmed.0050068. Accessed 26 May 2017.

Moses, S. (2016, December 6). Philanthropy vs. Mosquitoes: The funders giving big money to fight a tiny insect. *Inside Philanthropy*. https://www.insidephilanthropy.com/home/2016/12/6/attacking-killer-mosquitoes-these-funders-are-throwing-big-money-to-fight-a-tiny-insect. Accessed 26 May 2017.

NALED insecticide fact sheet. (2016). *Np spray*. http://nospray.org/naled-insecticide-fact-sheet/. Accessed 15 May 2017.

Nickerson, R. (1998, June). Confirmation bias: A ubiquitous phenomenon in many guises. *Review of General Psychology, 2*(2), 165–220.

Oxitec. (2016, August 5). *Environmental assessment for investigational use of Aedes aegypti OX513A*. https://www.fda.gov/downloads/AnimalVeterinary/DevelopmentApprovalProcess/GeneticEngineering/GeneticallyEngineeredAnimals/UCM514698.pdf. Accessed 12 May 2017.

PRNewswire. (2018, May 24). Oxitec launches field trial in Brazil for next generation addition to friendly™ mosquitoes platform. *PRNewswire*. https://www.oxitec.com/oxitec-launches-field-trial-in-brazil-for-next-generation-addition-to-friendly-mosquitoes-platform/. Accessed 25 Jul 2018.

Pearlman, A. (2017, November 16). Biohackers are using CRISPR on their DNA and we can't stop it. *New Scientist*. https://www.newscientist.com/article/mg23631520-100-biohackers-are-using-crispr-on-their-dna-and-we-cant-stop-it/. Accessed 25 Jul 2018.

Pew Research Center. (2018). Public Views of Gene Editing for Babies Depend on How It Would Be Used. July 26. https://www.pewresearch.org/science/2018/07/26/public-views-of-gene-editing-for-babies-depend-on-how-it-would-be-used/. Accessed September 30, 2019.

Polansky, R. (2016, September 15). Fla. Reps. urging feds to use GMO mosquitoes in Zika fight. *NBC2.com WBBH News*. http://www.nbc-2.com/story/33106339/fla-reps-urging-feds-to-use-gmo-mosquitoes-in-zika-fight#.V96xsygrKUk. Accessed 18 Sept 2016.

Polizzi, K. M., Stanbrough, L., & Heap, J. T. (2018). *A new lease of life: Understanding the risks of synthetic biology.* An emerging risks report published by Lloyd's of London. https://www.lloyds.com/news-and-risk-insight/news/lloyds-news/2018/07/a-new-lease-of-life. Accessed 4 Sept 2018.

Regalado, A. (2015, October 19). First gene-edited dogs reported in China. *MIT Technology Review*. https://www.technologyreview.com/s/542616/first-gene-edited-dogs-reported-in-china/. Accessed 10 Aug 2018.

Rose, J. (2016, July 6). Is Zika virus still around in the US? A CDC report suggests the threat is still very real. *Romper*. https://www.romper.com/p/is-zika-virus-still-around-in-the-us-a-cdc-report-suggests-the-threat-is-still-very-real-68398. Accessed 21 Jul 2017.

Servick, K. (2016, October 13). Brazil will release billions of lab-grown mosquitoes to combat infectious disease. Will it work? *Science*. http://www.sciencemag.org/news/2016/10/brazil-will-release-billions-lab-grown-mosquitoes-combat-infectious-disease-will-it. Accessed 31 May 2017.

Sikka, V., et al. (2016). The emergence of zika virus as a global health security threat: A review and a consensus statement of the INDUSEM Joint working Group (JWG). *Journal of Global Infectious Diseases, 8*, 1. http://www.jgid.org/article.asp?issn=0974-777X;year=2016;volume=8;issue=1;spage=3;epage=15;aulast=Sikka. Accessed 18 Sept 2018.

Tome, K. (2017, March 29). Gene drive in mosquitoes to drive away malaria. *Decoded Science*. https://www.decodedscience.org/gene-drive-mosquitoes-drive-away-malaria/61045. Accessed 4 Jun 2017.

Trump, B. D., Cegan, J., Wells, E., Poinsatte-Jones, K., Rycroft, T., Warner, C., et al. (2019). Co-evolution of physical and social sciences in synthetic biology. *Critical Reviews in Biotechnology, 39*(3), 351–365.

Trump, B. D., Cegan, J. C., Wells, E., Keisler, J., & Linkov, I. (2018). A critical juncture for synthetic biology: Lessons from nanotechnology could inform public discourse and further development of synthetic biology. *EMBO Reports, 19*(7), e46153.

Wilcox, C. (2016, January 31). No, GM mosquitoes didn't start the Zika outbreak. *Discovery Magazine*. http://blogs.discovermagazine.com/science-sushi/2016/01/31/genetically-modified-mosquitoes-didnt-start-zika-ourbreak/#.V82RpSgrKUk. Accessed 5 Sept 2016.

Xi, Z., Khoo, C., & Dobson, S. (2005, October 14). WoZbachia establishment and invasion in an Aedes aegypti laboratory poputation. *Science, 310*(5746). https://www.ncbi.nlm.nih.gov/pubmed/16224027. Accessed 2 Jul 2017.

Yeoh, O. (2017, November 19). SAVVY: Mosquitoes begone! *New Straight Times*. https://www.nst.com.my/lifestyle/sunday-vibes/2017/11/304995/savvy-mosquitoes-begone. Accessed 10 Jul 2018.

Synthetic Biology Industry: Biosafety Risks to Workers

Vladimir Murashov, John Howard, and Paul Schulte

Introduction

Synthetic biology involves two closely related capabilities: (1) the design, assembly, synthesis, or manufacture of new genomes, biological pathways, devices, or organisms not found in nature for use in agriculture, bio-manufacturing, health care, energy, and other industrial sectors and (2) the redesign of existing genes, cells, or organisms for the purpose of drug discovery and gene therapy. Synthetic biology has accelerated the growth of the biotechnology sector of the US economy. This rapid growth has been accompanied by a corresponding increase in the number of workers employed in the synthetic biology industry.

Despite the present and anticipated growth of synthetic biology, workplace health and safety considerations for the synthetic biology workplace have not kept pace with the rapid introduction of this enabling industrial technology. Specifically, biosafety practices for the laboratory need to be adapted to the more complex industrial-scale processes. Therefore, updated health and safety guidance specific to the synthetic biology industry is urgently needed.

Background

Synthetic biology is an emerging interdisciplinary field of biotechnology that applies the principles of engineering and chemical design to biological systems (Ball 2005). The earliest use of the term was by French biologist Stéphane Leduc in 1910. By 1974, synthetic biology was being viewed as the next phase in molecular

V. Murashov (✉) · J. Howard · P. Schulte
National Institute for Occupational Safety and Health, Washington, DC, USA
e-mail: vem8@cdc.gov

© Springer Nature Switzerland AG 2020
B. D. Trump et al. (eds.), *Synthetic Biology 2020: Frontiers in Risk Analysis and Governance*, Risk, Systems and Decisions, https://doi.org/10.1007/978-3-030-27264-7_8

biology by which scientists would "devise new control elements and add these new modules to the existing genomes or build up wholly new genomes" (Szybalski 1974). Such capabilities did not become practical until the turn of the millennium, as illustrated by the sequencing of the first human genome in 2001 (National Research Council 2015). With the emergence of the synthetic biology industry, in 2015 the Ad Hoc *Technical Expert Group on Synthetic Biology* of the United Nations Convention on Biological Diversity proposed defining synthetic biology as "a further development and new dimension of modern biotechnology that combines science, technology, and engineering to facilitate and accelerate the understanding, design, redesign, manufacture and/or modification of genetic materials, living organisms and biological systems" (Convention on Biological Diversity 2015).

Although it is often difficult to demarcate synthetic biology from other biotechnology research areas, synthetic biology can be understood to involve two closely related capabilities, both of which may have wide utility in commerce and health care and also create unique occupational safety and health risks (Howard et al. 2017). First, while the transfer of already existing genes from one cell to another characterized an earlier phase of the field of biotechnology, synthetic biology involves the design, assembly, synthesis, or manufacture of new genomes, biological pathways, devices, or organisms not found in nature. These operations are made possible by recent advances in DNA synthesis and DNA sequencing, providing standardized DNA "parts," modular protein assemblies, and engineering models (Baker et al. 2006; Eisenstein 2016). Recent milestones in constructing new genomes from DNA sequences include synthesis of a completely new chromosome (Smith et al. 2003); a new bacterial genome (Gibson et al. 2008); the first synthetic life form, a single-celled organism based on an existing bacterium (Gibson et al. 2010); and living protocells assembled entirely from nonliving, individual biological components (Miller and Gulbis 2015; Kurihara et al. 2015). This capability to construct new genomes is the core use of synthetic biology in advanced chemical manufacturing, agriculture, health care, energy, and other industrial sectors (National Research Council 2015). Such capability could also give rise to increased biological hazards for workers in synthetic biology industry.

The second capability of synthetic biology involves the redesign of existing genes, cells, or organisms for the purpose of drug discovery and gene therapy (Scott 2018) and is utilized primarily in the health-care industry (Synthetic Biology Project 2018). Modification of existing genes in animal and human cells is enabled by four major genome editing platforms: (1) meganucleases, (2) zinc-finger nucleases (ZFNs), (3) transcription activator-like effector nucleases (TALENs), and (4) the clustered regularly interspaced short palindromic (CRISPR)-associated system (Cas) (Yin et al. 2017). Progress in this branch of synthetic biology has yielded remarkable therapeutic advances in gene therapy well beyond the achievements of conventional drugs and biologic agents (Naldini 2015) and has led to the first FDA-approved gene therapy (for acute lymphoblastic leukemia) on August 30, 2017 (FDA 2017). As of March 15, 2018, there were 709 gene therapy clinical trials underway according to the NIH Clinical Trials database (https://clinicaltrials.gov/search?term=%22gene+therapy%22). These trials address a broad range of conditions from cancers to inherited disorders. Gene therapies could also bring harmful genetic changes and infections with replication-competent gene carriers and lead to adverse health effects among health-care workers upon unintentional exposures to gene therapy agents.

Due to these advances in underlying science and technology, synthetic biology is becoming a widespread enabling technology whose range of applications will only increase in the near future (Center for Biosecurity of UPMC 2012). The biotechnology sector of the US economy has grown on average greater than 10% each year over the past 10 years, and the sector is growing much faster than the rest of the economy (Carlson 2016). Workforce development and workforce protection are both critical in driving the national bioeconomy, where economic activity is powered by innovation in the biosciences (White House 2012). Synthetic biology is playing an increasing role in the commercial bioeconomy as providers of biological designs and optimized biological molecules and laboratory suppliers of customer-specified DNA, RNA, enzymes, and cell-cloning services and in drug development (Lokko et al. 2018). A 2015 analysis of the private sector landscape shows that 162 US companies were engaged in substantial activity in the synthetic biology area, drawing about $6 billion in investments from venture capital individuals and firms (Department of Defense 2015). In 2016 alone, over $1 billion was invested globally in synthetic biology companies (SynBioBeta 2017). The United Kingdom expects to achieve a market in synthetic biology products equivalent to $10 billion in British pound sterling by 2030 (Synthetic Biology Leadership Council 2015). To further facilitate venture capital investments in synthetic biology companies, business risks should be assessed and mitigated proactively. These risks include future liabilities stemming from adverse health and environmental effects such as occupational injuries and ill health that can increase insurance premiums and provoke legal actions (Murashov and Howard 2009).

Provided that this nascent technology is demonstrated as safe, synthetic biology is expected to continue expanding into new application domains. It has been touted as an enabling technology to solve problems not only on earth but also in space. For example, remarkable resilience and adaptation of fungi and yeast to space conditions including high levels of radiation and microgravity conditions has led to proposals for using synthetic biology based on these organisms to synthesize useful materials such as essential nutrients, medicine, and polymers for a range of applications during space travel and eventual colonization of other planets (deGrandpre 2017; Phelan 2018).

Workers participate in all phases of synthetic biology, from laboratory research and development through start-up and pilot operations to production, manufacturing, and end-of-product-life activities. There are hundreds of companies and laboratories worldwide engaged in synthetic biology activities (Department of Defense 2015; Carlson 2016). Workers are involved with each activity. The expanding scope of synthetic biology in the new bioeconomy both nationally and internationally marks an opportune time to review existing risk assessment and risk management measures to better protect current and future synthetic biology workers from harm.

Discussion

Industrial Synthetic Biology and Occupational Risks

Synthetic biology promises both tremendous societal benefits in treating human genetic disease (Lander 2015) and huge commercial market potential for technology investors (Hayden 2015). At the same time, synthetic biology has raised concerns about potential biosafety risks to workers and to the society in general (Trump et al. 2018; Howard et al. 2017).

The biosafety concerns about synthetic biology and its gene editing tools are similar to the concerns lodged about recombinant DNA technology when genetic engineering was first introduced in the early 1980s (Kuzma 2016). Those concerns include whether products resulting from the recombinant DNA technology would pose greater risks than those achieved through traditional manipulation techniques. For example, concerns were raised about potential biological hazards to workers in the field of genetic engineering (Berg et al. 1975; OSHA 1985). Ongoing reports of potential exposure incidents to workers at Level 3 and Level 4 containment laboratories only serve to increase biosafety concerns about synthetic biology (Young and Penzenstadler 2015; Young 2016; Grady 2017). A strong perception exists that biosafety rules cannot keep up with practices in the modern biotechnology laboratory (Pollack and Wilson 2010).

To manage the risk of biotechnology, in 1986, the White House Office of Science Technology and Policy (OSTP) developed the US government's *Coordinated Framework for the Regulation of Biotechnology* ("CF") (Office of Science Technology and Policy 1986). As a result of the CF, the Food and Drug Administration (FDA), the Environmental Protection Agency (EPA), the US Department of Agriculture (USDA), the National Institutes of Health (NIH), and the Occupational Safety and Health Administration (OSHA) outlined their respective roles in ensuring the safety of biotechnology research and products.

OSHA determined that the general duty clause, together with a set of existing occupational safety and health standards, provided an adequate and enforceable basis for protecting biotechnology workers and that no new synthetic biology-specific standards were necessary. OSHA also provided a set of guidelines for biotechnology laboratory worker safety based on existing OSHA standards. The specific OSHA standards that may be applicable to biotechnology laboratories include (1) blood-borne pathogens (OSHA 2010); (2) toxic and hazardous substances (OSHA 2005a); (3) access to employee exposure and medical records (OSHA 2005b); (4) hazard communication (OSHA 2014a); (5) exposure to toxic chemicals in laboratories (OSHA 2014b); (6) respiratory protection (OSHA 2014c); and (7) safety standards of a general nature (e.g., general environmental, walking and working surfaces, fire protection, compressed gases, electrical safety, and material handling and storage contained in 29 CFR Part 1910 Subparts J, D, E and L, H, S and N). As a part of the 1986 CF comment process, the National Institute for Occupational Safety and Health (NIOSH) recommended increased injury and illness surveillance of biotechnology workers given the gap in information about occupational health and safety risks to such workers (National Archives and Records Administration 1992). That recommendation has become even more salient with the rise of the industrial phase synthetic biology.

The Coordinated Framework was most recently updated in 2017 (OSTP 2017). The update aims to clarify roles of the three main agencies regulating the products of biotechnology: EPA, FDA, and USDA. While it does not reference OSHA explicitly, the update states that some occupational risks are addressed by EPA under the 1976 Toxic Substances Control Act (TSCA) and the 1972 Federal Insecticide, Fungicide, and Rodenticide Act (FIFRA). The update clarifies that under TSCA, new microorganisms utilized by synthetic biology would be considered "new chemicals." Specifically it states that "microorganisms formed by the deliberate combination of genetic material from synthetic genes that are not identical to DNA that would be derived from the same genus as the recipient, are considered 'intergeneric' (i.e., 'new') microorganisms, and so would be subject to the pre-manufacturing review provisions" (OSTP 2017). The Frank R. Lautenberg Chemical Safety for the 21st Century Act amended TSCA in 2016. Under this amendment, "EPA must make an affirmative finding on the safety of new chemical substances, including intergeneric organisms, before they are allowed into the marketplace" (OSTP 2017). The amendment did not change EPA authority to issue significant new use rules (SNURs) under the TSCA Section 5(a)(2) and consent orders under the TSCA Section 5(e). Under these sections, EPA has the authority to require implementation of exposure mitigation measures in the workplace (Murashov et al. 2011).

Under FIFRA, EPA regulated the sale, distribution, and use of all pesticides, including those produced through synthetic biology (OSTP 2017). The FIFRA registration process requires pesticide data submission which includes, among other data, information about worker exposure and a copy of the proposed labeling containing directions for use, storage, and disposal, as well as warnings, restrictions, and other information. Through FIFRA authorities, EPA developed a regulatory standard aimed specifically at worker protection. EPA's 2015 Worker Protection Standard for Agricultural Pesticides (WPS) is a regulation to reduce the risk of pesticide poisonings and injuries among agricultural workers and pesticide handlers (EPA 2015). The WPS contains requirements for pesticide safety training, notification of pesticide applications, use of personal protective equipment, restricted-entry intervals after pesticide application, decontamination supplies, and emergency medical assistance (Murashov et al. 2011).

Separately, OSHA considered the need to support a comprehensive employer-established infection control program and control measures to protect employees from aerosol exposures to infectious disease agents. OSHA published an Infectious Diseases Request for Information (RFI), held stakeholder meetings, conducted site visits, and completed the Small Business Regulatory Enforcement Fairness Act (SBREFA) process in support of rulemaking. However, OSHA froze this rulemaking process in 2017 and placed the Regulatory Agenda for the Infectious Diseases Notice of Proposed Rulemaking under a "long-term action" (OSHA 2018). It remains unclear whether the scope of the final ruling for such a standard would include industrial processes in synthetic biology.

Private sector groups have also called for improvements in the regulatory infrastructure to address the implications of new synthetic biology products (Ledford 2016; Bergeson et al. 2015; Carter et al. 2014; Lowrie and Tait 2010). Public interest groups have recommended applying the precautionary principle to any further research and commercialization of synthetic products until specific biosafety mechanisms can be developed to keep pace with synthetic biology advances (Friends of the Earth et al. 2010). Other groups have proposed detailed risk governance policies for commercial entities, users, and organizations engaging in synthetic genomics research, including compiling a manual specifically addressing biosafety in synthetic biology laboratories (Garfinkel et al. 2007; Cummings and Kuzma 2017).

Biosafety guidance specific to scientific advances in synthetic biology is necessary to fill the gap in safety oversight. Currently, the World Health Organization's *Laboratory Biosafety Manual* (WHO 2004) and *Biosafety in Microbiological and Biomedical Laboratories* (DHHS 2009), jointly co-authored by the US Centers for Disease Control and Prevention and the National Institutes of Health in the US Department of Health and Human Services, address an earlier phase of biotechnology aimed primarily at prevention of exposure to already existing pathogens in traditional biology laboratories. However, synthetic biology is using newly designed organisms and viral vectors (i.e., tools used to deliver genetic material inside a cell), not unmodified existing pathogens. Furthermore, industrialization of synthetic biology requires translating well-established safety guidelines for pathogenic organisms and recombinant DNA in laboratory research to industrial-scale manufacturing. The laboratory guidance should be also adapted to Do It Yourselfers (DIYers) as synthetic biology often is not being performed solely by "biologists," but by engineers, physical scientists, and others who are not familiar with fundamental biosafety measures such as biocontainment protections.

Health care and occupational risks Gene therapy is one of the most promising applications of synthetic biology in health care. Since the first gene therapy trial in 1990 (Blaese et al. 1995), various non-viral and viral delivery strategies of functional genes have been developed (Yin et al. 2017). In non-viral delivery, physical methods (electroporation and microfluidic technologies) and nanomaterial-based methods (lipid- and polymer-based nanoparticles, cell-penetrating peptides) are utilized (Yin et al. 2017). In viral delivery, viruses are used as gene transfer devices or "vectors," such as retrovirus, adenovirus, adeno-associated virus, and herpes simplex virus (Cross and Burmester 2006; Merten and Al-Rubeai 2011). For example, vectors made from members of the *Retroviridae* (retroviruses) family have gained attention as efficient gene transfer vehicles (Robbins and Ghivizzani 1998; Levine et al. 2006). All retroviruses have the ability to transcribe their single-stranded RNA genome into double-stranded DNA by the reverse transcriptase enzyme. Transcribed RNA-DNA can then be integrated into the host cell genome, producing permanent genetic change in the organism (Maetzig et al. 2011). Among retroviruses, lentiviruses, a subgroup of retroviruses, such as the human immunodeficiency virus (HIV), found in humans and animals, are most widely used in gene therapy (Schlimgen et al. 2016; Tomas et al. 2013).

To increase the range of cell types that viral vectors can infect, envelope glycoproteins responsible for cellular attachment are modified through "pseudotyping" with glycoproteins from another virus (Cronin et al. 2005). However, this increased tropism, or specificity of a viral vector for a particular host tissue, can also result in the unintended transduction of "off-target" cell types in a worker who becomes occupationally exposed to viral vectors used in genetic therapies. Pseudotyping viral vectors illustrates just one of the hazards that synthetic biology researchers, clinicians, and ancillary workers face when they are occupationally exposed to viral vectors. Other hazards associated with unintentional viral vector worker exposure include the generation of replication-competent viruses (Schambach et al. 2013), insertional mutagenesis, and transactivation of neighboring genome sequences which could lead to cancer and other diseases (Mosier 2004).

Since 1974, the biological safety of federally funded research involving recombinant DNA molecules and targeting medical applications has been addressed through the *NIH Guidelines for Research Involving Recombinant or Synthetic Nucleic Acid Molecules* (NIH 2016). These guidelines focus on biological hazards only and do not cover other hazards such as physical, mechanical, and chemical hazards. They outline general biosafety requirements and specific requirements for selected biological agents (e.g., influenza viruses) and for large-scale uses and production of organisms. The 2016 update of the NIH Guidelines streamlined the NIH protocol review process in light of decades of safety data, increased experience with recombinant DNA technology, and concurrent oversight from the US FDA, institutional review boards, and institutional biosafety committees. Under the NIH Guidelines, investigators must initially assess the risk of the agent to cause disease in laboratory workers or others if a release occurs. After the disease risk is assessed, a decision must be made as to the level of containment to control potential exposure. In determining the level of containment, factors such as virulence, pathogenicity, infectious dose, environmental stability, transmissibility, quantity, availability of treatment, and gene product effects such as toxicity, physiological activity, and allergenicity should be considered (NIH 2016).

For an organism containing genetic sequences from multiple sources, the 2016 NIH Guidelines require assessing the potential for causing human disease based on the source(s) of the DNA sequences and on the virulence and transmissibility functions encoded by these sequences. Combining sequences in a new biological organism may produce an organism whose risk profile could be higher than that of the contributing organisms or sequences. Using these considerations, the appropriate biosafety level (BSL) containment conditions (Levels 1 through 4) can be selected (NIH 2016). The NIH Guidelines highlight three complementary means of containment: (1) administrative containment, a set of standard practices that are generally used in microbiological laboratories; (2) physical containment, special procedures, equipment, and laboratory installations that provide physical barriers that are applied in varying degrees according to the estimated biohazard; and (3) biological containment, the application of highly specific biological barriers that limit either the infectivity of a vector or vehicle (plasmid or virus) for specific hosts or its dissemination and survival in the environment.

The 2016 NIH Guidelines require that any significant problems, violations, or any significant research-related accidents and illnesses are reported to the NIH Office of Science Policy within 30 days. Specifically, they prescribe that spills and accidents in BSL2 laboratories resulting in an overt exposure and spills and accidents in high containment (BSL3 or BSL4) laboratories resulting in an overt or potential exposure are immediately reported to the NIH Office of Science Policy.

In addition to the 2016 NIH Guidelines, enforced for federally funded research, private sector research laboratories generate site-specific safety guidance for working around viral vectors (Stanford University 2018; University of Cincinnati 2014; Gray 2011; Byers 2015).

The risks to workers may increase in the future as synthetic biology is commercialized. Proactive steps should be taken now to ensure worker health and safety protection as the new field of synthetic biology advances. Worker safety has been a guiding principle of biotechnology since the risks of recombinant DNA were first considered at the 1974 *Asilomar Conference on Recombinant DNA Molecules* (Berg et al. 1975), which led to the issuance of the first NIH Recombinant DNA Guidelines for federally funded research in 1976 (Fredrickson 1980). The risk control measures identified then—the use of biological and physical barriers to contain potentially hazardous organisms—remain the mainstay of the largely self-regulated, voluntary approach to worker protection in synthetic biology today.

Risk mitigation in synthetic biology The maturation of synthetic biology from laboratory experiments to industrial biofabrication processes requires enhanced risk governance strategies. These strategies, described below, include health surveillance, proactive risk management, prevention-through-design principles, dynamic guidance for synthetic biology processes, and attention and involvement by occupational health professionals and government officials.

Health Surveillance Synthetic biology risk assessment can be enhanced by adding health surveillance capabilities to current efforts, which NIOSH has been raising since its comments in response to OSHA announcement of Guidelines on Biotechnology in 1985 (OSHA 1985). Such surveillance includes recording, collecting, and analyzing injury and disease experience of the populations of workers exposed to synthetic biology laboratory, therapeutic, and industrial processes. Injury and disease surveillance efforts can be used to minimize potential worker harm. A temporal challenge to disease surveillance in synthetic biology is that many of the long-term adverse health effects such as adverse oncogenic effects of viral vectors may not be detectable for years or decades following exposure (Howard et al. 2017). Since workers move from job to job, a long-term exposure registry of synthetic biology workers should be considered. Registries have been successfully used and recommended for other hazardous agents and emerging technologies (Schulte et al. 2011).

Proactive Risk Management As synthetic biology emerges from the research laboratory into the bioeconomy, a greater number of occupational safety and health professionals will be involved in ensuring worker health and safety protections. The use of synthetic biotechnology in advanced manufacturing requires educating occupational safety and health professionals not currently involved in biosafety about risks to workers associated with synthetic biology. More professionals will have to take a role in proactively assessing the potential risks to workers as synthetic biology products become increasingly used in advanced biological manufacture and in routine clinical care delivery settings. Additionally, as the synthetic biology workforce expands, worker training tailored to safe approaches to commercial synthetic biology will be needed.

Proactive risk management approaches developed for other emerging technologies such as nanotechnology could be useful in synthetic biology risk assessment and risk management (Murashov and Howard 2009). Workers must be free to report deviations from high-reliability safety procedures without fear of reprisal, and employers should conduct detailed investigations of near-misses and other potential safety failures (Weick and Sutcliffe 2015; Trevan 2015). As applied to other emerging technologies, the proactive approach in synthetic biotechnology provides an opportunity to address occupational health and safety risks at the design stage of synthetic biotechnology workplaces, processes, and products prior to widespread dissemination in commercial arenas (Schulte et al. 2008).

Other aspects of proactive risk management include identifying industrial scenarios where workers could be exposed to synthetic biological products. Industrial synthetic biology is already a growing field, and several industries anticipate using the commercial applications: energy, chemicals, materials, pharmaceuticals, food, and agriculture as well as in the medical diagnostic and therapeutic areas (Erickson et al. 2011; Schmidt 2012; Rohn 2013). By identifying the processes where synthetic biology can be used in these industries, risk managers can make proactive assessments of possible risks to workers and what controls need to be put in place. Transportation and warehousing workers handling synthetic biological products and first responders to unplanned releases would also have potential exposure and should be included in the proactive risk assessments. Little is known about occupational exposures that could occur when synthetic biological products, vectors, or organisms are used in industrial scenarios. Exposure assessments of synthetic biology will depend on whether the area of interest is in or around containment structures or in the external environment.

Worker hazards that are unique to synthetic biology are not well defined. Some information on potential types of laboratory hazards may be drawn from laboratory-acquired infections (LAIs) that occurred over the last 60 years in clinical, research, and industrial microbiological laboratories (Sewell 1995; Wurtz et al. 2016). The history of LAIs generally has involved organisms (or toxins of these organisms) that have been known to cause disease or reasonably believed to cause diseases in people and animals. The extent to which LAIs have occurred in synthetic biology laboratories is not known because there is no plan for data collection. A recent global survey of LAIs in biosafety Level 3 and 4 laboratories found infrequent occurrence and identified "human error" as the causal factor in "a very high percentage" of the cases (Wurtz et al. 2016).

There is still uncertainty surrounding the hazards associated with the "construction in organisms that may contain genes or proteins that never existed together in a biological organism or that contain newly designed biological functions that do not exist in nature" (NIH 2016). It is not reasonable to characterize the hazards of synthetic biological organisms, processes, or products with a single overarching descriptor. There are and will be many different hazards. There will be a range of hazard severities. The hazard of a specific synthetic biological organism is a function of its pathogenicity or immunogenicity. Synthetic organisms are designed to reproduce and will evolve. The hazards associated with them may change as well. The nature of a hazard will drive management and control measures.

Prevention-Through-Design Worker protections in synthetic biology may benefit from applying prevention-through-design principles to promote further risk control research on physical and biological containment (NIOSH 2014). The mainstay of risk management for genetic engineering, including for synthetic biology, has been containment. Biosafety containment can be categorized as physical (or extrinsic) and biological (or intrinsic) containment.

Extrinsic containment was developed in the late 1940s and early 1950s chiefly at the US Army Biological Warfare Laboratories at Fort Detrick, Maryland. Extrinsic containment was designed to provide physical containment of highly infectious organisms in secure rooms or cabinets. Biological safety cabinets (BSCs) provide three classes of protection: (1) personal and environmental protection (Class I); (2) personal and environmental protection as well as product protection (Class II); and (3) maximal protection through a gas-tight enclosure where gloves are attached to the front of the BSC to prevent direct contact with hazardous materials (Class III), often referred to as glove boxes (DHHS 2009). BSCs are now the mainstay of extrinsic containment in laboratories around the world. The effectiveness of the personal protective equipment such as gloves and coveralls to reduce potential for exposures to biological agents is under active investigation (Villano et al. 2017). A 2016 systematic review of reported studies used very low quality evidence to conclude that (1) more breathable types of personal protective equipment (PPE) may not lead to more contamination but may have greater user satisfaction and (2) double gloving and protective clothing "doffing" guidance from the Centers for Disease Control and Prevention appear to decrease the risk of contamination and that active training in PPE use may reduce PPE and doffing errors more than passive training (Verbeek et al. 2016).

Intrinsic containment is a more recent type of containment, which leverages the fact that synthetic biology is chiefly an engineering discipline in the life sciences. Organisms live or die through a variety of processes, some of which can be interrupted. Intrinsic containment is still under active development in the field of synthetic biology. The aims of intrinsic containment include (1) controlling growth of the engineered organism in the research laboratory or after an unintentional environmental release; (2) preventing the horizontal flow of genetic material from a synthetic organism to a natural one (gene flow); (3) preventing the use of engineered microbes as bioterror agents; and (4) protecting the intellectual property of biotechnology companies (Cai et al. 2015). Genetic safeguards intrinsic to the synthetic organism can restrict its viability in defined environments (Schmidt and de Lorenzo 2012). Designing these safeguards into synthesized organisms can protect workers and supports designing out hazards and preventing the occurrence of harmful exposures.

Since the 1980s, the field of intrinsic containment has grown rapidly to encompass a number of different strategies. Control of cell growth by engineered auxotrophy, i.e., the inability of an organism to synthesize a particular organic compound required for its growth, protects against it surviving environmental release (Steidler et al. 2003). Intrinsic containment methods to produce safer viral vectors involve splitting gene vector components into three plasmids; using vectors without viral accessory proteins that are important for a natural virus as a pathogen, but not as a vector (Sakuma et al. 2012); and using self-inactivating (SIN) vectors which can help mitigate the risk of insertional gene activation (Cockrell et al. 2006; Zufferey et al. 1998). Other methods include designing engineered regulators that control the expression of essential genes (Gallagher et al. 2015), transcriptional and recombinational strategies to control essential gene functions (Gallagher et al. 2015), the use of microbial kill switches (Chan et al. 2016) and other vector suicide strategies (Bej et al. 1988). Developing a quantitative assay for insertional mutagenesis can help produce safer viral vectors (Bokhoven et al. 2009). Finally, a largely theoretical intrinsic containment method involves engineering organisms with chromosomes made not from DNA and RNA, but from xeno ("stranger") nucleic acids or XNA (Schmidt 2010). Unrealistic as such "organisms" seem to us today, their utility could prevent "gene flow" with DNA-based organisms, serving as a genetic "firewall."

Rigorous effectiveness studies should assess these intrinsic containment methods and others that emerge as synthetic biotechnology becomes more commercial. Although funding in this area is not customarily a high priority for governmental biomedical funders or for entrepreneurs (Garfinkel 2012), it will likely advance worker protections.

Dynamic Guidance Safety guidance that is specific to synthetic biology should be developed in an electronically updatable format that reflects advances in risk science. NIOSH's *Approaches to Safe Nanotechnology* serves as an example of safety guidance for an emerging technology (NIOSH 2009). This safety guidance should include steps to foster a robust safety culture characterized by employer commitment to and worker involvement in safe synthetic biology. Model guidance could be adopted by national governments as a mandatory standard or used as the basis for a national or international consensus standard (Murashov and Howard 2008; Murashov et al. 2011).

Greater Awareness and Involvement Medical professionals and government agencies contributing to protecting worker safety should have an understanding of the complex risk assessments and risk management issues inherent in synthetic biotechnology. During the research-only phase of synthetic biology, biosafety professionals have worked diligently to keep workers safe. The increases in biosafety research laboratories, the number of workers potentially exposed to synthetic biology products, the use of gene transfer-viral vectors in synthetic biology, and the emerging commercialization of synthetic biology oblige the involvement of the occupational safety and health practice community and governmental occupational safety and health research and regulatory agencies.

Conclusion

As the *Presidential Commission for the Study of Bioethical Issues* recommended in 2010, maximizing synthetic biology's benefits and minimizing its harms will benefit from risk assessment and risk reduction strategies (Presidential Commission for the Study of Biomedical Issues 2010). NIOSH has identified synthetic biology as a possible hazard to workers; however, where and how exposure to these hazards could occur is not well defined. NIOSH is establishing an Emerging Technologies branch that will work to identify where hazards to workers might occur and how to mitigate exposures in emerging technologies such as synthetic biology.

As synthetic biology enters more and more industrial workplaces, engagement from the entire occupational safety and health practice community is needed for the responsible development of commercial synthetic biology while protecting the health and safety of its workers (Moe-Behrens et al. 2013). Proven risk mitigation approaches for emerging technologies including health surveillance, proactive risk management, prevention-through-design principles, and dynamic guidance should be implemented to ensure that no worker suffers adverse health effects in the emerging synthetic biology workplace and that the synthetic biology realizes its full potential in improving quality of life for all.

Disclaimer The findings and conclusions in this report are those of the authors and do not necessarily represent the views of the National Institute for Occupational Safety and Health, the Centers for Disease Control and Prevention, or the US Department of Health and Human Services.

References

Baker, D., Church, G., Collins, J., et al. (2006). Engineering life: Building a FAB for biology. *Scientific American, 294*, 44–51.

Ball, P. (2005). Synthetic biology for nanotechnology. *Nanotechnology, 16*, R1–R8.

Bej, A. K., Perlin, M. H., & Atlas, R. M. (1988). Model suicide vector for containment of genetic engineered microorganisms. *Applied and Environmental Microbiology, 54*(10), 2472–2477.

Berg, P., Baltimore, D., Brenner, S., Roblin, R. O., III, & Singer, M. F. (1975). Summary statement of the Asilomar Conference on Recombinant DNA Molecules. *PNAS, 72*(6), 1981–1984.

Bergeson, L. L., Campbell, L. M., Dolan, S. L., et al. (2015). *The DNA of the U.S. regulatory system: Are we getting it right for synthetic biology?* Washington, D.C.: Wilson Center. Retrieved from http://www.synbioproject.org/publications/dna-of-the-u.s-regulatory-system/. Accessed on 16 Mar 2018.

Blaese, R. M., Culver, K. W., Miller, A. D., et al. (1995). T lymphocyte-directed gene therapy for ADA-SCID: Initial trial results after 4 years. *Science, 270*, 475–480.

Bokhoven, M., Stephen, S. L., Knight, S., et al. (2009). Insertional gene activation by lentiviral and gammaretroviral vectors. *Journal of Virology, 83*(1), 283–294.

Byers, K. B. (2015). Biosafety tips. *Applied Biosafety, 20*, 250–252.

Cai, Y., Agmon, N., Choi, W. J., et al. (2015). Intrinsic biocontainment: Multiplex genome safeguards combine transcriptional and recombinational control of essential yeast genes. *PNAS, 112*(6), 1803–1808.

Carlson, R. (2016). Estimating the biotech sector's contribution to the US economy. *Nature Biotechnology, 34*(3), 247–255.

Carter, S. R., Rodemeyer, M., Garfinkel, M. S., & Friedman, R. (2014). *Synthetic biology and the U.S. biotechnology regulatory system: Challenges and options.* J. Craig Venter Institute. Retrieved from http://www.jcvi.org/cms/fileadmin/site/research/projects/synthetic-biology-and-the-us-regulatory-system/full-report.pdf. Accessed on 16 Mar 2018.

Center for Biosecurity of UPMC. (2012). The industrialization of biology and its impact on national security. Retrieved from http://www.upmchealthsecurity.org/our-work/pubs_archive/pubs-pdfs/2012/2012-06-08-industrialization.pdf. Accessed on 13 Mar 2018.

Chan, C. T. Y., Lee, J. W., Cameron, E., Bashor, C. J., & Collins, J. J. (2016). 'Deadman' and 'passcode' microbial kill switches for bacterial containment. *Nature Chemical Biology, 12*, 82–86.

Cockrell, A. S., Ma, H., Fu, K., McCown, T. J., & Kafri, T. (2006). A trans-lentiviral packaging cell line for high-titer conditional self-replicating HIV-1 vectors. *Molecular Therapy, 14*(1), 276–284.

Convention on Biological Diversity. (2015). *Report of the Ad Hoc technical expert advisory group on synthetic biology. UNEP/CBD/SYNBIO/AHTEG/2015/1/3.* Retrieved from https://www.cbd.int/doc/meetings/synbio/synbioahteg-2015-01/official/synbioahteg-2015-01-03-en.pdf. Accessed on 19 Mar 2018.

Cronin, J., Zhang, X.-Y., & Reiser, J. (2005). Altering the tropism of lentiviral vectors through pseudotyping. *Current Gene Therapy, 5*(4), 387–398.

Cross, D., & Burmester, J. K. (2006). Gene therapy for cancer treatment: Past, present and future. *Clinical Medicine & Research, 4*(3), 218–227.

Cummings, C. L., & Kuzma, J. (2017). Societal risk evaluation scheme (SRES): Scenario-based multi-criteria evaluation of synthetic biology applications. *PLoS One, 12*(1), e0168564.

deGrandpre, A. (2017). The survival of a Mars mission could depend on astronaut urine. *Washington Post* August 23, 2017. Retrieved from https://www.washingtonpost.com/news/speaking-of-science/wp/2017/08/23/the-survival-of-a-mars-mission-could-depend-on-astronaut-urine/?utm_term=.903e0a6953c8. Accessed 19 Mar 2018.

Department of Defense (DOD). (2015). *Technical assessment: Synthetic biology.* Office of Technical Intelligence, Office of the Assistant Secretary of Defense for Research & Engineering. Retrieved from http://defenseinnovationmarketplace.mil/resources/OTI-SyntheticBiologyTechnicalAssessment.pdf. Accessed 19 Mar 2018.

Department of Health and Human Services (DHHS). (2009). *Biosafety in microbiological and biomedical laboratories* (DHHS Publication No. (CDC) 21–1112) (5th ed.). U.S. Department of Health and Human Services. Retrieved from https://www.cdc.gov/biosafety/publications/bmbl5/bmbl.pdf. Accessed on 16 Mar 2018.

Eisenstein, M. (2016). Living factories of the future. *Nature, 531*, 401–403.

Environmental Protection Agency. (2015). Pesticides; Agricultural Worker Protection Standard Revisions, November 2, 2015. Retrieved from https://www.gpo.gov/fdsys/pkg/FR-2015-11-02/pdf/2015-25970.pdf. Accessed on April 2, 2018.

Erickson, R., Sing, H. R., & Winters, P. (2011). Synthetic biology: Regulating industry uses of new biotechnologies. *Science, 333*, 1254–1256.

Food and Drug Administration (FDA). (2017). FDA approval brings first gene therapy to the United States: CAR T-cell therapy approved to treat certain children and young adults with B-cell acute lymphoblastic leukemia. *FDA News Release*. Retrieved from https://www.fda.gov/NewsEvents/Newsroom/PressAnnouncements/ucm574058.htm. Accessed on 15 Mar 2018.

Fredrickson, D. S. (1980). A history of the recombinant DNA guidelines in the United States. Retrieved from https://profiles.nlm.nih.gov/FF/B/B/K/C/_/ffbbkc.pdf. Accessed on 2 Apr 2018.

Friends of the Earth U.S., International Center for Technology Assessment, & ETC Group. (2010). The principles for the oversight of synthetic biology. Retrieved from http://www.etcgroup.org/sites/www.etcgroup.org/files/The%20Principles%20for%20the%20Oversight%20of%20Synthetic%20Biology%20FINAL.pdf. Accessed 16 Mar 2018.

Gallagher, R. R., Patel, J. R., Interiano, A. L., Rovner, A. J., & Isaacs, F. J. (2015). Multilayered genetic safeguards limit growth of microorganisms to defend environments. *Nucleic Acids Research, 43*(3), 1945–1954.

Garfinkel, M. (2012). Biological containment of synthetic microorganisms: Science and policy. European Science Foundation/Standing Committee for Life, Earth, and Environmental Sciences Strategic Workshop, November 13–14, 2012. Retrieved from http://www.embo.org/documents/science_policy/biocontainment_ESF_EMBO_2012_workshop_report.pdf. Accessed on 29 Mar 2018.

Garfinkel, M. S., Endy, D., Epstein, G. L., & Friedman, R. M. (2007). *Synthetic genomics: Options for governance*. Retrieved from http://www.synbiosafe.eu/uploads/pdf/Synthetic%20Genomics%20Options%20for%20Governance.pdf. Accessed on 16 Mar 2018.

Gibson, D. G., Benders, G. A., Andrews-Pfannkoch, C., et al. (2008). Complete chemical synthesis, assembly, and cloning of a Mycoplasma genitalium genome. *Science, 319*, 1215–1219.

Gibson, D. G., Glass, J. I., Lartigue, C., et al. (2010). Creation of a bacterial cell controlled by a chemically synthesized genome. *Science, 329*, 52–56.

Grady, D. (2017). Work stops at C.D.C.'s top deadly germ lab over air hose safety. *The New York Times*, February 17, 2017. Retrieved from https://www.nytimes.com/2017/02/17/health/cdc-germ-lab-safety.html. Accessed on 20 Mar 2018.

Gray, J. T. (2011). Laboratory safety for oncogene-containing retroviral vectors. *Applied Biosafety, 16*(4), 218–222.

Hayden, E. C. (2015). Tech investors bet on synthetic biology. *Nature, 527*, 19.

Howard, J., Murashov, V., & Schulte, P. (2017). Synthetic biology and occupational risk. *Journal of Occupational and Environmental Hygiene, 14*(3), 224–236.

Kurihara, K., Okura, Y., Matsuo, M., Toyota, T., Suzuki, K., & Sugawara, T. (2015). A recursive vesicle-based model protocell with a primitive model cell cycle. *Nature Communications, 6*, 1–7.

Kuzma, J. (2016). Reboot the debate on genetic engineering. *Nature, 531*, 165–167.

Lander, E. S. (2015). Brave new genome. *The New England Journal of Medicine, 373*(1), 5–8.

Ledford, H. (2016). Gene editing surges as U.S. rethinks regulations. *Nature, 532*, 158–159.

Leduc, S. (1910). *Theorie physico-chimique de la vie et generations spontanees* (p. 202). Paris: A. Poinat.

Levine, B. L., Humeau, L. M., Boyer, J., et al. (2006). Gene transfer in humans using a conditionally replicating lentiviral vector. *PNAS, 103*(46), 17372–17377.

Lokko, Y., Hejde, M., Schebesta, K., Scholtes, P., Van Montagu, M., & Giacca, M. (2018). Biotechnology and the bioeconomy – Towards inclusive and sustainable industrial development. *New Biotechnology, 40*, 5–10.

Lowrie, H., & Tait, J. (2010). *Policy brief: Guidelines for the appropriate risk governance of synthetic biology*. Lausanne: International Risk Governance Council. Retrieved from http://www.irgc.org/IMG/pdf/irgc_SB_final_07jan_web.pdf. Accessed on 16 Mar 2018.

Maetzig, T., Galla, M., Baum, C., & Schambach, A. (2011). Gammaretroviral vectors: Biology, technology and application. *Viruses, 3*, 677–713.

Merten, O.-W., & Al-Rubeai, M. (2011). *Viral vectors for gene therapy*. New York: Human Press.

Miller, D. M., & Gulbis, J. M. (2015). Engineering protocells: Prospects for self-assembly and nanoscale production lines. *Life, 5*, 1019–1053.

Moe-Behrens, G. H. G., Davis, R., & Haynes, K. A. (2013). Preparing synthetic biology for the world. *Frontiers in Microbiology, 4*(5), 1–10.

Mosier, D. E. (2004). Introduction for 'safety considerations for retroviral vectors: A short review'. *Applied Biosafety, 9*(2), 68–75.

Murashov, V., & Howard, J. (2008). The U.S. must help set international standards for nanotechnology. *Nature Nanotechnology, 3*, 635–636.

Murashov, V., & Howard, J. (2009). Essential features of proactive risk management. *Nature Nanotechnology, 4*, 467–470.

Murashov, V., Schulte, P., Geraci, C., & Howard, J. (2011). Regulatory approaches to worker protection in nanotechnology industry in the USA and European Union. *Industrial Health, 49*, 280–296.

Naldini, L. (2015). Gene therapy returns to centre stage. *Nature, 526*, 351–360.

National Archives and Records Administration. (1992). Exercise of federal oversight within scope of statutory authority: Planned introductions of biotechnology products into the environment. *Federal Register, 57*(39), 6753–6762.

National Research Council of the National Academies. (2015). *Industrialization of biotechnology: A roadmap to accelerate the advanced manufacturing of chemicals*. Washington, D.C.: The National Academies Press.

NIH. (2016). *NIH guidelines for research involving recombinant or synthetic nucleic acid molecules (NIH guidelines)*. U.S. Department of Health and Human Services, National Institutes of Health, Office of Science Policy, April, 2016. Retrieved from https://osp.od.nih.gov/wp-content/uploads/NIH_Guidelines.html. Accessed on 15 May 2018.

NIOSH. (2009). *Approaches to safe nanotechnology: Managing the health and safety concerns associated with engineered nanomaterials* (DHHS (NIOSH) Publication No. 2009–125). Cincinnati: U.S. Department of the Health and Human Services, Centers for Disease Control and Prevention, National Institute for Occupational Safety and Health. Retrieved from http://www.cdc.gov/niosh/docs/2009-125/pdfs/2009-125.pdf. Accessed on 29 Mar 2018.

NIOSH. (2014). *The state of the national initiative on prevention through design* (DHHS (NIOSH) Publication No. 2014–123). Cincinnati: U.S. Department of the Health and Human Services, Centers for Disease Control and Prevention, National Institute for Occupational Safety and Health. Retrieved from http://www.cdc.gov/niosh/docs/2014-123/pdfs/2014-123.pdf. Accessed on 29 Mar 2018.

Occupational Safety and Health Administration (OSHA). (2018). Infectious diseases rulemaking. Retrieved from https://www.osha.gov/dsg/id/index.html. Accessed on 14 Mar 2018.

Office of Science Technology and Policy (OSTP). (1986). Coordinated framework for regulation of biotechnology. *Executive Office of the President*, June 26, 1986. Retrieved from http://www.aphis.usda.gov/brs/fedregister/coordinated_framework.pdf. Accessed on 20 Mar 2018.

Office of Science Technology and Policy (OSTP). (2017). Modernizing the regulatory system for biotechnology products: Final version of the 2017 update to the coordinated framework for the regulation of biotechnology. Retrieved from https://obamawhitehouse.archives.gov/sites/default/files/microsites/ostp/2017_coordinated_framework_update.pdf. Accessed on 14 Mar 2018.

OSHA. (1985). Agency guidelines on biotechnology. *Federal Register, 50*(71), 14468–14469.

OSHA. (2005a). Toxic and hazardous substances, *Code of Federal Regulations, Title 29*, Part 1910.1000. 2005. pp. 7–18. Retrieved from https://www.osha.gov/pls/oshaweb/owadisp.show_document?p_id=9991&p_table=STANDARDS. Accessed on 20 Mar 2018.

OSHA. (2005b). Access to employee exposure and medical records, *Code of Federal Regulations Title 29*, Part 1020. 2005. pp. 93–102. Retrieved from https://www.osha.gov/pls/oshaweb/owadisp.show_document?p_table=STANDARDS&p_id=10027. Accessed on 20 Mar 2018.

OSHA. (2010). Bloodborne pathogens, *Code of Federal Regulations, Title 29,* Part 1910.1030. 2010. pp. 265–278. Retrieved from https://www.osha.gov/pls/oshaweb/owadisp.show_document?p_table=STANDARDS&p_id=10051. Accessed on 20 Mar 2018.

OSHA. (2014a). Hazard communication, *Code of Federal Regulations, Title 29,* Part 1910.1200. 2014. pp. 463–591. Retrieved from https://www.osha.gov/pls/oshaweb/owadisp.show_document?p_table=standards&p_id=10099. Accessed on 20 Mar 2018.

OSHA. (2014b). Occupational exposure to hazardous chemicals in laboratories, *Code of Federal Regulations, Title 29,* Part 1910.1450. 2014. pp. 591–651. Retrieved from https://www.osha.gov/pls/oshaweb/owadisp.show_document?p_table=STANDARDS&p_id=10106. Accessed on 20 Mar 2018.

OSHA. (2014c). Respiratory protection, *Code of Federal Regulations, Title 29,* Part 1910.134. 2014. pp. 426–452. Retrieved from https://www.osha.gov/pls/oshaweb/owadisp.show_document?p_table=STANDARDS&p_id=12716. Accessed on 20 Mar 2018.

Phelan, M. (2018). Why fungi adapt so well to life in space. *Scienceline,* March 7, 2018. Retrieved from http://scienceline.org/2018/03/fungi-love-to-grow-in-outer-space/. Accessed on 29 Mar 2018.

Pollack, A., & Wilson, D. (2010). Safety rules can't keep up with biotech. *The New York Times,* May 27, 2010. Retrieved from http://www.nytimes.com/2010/05/28/business/28hazard.html?_r=0. Accessed on 20 Mar 2018.

Presidential Commission for the Study of Biomedical Issues. (2010). *New directions: The ethics of synthetic biology and emerging technologies.* Washington, D.C.: Presidential Commission for the Study of Biomedical Issues. Retrieved from https://bioethicsarchive.georgetown.edu/pcsbi/sites/default/files/PCSBI-Synthetic-Biology-Report-12.16.10_0.pdf. Accessed on 29 Mar 2018.

Robbins, P. D., & Ghivizzani, S. C. (1998). Viral vectors for gene therapy. *Pharmacology & Therapeutics, 80*(1), 35–47.

Rohn, J. (2013). Synthetic biology goes industrial. *Nature Biotechnology, 31,* 773.

Sakuma, T., Barry, M. A., & Ikeda, Y. (2012). Lentiviral vectors: Basic to translational. *The Biochemical Journal, 443,* 603–618.

Schambach, A., Zychlinski, D., Ehrnstroem, B., & Baum, C. (2013). Biosafety features of lentiviral vectors. *Human Gene Therapy, 24,* 132–142.

Schlimgen, R., Howard, J., Wooley, D., Thompson, M., Baden, L. R., Yang, O. O., Christiani, D. C., Mostoslavsky, G., Diamond, D. V., Gilman Duane, E., Byers, K., Winters, T., Gelfand, J. A., Fujimoto, G., Hudson, W., & Vyas, J. M. (2016). Risk associated with lentiviral vector exposures and prevention strategies. *Journal of Occupational and Environmental Medicine, 58*(12), 1159–1166.

Schmidt, M. (2010). Xenobiology: A new form of life as the ultimate biosafety tool. *BioEssays, 32*(4), 322–331.

Schmidt, M. (2012). Introduction. In M. Schmidt (Ed.), *Synthetic biology: Industrial and environmental applications* (pp. 1–6). Wiley-Blackwell: Weinheim.

Schmidt, M., & de Lorenzo, V. (2012). Synthetic constructs in/for the environment: Managing the interplay between natural and engineered biology. *FEBS Letters, 586,* 2199–2206.

Schulte, P., Rinehart, R., Okun, A., Geraci, G. L., & Heidel, D. S. (2008). National prevention through design (PtD) initiative. *Journal of Safety Research, 39*(2), 115–121.

Schulte, P., Mundt, D. J., Nasterlack, M., Mulloy, K. B., & Mundt, K. A. (2011). Exposure registries: Overview and utility for nanomaterial workers. *Journal of Occupational and Environmental Medicine, 53*(6 Suppl), S42–S47.

Scott, A. (2018). A CRISPR path to drug discovery. *Nature, 555,* S10–S11.

Sewell, D. L. (1995). Laboratory-associated infections and biosafety. *Clinical Microbiology Reviews, 8,* 389–405.

Smith, H. O., Hutchison, C. A., Pfannkoch, C., & Venter, J. C. (2003). Generating a synthetic genome by whole genome assembly: φX174 bacteriophage from synthetic oligonucleotides. *PNAS, 100*(26), 15440–15445.

Stanford University. (2018). Biosafety manual. Retrieved from https://ehs.stanford.edu/manual/biosafety-manual. Accessed 20 Mar 2018.

Steidler, L., Neirynck, S., Huyghebaert, N., et al. (2003). Biological containment of genetically modified Lactococcus lactis for intestinal delivery of human interleukin 10. *Nature Biotechnology, 21,* 785–789.

SynBioBeta. (2017). The synthetic biology industry: Annual growth update. Retrieved from http://synbiobeta.com/wp-content/uploads/sites/4/2017/03/Synthetic-Biology-Industry-Annual-Growth-Update-2017.pdf. Accessed 20 Mar 2018.

Synthetic Biology Leadership Council. (2015). Biodesign for the bioeconomy: UK synthetic biology strategic plan 2016. Retrieved from http://www.synbio.cam.ac.uk/news/synbio-strategic-plan-2016. Accessed 19 Mar 2018.

Synthetic Biology Project. (2018). *What is synthetic biology?* Washington, D.C.: Woodrow Wilson Center. Retrieved from http://www.synbioproject.org/topics/synbio101/definition/. Accessed on 19 Mar 2018.

Szybalski, W. (1974). In vivo and in vitro initiation of transcription. In A. Kohn & A. Shatkay (Eds.), *Control of gene expression* (pp. 23–24). New York: Plenum Press, 404–405, 411–412, and 415–417.

Tomas, H. A., Rodriquez, A. F., Alves, P. M., & Corodinha, A. S. (2013). Chapter 12: Lentiviral gene therapy vectors: Challenges and future directions. In F. M. Molina (Ed.), *Gene therapy: Tools and potential applications*. Open Source, InTech, 2013. Retrieved from http://www.intechopen.com/books/gene-therapy-tools-and-potential-applications/lentiviral-gene-therapy-vectors-challenges-and-future-directions. Accessed on 19 Mar 2018.

Trevan, T. (2015). Biological research: Rethink biosafety. *Nature, 527,* 155–158.

Trump, B. D., Foran, C., Rycroft, T., Wood, M. D., Bandolin, N., Cains, M., et al. (2018). Development of community of practice to support quantitative risk assessment for synthetic biology products: Contaminant bioremediation and invasive carp control as cases. *Environment Systems and Decisions, 38*(4), 517–527.

University of Cincinnati. (2014). Viral vectors: Web training. Biosafety Office. Retrieved from http://researchcompliance.uc.edu/biosafety/Training/ViralVectorWebtraining.aspx. Accessed on 20 Mar 2018.

Verbeek, J. H., Ijaz, S., Mischke, C., Ruotsalainen, J. H., Mäkelä, E., Neuvonen, K., Edmond, M. B., Sauni, R., Kilinc Balci, F. S., & Mihalache, R. C. (2016). Personal protective equipment for preventing highly infectious diseases due to exposure to contaminated body fluids in healthcare staff. *Cochrane Database of Systematic Reviews,* (4), CD011621. https://doi.org/10.1002/14651858.CD011621.pub2.

Villano, J. S., Follo, J. M., Chappell, M. G., & Collins, M. T., Jr. (2017). Personal protective equipment in animal research. *Comparative Medicine, 67*(3), 203–214.

Weick, K. E., & Sutcliffe, K. M. (2015). *Managing the unexpected: Sustained performance in a complex world.* Hoboken: Wiley.

White House. (2012). *National bioeconomy blueprint.* Washington, D.C.: Executive Office of the President. Retrieved from https://obamawhitehouse.archives.gov/sites/default/files/microsites/ostp/national_bioeconomy_blueprint_april_2012.pdf. Accessed on 19 Mar 2018.

World Health Organization (WHO). (2004). *Laboratory biosafety manual* (3rd ed.). Geneva: World Health Organization. Retrieved from http://www.who.int/csr/resources/publications/biosafety/Biosafety7.pdf. Accessed on 16 Mar 2018.

Wurtz, N., Papa, A., Hukic, M., DiCaro, A., Lepare-Goffart, I., Leroy, E., et al. (2016). Survey of laboratory-acquired infections around the world in biosafety level 3 and 4 laboratories. *European Journal of Clinical Microbiology & Infectious Diseases, 35*(8), 1247–1258.

Yin, H., Kauffman, K. J., & Anderson, D. G. (2017). Delivery technologies for genome editing. *Nature Reviews: Drug Discovery, 16,* 387–399.

Young, A. (2016). CDC labs repeatedly faced secret sanctions for mishandling bioterror germs. *USA Today,* May 10, 2016. Retrieved from http://www.usatoday.com/story/news/2016/05/10/cdc-lab-secret-sanctions/84163590/. Accessed 20 Mar 2018.

Young, A., & Penzenstadler N. (2015). Inside America's secretive biolabs. *USA Today*, May 28, 2015. Retrieved from http://www.usatoday.com/story/news/2015/05/28/biolabs-pathogens-location-incidents/26587505/. Accessed 20 Mar 2018.

Zufferey, R., Dull, T., Mandel, R. J., et al. (1998). Self-inactivating lentivirus vector for safe and efficient in vivo gene delivery. *Journal of Virology, 72*(12), 9873–9880.

Designing a "Solution-Focused" Governance Paradigm for Synthetic Biology: Toward Improved Risk Assessment and Creative Regulatory Design

Adam M. Finkel

Introduction

Any commercially available, bench-scale, or even proposed "back-of-envelope" application of synthetic biology (SynBio) will prompt discussion and debate—perhaps highly philosophical, perhaps highly practical and legalistic—both about how to *think* about the application and what, if anything, to *do* about it, pro or con. The former kind of discussion is the domain of precautionary or "permissionless" (Thierer 2016) rhetoric, of quantitative risk assessment, and of cost-benefit analysis; the latter is the domain of risk management, regulation, information disclosure, industrial policy, and other interventions.

SynBio applications are controversial because of their promise and their peril—in short, because they can greatly reduce risks and also because they threaten to expose humans and the natural environment to new or increased risks. I will discuss these issues throughout this chapter, but a working introductory definition of SynBio is "using tools of molecular biology to engineer new or improved cellular products or processes" (see Cameron et al. 2014). A working definition of quantitative risk assessment (QRA) is "a method that synthesizes information from basic sciences (e.g., toxicology, epidemiology, chemistry, statistics) to explore the probability that one or more adverse outcomes will occur from a product or process, and to gauge the severity of each outcome" (see Kaplan and Garrick 1981). As I will discuss below in the section "Risk Assessment Methodologies for SynBio", the output of a *useful* QRA is not a yes/no pronouncement about the existence of a risk, or even a quantitative estimate both of its likelihood and its consequence. It is, instead, a "characterization" of risk (NAS 1983) that offers information about (1) the extent of

A. M. Finkel (✉)
Environmental Health Sciences, University of Michigan School of Public Health,
Ann Arbor, MI, USA
e-mail: adfinkel@umich.edu

© Springer Nature Switzerland AG 2020
B. D. Trump et al. (eds.), *Synthetic Biology 2020: Frontiers in Risk Analysis and Governance*, Risk, Systems and Decisions, https://doi.org/10.1007/978-3-030-27264-7_9

scientific uncertainty that, if analysts are honest and humble, precludes them from pinning down the probability or severity with precision and (2) the extent and nature of interindividual variability in the risk, so that different populations can appreciate that probabilities and severities also depend on who is facing the hazardous condition(s).

This chapter, the capstone product of a project supported by the Alfred P. Sloan Foundation, breaks new ground in two fundamental and complementary ways—one dealing with analysis of evidence and one with evidence-based action. With respect to analysis, many thought processes about and formal assessments of possible harms to human health, safety, and environment (HSE) begin and end with the most simple of questions: is it "safe"? Or, slightly more broadly, "does it promise economic benefits in excess of the (monetized) harms it presents?" Such questions allow (or at least encourage) only dichotomous answers, but worse, *they crowd out more sophisticated, sweeping, and bold risk assessment questions.* Similarly, many interventions (risk management) to control possible HSE harms consider only the narrowest range of actions: should we ban the new process/product/activity, or should we declare "nothing to see here; let's move on?" Here my concern is that the "poverty of choices" can lead to poor decisions, akin to how a "poverty of questions" can lead to poor analyses.

I instead start from the premise that asking a wider range of questions and considering a wider range of actions are both unambiguous virtues. This is not to say that expansive and protracted analyses always outperform simpler ones or that circumscribing the choices to "go/no-go" is always mediocre—*only that simplifying the analysis and narrowing the range of options should be done consciously and at least somewhat reluctantly.*

These twin considerations apply in spades to the new arena of SynBio (for an excellent primer on the issues raised, see Moe-Behrens et al. 2013, or Rodemeyer 2009). First, to a greater extent than is true for most other new and continuing sources of HSE risks, the dangers posed by SynBio applications are offset (sometimes partially, completely, or "more than completely") by their direct and often unprecedented power to *reduce other risks.* Hence, I argue that traditional "is it safe?" risk assessment questions are particularly myopic here, as they ignore the real possibility that the new application *is at the same time both objectively dangerous and yet a risk-reducing improvement over the status quo.* Traditional "go/no-go" risk management choices are also particularly inappropriate for new SynBio technologies, because of their novelty, the rapidity with which unforeseen risks or unforeseen risk-reducing benefits may be realized soon after their deployment, their ethically controversial nature, and their dependence on a social license to operate and perhaps even public sector funding. For decisions like these, society has the opportunity (arguably the responsibility to itself and to posterity) to consider many shades of gray between draconian regulation and laissez-faire—*as well as various creative options that are actually either more stringent than even an outright ban or more encouraging than even a hands-off posture.*

To introduce the rich range of assessment questions and management options that I urge should be posed and considered in the analysis and governance of SynBio,

I offer a hierarchical ordering of each; the risk assessment questions ranging from the most rudimentary to the most nuanced and expansive and the control options ranging from the most favorable to the SynBio application to the most restrictive.

Table 1 (adapted from Finkel 2018a) presents ten distinct levels of analytic complexity, several of which I will highlight here. Level 1 represents the most qualitative appraisal possible: the "is it safe?" (or "is it costly to avoid?") question. Level 5 offers the traditional cost-benefit question: are the expected risks reduced by the policy greater than its expected costs? As we will see in detail below, this question can easily be recast as an appraisal of the net benefit profile of a new application or technology: on average, does it reduce risk by more than it exacerbates it? The remainder of the hierarchy basically enriches the simple cost-benefit (or risk-risk) estimate with considerations of the two most fundamental phenomena surrounding all risks—the uncertainty impeding our ability to precisely quantify risk and the interindividual variability that makes any risk estimate uniquely applicable to only one individual or subgroup within the affected population. A cost-benefit (or risk-risk) analysis that fully considered both phenomena would ask questions of the form "for these particular citizens, what is the range of possible outcomes (the new technology reduces, leaves unchanged, or increases risks, by how much?), and what is the probability of each outcome?"

A "Solution-Focused" Partnership between Analysis and Action

Armed with the answers to one or more assessment questions about a SynBio application, society could then consider whether they justify a response at or near the highly "bullish" left end of the spectrum, the highly restrictive right-hand end, or somewhere in between. Figure 1 displays a very broad range of possible responses to a SynBio application; of particular note, the right-hand tail of the range offers somewhat more ambitious prospects for SynBio control than are generally contemplated, while the left-hand region offers various gradations of incentives and support for SynBio that go far beyond merely permissive responses. The other unusual feature of this schema is that it explicitly construes the SynBio application as competing with existing materials or applications—therefore, options exist to constrain the SynBio application indirectly (by subsidizing the competing application or loosening regulations on it) or to promote SynBio indirectly (by regulating or banning the existing application).

The main contribution of this chapter, however, comes *in the space between risk assessment questions and risk management control options*, simply because analysis should not result in a specific action when the dots are connected poorly and with little forethought. Knowing only the net risk (or net benefit) of a SynBio application, however comprehensively that risk is assessed (see Table 1), one can certainly take some action somewhere along the spectrum in Fig. 1, *but this is far from the*

Table 1 Characteristics of a ten-rung ladder of complexity in policy decision-rules

Level	Verbal description	Question posed	What analysis does it demand from		
			Risk assessment?	Cost assessment?	Benefits valuation?
1	Prevent or eliminate a risk (or a cost) if it is "real"	Is $R > 0$ OR $C > 0$?	That the risk exists or could plausibly exist	That taking action would be costly or could plausibly be costly	Nothing
2	Attain a "bright line" of safety	Is $R < R*$?	Compare risk's magnitude to the bright line	Nothing	Nothing
3	Pass a "double bright line" test	Is $R < R*$ *and* is $C < C*$?	Showing that risk is "unacceptable" or "significant" without action	Showing that risk can be reduced without unacceptably high costs	Nothing
4	Compare arbitrary point estimates of total benefit and total cost	Is $[(P \times \Delta R_? \times \mathrm{VSL}) - C_?] > 0$?	"guesstimate" (low, central, conservative?) of risk reduction	"guesstimate" (low, central, conservative?) of cost	Point estimate of VSL
5	Compare reconciled point estimates of total benefit and total cost (expected values, in this example)	Is $\left[\left(P \times \overline{\Delta R} \times \mathrm{VSL}\right) - \overline{C}\right] > 0$?	Propagate mean estimates of input variables for risk reduction	Propagate mean estimates of input variables for cost	Point estimate of VSL
6	Develop a probability density function (pdf) for net benefit	With what probability is $[pdf(\Delta R)x\ pdf(\mathrm{VSL}\ x\ P) - pdf(C)] > 0$?	Monte Carlo or other methods to generate pdf of risk reduction	Monte Carlo or other methods to generate pdf for cost	Distribution of mean VSLs across studies
7	Develop a pdf for net risk minus net cost	With what probability is $[pdf(\Delta R_{\mathrm{NET}})\ x\ pdf(\mathrm{VSL}\ x\ P) - pdf(C_{\mathrm{NET}})] > 0$?	As in #6, plus pdfs for significant co-risks or "anti-risks" (co-benefits)	As in #6, plus pdfs for significant "co-costs" (innovation drag, etc.) and "anti-costs" (general equilibrium effects, employment in "green" sectors, etc.)	As in #6

			Focus on net risk to a highly affected (or highly favored) subpopulation	Focus on net cost to a highly affected (or highly favored) subpopulation	Point estimate of VSL (or nothing, if risk or cost is deemed unacceptable in its natural units)
8	Compare pdfs of net risk and net cost with some attention to interindividual variability in each	Which is larger, NB (or net risk) to subgroup i or NB (or net risk) to subgroup j?			
9	Assess the uncertain net benefit to each individual (or subpopulation) separately, and then aggregate	Is $\sum_{i=1}^{P}\left\{\left[pdf\left(\Delta r_i\right) \times v_i\right] - pdf\left(c_i\right)\right\} > 0$?	Assess net risk (and its uncertainty) separately for each subpopulation (stratified by exposure, susceptibility, or both)	Assess net cost (and its uncertainty) separately for each subpopulation (stratified by industrial sector, income group of consumers, or both) Assess price elasticity of demand so as to estimate apportionment of costs between producers and consumers	Use individual data points in VSL stated preference studies to develop subgroup VSL estimates
10	Combine estimates of individual net benefit via any non-trivial social welfare function	Is $\sum_{i=1}^{P} w_i\left\{\left[pdf\left(\Delta r_i\right) \times v_i\right] - pdf\left(c_i\right)\right\} > 0$?	As in #9	As in #9	Develop weights (the w_i) via some principled criteria

Key to abbreviations: *NB* net benefit, *P* number of persons in affected population, *VSL* value of a statistical life, ΔR (ΔC) change in total risk (total cost) that a policy affects, Δr (Δc) change in risk (cost) for a particular *individual*

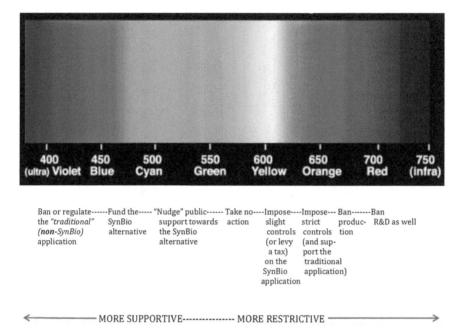

Fig. 1 Spectrum of possible governance responses to a synthetic biology application

only paradigm for linking the results of assessment to the form, ambition, and stringency of control. When we look solely to risk assessment to inform and guide action, we implicitly assume that the amount of concern or worry proportionally dictates the amount of resources we should expend to reduce the given hazard. Instead, I have proposed (Finkel 2011; see also Natl Acad Sci 2009; Goldstein 2018) that a host of questions should intervene between the two parts of the "big risks, large controls" mantra:

> Risk assessment for its own sake is an inherently valuable activity but, at best, a risk assessment can illuminate what we should *fear*, and tap into our inexhaustible supply of worry—whereas a good solution-focused analysis can illuminate what we should *do*, and mobilize our precious supply of resources. (Finkel 2011, p. 781)

By "solution-focused," I mean a decision process that eschews risk assessment performed in one-risk-at-a-time isolation and disconnected from the appraisal of what solutions are or may be available to control the risks being compared. "Solution-focused risk assessment," or SFRA, seeks above all to resist the temptation to declare victory when a risk has been quantified and a lower level of risk deemed "acceptable." Such a mindset suffers from two fundamental flaws: it defines success as an isolated risk reduction, rather than a more comprehensive solution, and it is often satisfied with the *aspirational* success of declaring an acceptable risk level, even if the risk actually never is reduced to that level. So SFRA instead emphasizes (1) that we must compare risk-reducing (welfare-enhancing) *opportunities*, not disembodied risks, and (2) that the earlier in the decision process we array the possible solutions, the less likely we will *define away* promising answers to risk-risk and cost-benefit dilemmas and end up with a course of action that is inferior to others we neglected to consider. For example, the US Environmental Protection Agency (EPA) has a vigorous research program concerned with the toxicity of various plasticizers (beginning with bisphenol A (BPA)) that can leach into drinking water provided in disposable plastic bottles. Eventually, this work may lead to regulatory limitations on the allowable concentrations of BPA in bottled water. If EPA assesses the risks more holistically, it might lead to a suite of concentration limits on the various substitutes for BPA as well.

But imagine posing the question not as "how many parts per billion of each substance is acceptable?" but as "how can the market deliver clean, cold drinking water at an affordable price and with the smallest environmental and human health footprint?" *That* linkage between analysis and action might prompt public discussion of the energy use and disposal issues associated with the current annual production of 49 billion plastic bottles in the USA (from a baseline of essentially zero several decades ago, a time when US consumers did not want for ready access to drinking water). And *that* question might lead to discussions of how governmental incentives, taxes, or investments in infrastructure might help reduce the runaway demand for plastic water bottles of any kind and increase the supply of "free" (funded by taxpayers) or low-cost drinking water provided as we remember it in the 1960s–1990s—available in public places and lobbies of private buildings, via fountains and water coolers.

This emphasis on solutions is not only the polar opposite of the way EPA and other US federal HSE regulatory agencies have largely construed their missions since their founding decades ago but is very different from recent "baby steps" EPA has taken to ground risk assessment in practical utility. In particular, EPA has highlighted, particularly in referring to its 1998 Guidelines for Ecological Risk Assessment, that it incorporates "problem formulation" into its planning as a way to make risk assessment more useful. Unfortunately, this semantic change only means that EPA sometimes asks up front the question "how can we limit the scope of our research and risk analysis to issues that can help set a risk reduction goal?"—and this is quite different from "how can we harness risk assessment to discriminate among possible ways to fulfill a human need effectively and with minimal imposition of new risks?" SFRA is not a wholly new concept by any means, however—it can be thought of as a marriage between QRA and a more impressionistic "innovation-based strategy for a sustainable environment" (Ashford 2000) that steers industrial policy toward solutions that minimize risks.[1]

---

[1] I also must acknowledge that after contributing to NAS (2009) and writing Finkel (2011), I realized that a prior report (Nelson and Banker 2007) introduced many of the same concepts as SFRA. I was led astray by the report's title, which began with "Problem Formulation," and I didn't appreciate that Nelson and Banker used the word "problem" in exactly the opposite way that EPA does and exactly the way I advocate—to them, the "problem" is the unfulfilled human need that competing technologies profess to supply (not the "problems" the technologies pose), and hence the goal of analysis is to solve *that* problem in a risk-decreasing manner.

SFRA is also particularly useful for emerging technologies such as SynBio, because it can reveal and supplant the *false choice* between risk and benefit. As Caruso (2008) pointed out near the inception of SynBio as a viable set of technologies, developers often advocate postponing risk-related inquiries until the benefits can be communicated (she quotes an official in Spain as saying "Let's first see what [the technology] is good for. If you first ask the question about risk, then you kill the whole field"). The central premise of SFRA, of course, is that *the acceptability of a new risk depends crucially on "what the technology is good for."* By exploring benefit and risk simultaneously (and by comparing the findings to benefit and risk analyses for current approaches to solving the same problem), SFRA can help avoid foolish actions (where small new risks are deemed intolerable despite massive risk reductions they can provide) and foolish inactions (where large new risks are permitted on account of small or phantom benefits they offer).

Bearing in mind these two premises—that risk-reducing opportunities should be compared, not simply "optimized" one at a time, and that creative questions about human needs can impel thoughtful discussion about fulfilling those needs in risk-reducing and welfare-enhancing ways—how might society grapple with new SynBio applications in a "solution-focused" paradigm? Here and in a recent article (Finkel et al. 2018a), I outline four different, and increasingly "solution-focused," ways to evaluate the merits of *any* SynBio application:

1. Does the application have positive net benefit? That is, does its risk reduction potential exceed its propensity to create additional risks?[2]
2. Compared to other ways to produce the same or similar material, does the SynBio application have *greater marginal net benefit* than the alternative(s)?[3]
3. Compared to other ways to fulfill the same human need, does the SynBio application have *greater marginal net benefit* than the alternative(s)?[4]
4. Does the existing dominant means of fulfilling (or failing to fulfill) a particular human need have a particularly *poor* risk profile, such that society might look to an *unmet* application of SynBio to displace it?

I emphasize that the nature of the new SynBio applications, as well as the stage of the product life cycle they occupy at the time of this writing, makes the application

---

[2] In the section below entitled "Risk Assessment Methodologies for SynBio", I will elaborate on what this question might mean given that both the risk-reducing and risk-increasing attributes of any technology are surrounded by uncertainty. At this point, one can certainly interpret the question to refer to the expected value (mean) risk reduction net of the expected risk increase.

[3] I will also elaborate later on the concept of "greater marginal net benefit." At this point, consider this term as shorthand for a case where a conventional way to produce a material has positive net benefit (reduces risks more strongly than it imposes them), but where a SynBio application has *even greater* net risk-reducing potential. But the term also applies to cases where the SynBio application has *negative* net benefit, but could replace a conventional application whose net benefit profile is even more strongly negative (a "lesser of two evils" case).

[4] Here as well I will elaborate below about the distinction between "ways to produce a material" and "ways to fulfill a function."

of the SFRA concept to this set of risks and benefits particularly timely, for three reinforcing reasons:

1. The risks and benefits involved are so different from most of what has come before that the substance-by-substance paradigm is simply a caricature of what is needed.
2. The applications are poised for completion but are largely not "out in the world"—*so we have an opportunity to start a revolution in technology with the simultaneous transformation of governance arrangements that are fit-for-purpose.* Such an approach will help minimize the need to "grandfather" a first generation of products and can help avoid untoward events that can both threaten human health or the environment and can fatally stigmatize this new technology before it can achieve successes.
3. As one of my colleagues has observed (Coglianese 2012), when a governance system waits for a tragic failure to occur (viz., the BP oil spill), it can be doubly unfortunate, because in addition to the tangible damage done, there is usually a rush to apply ill-conceived policy band-aids that can actually make future failures even more likely or more severe. *The "first failure" of SynBio could wipe away most hope for a proactive system of governance, one that we have time to craft now.*

This report will describe and evaluate the various linkages among the design of risk assessments, the use of risk and benefit information to make solution-focused comparisons among technologies and materials, and the risk-informed governance of SynBio applications. In turn, I will discuss:

- The crucial components of a risk assessment method that, when adapted to the special challenges of SynBio risks, can provide reliable, transparent, and "humble" (Andrews 2002) information. Here I will emphasize the extent to which existing risk assessment methods can be sensibly ported over to the SynBio context. While I will also highlight areas where new methods will have to be developed, this report will not per se generate any new risk assessment algorithms.
- The attributes of various solution-focused risk management questions that might allow for the reasoned expansion of some SynBio applications, the restriction of others, and the imposition of "prudent vigilance" (PCSBI 2010) on still others.
- The importance of revealing the many hidden value judgments that permeate the process of risk and cost-benefit analysis, so that governance decisions can be made with a fuller appreciation of their ethical implications.
- The current state of risk communication (and "benefit communication") for SynBio, as reflected in written pronouncements on these matters by leading pioneers in the field.
- A table summarizing tentative conclusions about how each broad class of SynBio application measures up, applying a solution-focused governance context.
- The potential to complement the solution-focused approach with a "solution-generating" mindset for SynBio.

Risk Assessment Methodologies for SynBio

Although this report is not intended to break new ground in quantifying the risks (or net risks) of SynBio applications, I hope here to jump-start a discussion of *how* society could do so. It is troubling that so much of the "risk assessment" dialogue and writing about SynBio contains little or no systematic, careful, or thorough estimation of *any* risks or benefits: rather, these discussions have introduced and perpetuated two of the most fundamental errors possible in risk assessment: (1) stating or implying that if an outcome (bad or good) is *possible*, it is likely or certain to transpire (this is insensitivity to probability[5]) or (2) stating or implying that one possible magnitude of the harm or benefit is its *expected* magnitude (insensitivity to uncertainty, or simply biased mis-estimation). Many conversations or pairs of opposing peer-reviewed articles about a SynBio application merely pit the claim that "this innovation will cure disease X" (or "clean up environmental problem Y," or "produce valuable product Z much more cheaply than any current method can"), against the counterclaim that "but it can spread a mutant protein throughout the human genome."[6] This is perhaps an example of a "risk-aware" conversation, but it is certainly not the basis for a sensible risk-benefit governance decision. For that latter—and vastly more useful—task, society needs at least a minimum set of raw materials with which to quantify risks and benefits, instead of a claim of good or harm that provides no information about its probability or magnitude.

This section of the report will sketch out such a core set of raw materials, useful for any of the risk management questions posed earlier and explored in more detail in the section "Implementing a Solution-Focused Management Regime". I will also elaborate on a richer set of risk assessment inputs that could help organize a more robust and intellectually honest analysis of the goods and harms of encouraging/discouraging any given SynBio application. Because the methods of QRA were first applied to the exposure and dose-response questions posed by synthetic chemicals in the environment and workplace, the discussions herein will use chemical risk assessment as a jumping-off point and template. QRA for SynBio will of course have to evolve to accommodate the challenges of estimating probability and severity for the novel risk (and risk-reducing) scenarios these applications pose, but QRA has previously risen, albeit fitfully, to similar challenges in other highly complex systems. Examples include risk assessment for pathogens (Mokhtari et al. 2006), the evolution of antibiotic resistance (Cox and Popken 2014), the adverse effects of molecules that can catalyze reactions (Hammitt 1990), the paradoxical dose-

---

[5] Of course, this insensitivity to probability can work in the reverse direction: stating or implying that if an outcome is highly unlikely, it *cannot* transpire.

[6] For one of many examples of a vague claim of massive harm, see Bunting (2007): "Creating fantastic bacteria in a contained laboratory is one thing, but what happens when they get out and cross with their wild cousins, mutating into organisms we had never foreseen?" For one of many examples of a vague claim of massive benefit, see Hylton (2012), quoting Craig Venter as saying that "Agriculture as we know it needs to disappear. We can design better and healthier proteins than we get from nature."

response relationships for immunotoxins such as beryllium (Willis and Florig 2002), the probability and consequences of contaminating an entire extraterrestrial environment with Terran microorganisms (NAS 2006), and the behavior of prions in the environment and in vivo (Schwermer et al. 2007).

Fundamental Concepts

Although there exist dozens of definitions and typologies of risk in the peer-reviewed and "gray" literatures (as well as in public discourse), no adequate definition of "risk" can fail to incorporate all of these three most fundamental questions (Kaplan and Garrick 1981): (1) what can happen?[7]; (2) with what probability can it happen?; and (3) how severe are the consequences if/when it happens? Once the "what?" question has been posed, any appraisal of "risk" must therefore integrate—perhaps via simple multiplication, perhaps via any more complicated function of the two— information about both probability and consequence; otherwise, *it is not a properly construed expression of risk.*

*In particular, two common "risk-like pronouncements" about some eventuality are **not** correct or useful expressions of risk.* To state (whether perfunctorily or as the culmination of a seemingly sophisticated technical analysis) that "exactly this consequence could happen" ignores or erases all of the powerful information that probability brings to the table. No matter how precisely one explains the exact *hazard* (e.g., a precise hazard statement would be "if the rope breaks, you will fall 500 feet to your certain death"), only by adding information about its probability can we reveal whether the *risk* is trivial, apocalyptic, or anywhere in between. Conversely, to state that "there is exactly one chance in 123.4 (a probability of 8.104×10^{-4}) that the rope will break" is useless without knowing whether the resulting fall will cover 5 inches, 5 feet, or 5 kilometers. Carelessness about probability (the former type of lapse mentioned above) often stems from the orientation that it is immoral to allow any non-zero probability of an involuntary harm to persist, but surely a tiny residual risk is at least *less* immoral than a large one.

Carelessness about severity is more insidious: when an outcome appears to be fully-described but is not, all kinds of value-laden conditions can be tacitly imposed upon the analysis or the decision. For example, consider the claim that more than 1000 Americans died "needlessly" (Gigerenzer 2006) in the year after September 11, 2001, when they chose to drive rather than to fly, because the per-mile *probability* of highway death is much greater than that of death in an aircraft. But despite the clear rank order of the probabilities, the *risk* of driving is only clearly greater than the risk of flying if the outcomes have identical severity—and it was far from

---

[7] Here I deliberately broaden Kaplan and Garrick's first principle (theirs was "What can go wrong?") to incorporate the notion that any policy, product, or activity can have either harmful and/ or salutary effects.

"irrational" to regard the specter of a protracted death-by-hijacking as qualitatively more dire than a sudden car crash (Finkel 2008; Assmuth and Finkel 2018).

Although [probability combined with severity] is the core of any meaningful expression of risk, properly considering both inputs may still yield an impoverished or an ambiguous risk estimate or risk-based conclusion, if one or more of these most basic definitional issues about risk are not considered:

- *"Pathway risk"* versus *total risk*. Any source of risk can present multiple conse-quences simultaneously, so it is important to consider all the major pathways or else explicitly highlight the partial nature of the thought process. For instance, a chemical in household water may be capable of causing several different acute health effects, and still other chronic effects, and can enter the body via inges-tion, dermal contact, or inhalation (as in showering with hot water). Each path-way, and each effect, may merit its own risk assessment, the panoply of which combine to yield a holistic estimate of overall risk.
- *Conditional probability* versus *unconditional probability*. Many risk appraisals involve two different kinds of probability: the chance that some untoward effect will occur and then the likelihood that the results of the event will proceed to cause harm. The former assessment of probability often involves "fault trees" or other means for estimating the odds of a discrete occurrence (such as an acciden-tal release of a particular chemical from a manufacturing plant or transportation system), while the latter may involve using toxicology data or epidemiologic studies to estimate the "potency" of the substance (the probability that a given concentration will cause a particular adverse health effect.) In such cases, the risk assessment must consider the joint probability both that the event will occur and that the health effect will occur conditional on such event.
- *Isolated risk* versus *aggregated risk*. A particular exposure may be the only con-tributor to a given health or environmental effect (e.g., beryllium is the only known cause of chronic beryllium disease), or it may add a small increment to an existing background risk of that effect (e.g., the amount of ionizing radiation emitted from nuclear power plants versus the natural background of radiation from Earth's crust and from cosmic rays). This does not mean that incremental involuntary exposures should be ignored merely because they may be small rela-tive to unavoidable background exposures, only that decision-makers and the public should know whether a policy would reduce a small or a large fraction of the aggregate risk.
- *Point estimate of risk* versus *acknowledging uncertainty in risk*. It is simply mis-leading to present probability or consequence estimates without providing confi-dence bounds (Finkel and Gray 2018) and preferably a probability density function (and note that uncertainty in risk-risk comparisons (Finkel 1995) is gen-erally X^2 as large as single-risk uncertainty).
- *Point estimate of risk* versus *acknowledging interindividual variability in risk*. QRA has suffered mightily from examples where population-average risks were deemed acceptable, when in fact risks varied dramatically depending upon the exposure, susceptibility, or other characteristics of subpopulations (Finkel 2008).

- *"Target risk"* versus *ancillary risk(s)*. There is a growing literature attesting to the folly of assuming that an intervention to reduce one risk will have no untoward consequences in another risk area (e.g., increasing highway mileage per gallon but failing to improve upon the safety performance of lighter-weight cars) (Graham and Wiener 1997). This literature, however, is tempered by a second round of scholarship (Rascoff and Revesz 2002; Finkel 2007) helping us distinguish between legitimate and sham trade-offs.
- *Life cycle orientation.* Ideally, the risks of a product or technology should be assessed from its production through its use and disposal, with an eye both toward general population risks and the disproportionate risks that workers usually bear (Powers et al. 2012).

This subsection concludes by emphasizing that when the stakes are high, QRA is unambiguously preferable to the three most commonly touted alternatives to it:

1. A "precautionary principle" that requires society to avert (or eschew) a single disfavored eventuality to the exclusion of others (Friends of the Earth, International Center for Technology Assessment, and ETC Group (2012)). Once precaution advocates realize that other advocates—for example, those insisting on the Iraq invasion of 2003 on the grounds that a "1% chance" of hidden chemical/biological weapons there should be regarded as a certainty (Suskind 2007) or those implicitly urging extreme precaution about economic costs rather than the harms caused by market failures—can define precaution to mean the opposite of what they do, the inadequacy of "pure precaution" is obvious (Montague and Finkel 2007).
2. "Scenario analysis" (Aldrich 2018), which commonly fails to discriminate between dire scenarios that are highly unlikely to occur and those that are far more plausible.
3. Qualitative risk assessment, in which hazards or scenarios are given color-coded severity rankings—this practice is often seen as intermediate between scenario analysis and full QRA, but various scholars have shown (e.g., Cox 2008) that following its dictates can be *worse* than choosing randomly without any risk information.

The Special Case of "Risk in the Name of Risk"

Analyzing the probabilities, severities, uncertainties, and other aspects of a SynBio application is rather more difficult even than ascertaining its downside risk(s), because many of the most interesting applications also promise to deliver significant *risk reduction* either as a prime mover or as incidental to it. Hence, the analysis needs to consider *net risk reduction* (or net increase) rather than the downside alone. Of course, a well-developed literature and set of practices exist for cost-benefit analysis (CBA), which can be thought of as the technique of comparing the risk of a product or practice to the benefits of producing it without constraints.[8] Here, I assume that the economic costs of reducing the risks of a SynBio application are small relative to the more fundamental question: do the risk reduction *benefits* the application offers exceed the novel risks the application poses? Net risk analysis, like a traditional CBA, requires two separate considerations, which could be termed "$R\downarrow$" (the decrease in risk that the application promises) and "$R\uparrow$" (the increased risk imposed by the application). If the SynBio application offers positive net benefit (does more good than harm), then the difference $[R\downarrow - R\uparrow]$ will be greater than zero.[9]

However, estimating the magnitudes of the two terms in a "risk in the name of risk" trade-off is rather different from, but in some ways easier than, the standard estimation problem in CBA. Standard CBA requires estimation of the economic costs of control, which can be surprisingly difficult (Finkel 2012). Standard CBA also requires that the benefits of control (aka. risk reduction) be "monetized" or converted from "natural units" (e.g., expected number of lives saved due to the controls or expected increase in biodiversity or other ecological indices) into dollars, in order that benefit can be compared to cost, and this is a highly controversial practice (Ackerman and Heinzerling 2005). In contrast, the estimation problem here does not require monetization, as both the risks imposed and the risks reduced are in "natural units," such as the expected number of lives lost or the estimated acres of habitat destroyed. When the risks on both sides of the ledger are in the same natural unit, there is no need to convert either to a dollar metric, although issues of commensurability will still persist if the natural units are different for risks reduced versus risks imposed. In considering the governance of an emerging SynBio or other technology, of course, we may have to consider that in order for the risk-superior application to be used, government may choose to subsidize it (hence accruing public costs that must be subtracted from *monetized* net benefits) or may have to regulate/tax/ban the riskier alternative (which would impose monetary costs in the form of reduced consumer surplus).

It is also quite possible that even if the risks reduced and risks imposed are in the same natural unit, the effects will accrue to different populations—see Graham and Wiener (1997) for a comprehensive treatment of the 2×2 different situations where either risks or populations (or both) can be identical or different. In such cases, simple subtraction may not yield a coherent net estimate or one that reveals important information about equity.

---

[8] This formulation is the obverse of how CBA is usually described—namely, as the benefits of *controlling* some risk or hazard less the economic costs of controlling it. Here, I deliberately reversed the description to make it more apt for a SynBio decision problem, where we might be comparing the risks posed by a product against the "savings" we obtain by *not* controlling it.

[9] More generally, if society is contemplating some controls on the SynBio application, then the net benefit of allowing the application to proceed, with controls, would be $[R\downarrow - R\uparrow* - C]$, where C is the economic cost of the controls. Presumably, in this case, the $R\uparrow*$ term would be smaller than $R\uparrow$ alone, because the effect of the controls would be to *decrease* the untoward risks of the application. So depending on the relationship between C and $R\uparrow*$, the net benefit of [approving with controls] could be larger or smaller than the "unfettered net benefit" estimate in the main text above.

This is not to say that estimating $[R\downarrow- R\uparrow]$ is by any means easy, only that it is conceptually straightforward. The first term could often be thought of as the baseline "toll" of some HSE problem, modified by the expected amount by which the SynBio application would effectively reduce that toll (see, e.g., Rooke 2013 for a catalog of SynBio advances that might reduce various human diseases). For example, suppose that the Oxitec hybrid mosquito (see summary of this case study in the section entitled "Broad/Tentative Observations about Comparative Risk Profiles of SynBio Categories") could, with 80% probability, reduce by 95% the number of mosquitoes capable of transmitting dengue fever in a region of the world where the disease was killing one million people annually (and with 20% probability would be ineffective). Then the expected amount of risk reduction the application would offer would be 760,000 statistical cases of disease averted per year (0.8 probability of reducing the death toll by 950,000).[10]

The $R\uparrow$ term is in many ways at the heart of this project, as it represents the untoward side effects of a SynBio application, and it is more difficult to estimate because almost by definition the raw materials of probability and severity of consequence are as yet unrealized (Dana et al. 2012). Conceptually, a useful estimate of $R\uparrow$ requires information on:

- The nature of each particular downside scenario (analogous to the "hazard identification" stage in classical human health risk assessment)
- The probability of each scenario manifesting itself
- The severity of the consequences if the scenario occurs
- How the consequences are actually experienced by the affected human population or ecological niche

With this raw material, the downside risk $R\uparrow$ is the sum of the [probability times experienced consequence] of each scenario, preferably with both probability and consequence expressed with the uncertainty in each. Once the risk is estimated, society could choose to treat very small risks as functionally equal to zero (Wareham and Nardini 2015) and, of course, could choose to *reduce* the probability and/or the severity of a risk by requiring developers to add additional safeguards to reduce the probability of an accidental release or to render an organism "inherently safe" even if released (Schmidt and de Lorenzo 2012; Wright et al. 2013).

The foregoing is, of course, a "much easier said than done" summary of how to arrive at a reasonable downside risk estimate for a SynBio application. Perhaps the most useful reference for understanding the tasks involved in estimating a downside SynBio risk is found in Bedau et al. (2009), which gives a "checklist" of how to think about scenarios. In looking for a template that could be improved upon for performing a state-of-the-art risk assessment for an emerging SynBio application (in this case, the risks of engineering hybrid mosquitoes to control dengue fever),

---

[10] There are, I hasten to add, good reasons not to combine mutually exclusive probabilities in this manner (NAS 1994, Chap. 9); one could certainly highlight rather than obscure the uncertainty in this example by summarizing the risk reduction as "an 80% chance of a reduction of 950,000 cases and a 20% chance of zero reduction."

Finkel, Trump et al. (2018a) recommended the assessment performed by Hayes et al. (2015), which offers a very complete risk assessment with respect to the probabilities of many downside risks, although it does not quantify the range of possible severities for any of the scenarios.

As QRA for SynBio improves, analysts can make greater use of existing techniques to cope with the particularly vexing problems inherent in estimating these probabilities and severities, including:

- Techniques for estimating the probabilities of unprecedented or "virgin" risks (Kousky et al. 2010)
- Techniques for bounding the probability of "surprise" (Shlyakhter 1994)
- Techniques for handling "deep uncertainty" (Cox 2012)
- Structured expert elicitation methods that force respondents to construct logically coherent scenarios (Cooke et al. 2007)

In contrast to the real need for additional complexity, it is also possible that SynBio risk analysts may be able to invoke some "first principles" for distinguishing high-concern scenarios from other ones, allowing for simpler assessments. For example, it *may* be the case that hybrid organisms designed to be less fit than the wild type cannot pose a significant risk to the ecosystem; if so, any scenarios involving mutations in which the hybrid organism remains less fit than before pose risks that might safely be ignored in a risk-risk analysis.

Risk Assessment in the Solution-Focused Regime

As I will discuss in the next section of this chapter, the bridge between net risk assessment and solution-focused governance is conceptually simple; it involves comparing the net risk profiles of various approaches to solving a human need or fulfilling a human want and using policy tools to support and encourage the solution(s) with the relatively most favorable profile, while discouraging, regulating, or banning solutions with inferior risk profiles. In comparing a new SynBio ("s") application to the most useful conventional ("c") solution to the same HSE problem, the question boils down to whether this equation is positive:

$$\left(R \downarrow_s - R \uparrow_s \right) - \left(R \downarrow_c - R \uparrow_c \right)$$

This equation symbolizes the incremental net benefit of the SynBio application over the conventional solution. Rearranging terms, the same equation can be expressed as:

$$\left(R \downarrow_s - R \downarrow_c \right) - \left(R \uparrow_s - R \uparrow_c \right),$$

which represents the incremental risk-reducing power of the SynBio alternative net of its incremental risk-increasing potential. In either case, if the equation yields a result greater than zero, the SynBio alternative can be said to have positive incremental net benefit over the "competition."

Alternatively, if we define the quantity RR_i, the "risk remaining" after a solution is implemented to partially eliminate a hazard (i.e., the status quo risk minus $R\downarrow_i$), then we could evaluate the equation:

$$\left(RR_c + R\uparrow_c \right) - \left(RR_s + R\uparrow_s \right),$$

which is the total risk (old plus new) for each solution. If this equation has a positive sum, then the SynBio application results in less total risk than the conventional solution it could supplant.

Implementing a Solution-Focused Management Regime

Armed with reliable methods to construct the risk-risk profiles (with attendant uncertainties) of a set of technologies, substances, or processes that includes one or more SynBio applications, how can government and the citizenry move from analysis to action? How can they/we decide what strictures, encouragements, outreach, taxes, subsidies, research, or other concerted actions are desirable or optimal? Although other orderings are possible, what follows is a chronological ordering of six tasks describing how SFRA maps onto this question of SynBio governance. Note that most of these elements are also described in a video in the "Risk Bites" series available on YouTube (Maynard and Finkel 2018).

(a) *Pose the fundamental question "which human need or want is unfulfilled?"* Unlike "problem formulation" as defined by EPA (see "A 'Solution-Focused' Partnership between Analysis and Action"), this mindset defines "the problem" not as a specific hazard presented by one product or process but essentially the opposite—as something one or more technologies might be able to solve. In other words, conventional risk assessment would ask "how much perchloroethylene can/should be emitted when dry-cleaning clothes?", while SFRA would ask "how can consumers clean their clothes most safely and effectively?" Here I will also distinguish between "needs" (e.g., humanity needs better methods to control disease-carrying mosquitoes without introducing new and untoward risks) and "wants" (consumers may benefit from a less-expensive or higher-quality artificial food-grade vanillin). SynBio applications can fulfill either needs or wants, but risk management governance may wish to consider these differently when balancing marginal risk increases against marginal risk reduction or other benefits.

(b) *Array a narrow or an expansive list of possible solutions to fulfill the need or satisfy the want.* Although the most fundamental distinction between SFRA and conventional risk assessment/management is that the former evaluates solutions rather than quantifies risks, the breadth and ambition of the solutions considered greatly distinguishes SFRA exercises from each other. It is possible to consider only "window dressing" responses to a human need (e.g., a medical professional advising a patient complaining of tight pants could suggest s/he get used to the discomfort or buy larger pants), or instead to emphasize "upstream" remedies that require much more expansive changes (in this case, advising the patient to change his/her diet or undergo bariatric surgery). In the chemical risk assessment arena, Finkel (2011) develops a case study contrasting narrow sets of possible solutions to the occupational and environmental health risks of chlorinated solvents for stripping paint from airplanes (one solvent versus another), somewhat more expansive sets (adding mechanical abrasives such as crushed walnut shells to the comparison), and very ambitious sets (including the option of leaving planes unpainted or even employing market mechanisms to reduce demand for business travel by plane). A good description of the correlation between the degree of upstream intervention and the "radicality" of the contemplated intervention is found in Løkke (2006), who notes that "the levels should not be mistaken as a grading of alternative solutions; most people would agree that preventive strategies are better than cleaning up, and increasing radicality will often but not necessary lead to better environmental solutions."

(c) *Estimate the net risk consequences (or non-risk benefit minus risk, for "wants")*
   of each solution. Using the risk assessment techniques and goals described in the section "Risk Assessment Methodologies for SynBio", SynBio and other means of solving a particular problem can be compared by deriving (with uncertainty) the extent to which each solution can reduce risks to the greatest extent, net of the new risks it poses. If there is no particular problem, but instead a set of ways to satisfy consumer wants, the comparison is similar, except that the "pro" term of every pro-net-of-con estimation would instead represent the benefits (perhaps using consumer surplus as a proxy) of each application or product. First, it is usually easy to reject outright solutions or products that have a negative profile (new risks exceed risk reduction or other benefits), as these are usually inferior to the status quo. Among the remaining choices, and because of uncertainty, there may well be no unique "winner" in any of these comparisons; often, the solution with the highest expected net risk reduction may not have the most favorable risk profile when the reasonable upper bound for the "con" term of the estimate is substituted (see the case study of dengue fever control in Finkel et al. 2018a). But choosing for or against an option that is "better on average but may be worse" (or "worse on average but may be better") is conceptually straightforward when the decision-maker openly chooses a degree of aversion to one unfortunate outcome or the other based on his/her own or on public attitudes toward regret (Lempert and Collins 2007).

(d) (1) *"Choose" the solution with the most favorable net risk profile—which is to*
   say, consider regulating, discouraging, or banning the less-favorable solution(s)
   and consider promoting, encouraging, or subsidizing the most favorable one.
   Depending on how intrusive these interventions are, when a government goes

beyond merely providing information about different products/technologies and implements regulations, taxes, subsidies, or the like to make it easier to sell and use some technologies and harder to use others, this may well smack of "picking winners and losers." This criticism must give decision-makers pause, especially if it is clear that by advancing a particular application, one monopoly producer will reap all the benefits (and in rarer cases, a single producer of a net riskier application will also bear all the costs of the other's "win"). But there is an element, perhaps a large one, of hypocrisy in denunciations of government's "picking winners." It is an article of faith that when the "free market" picks winners and losers, as happens constantly and relentlessly, *those* decisions stem from adequate information and by definition increase net economic benefit. So it is the case that *governance* decisions that advantage some producers over others can similarly be evidence-based and can provide net economic benefits as well as reducing externalities. The related claim that regulation should generally avoid specifying the means of compliance (technology-based standards) and instead set performance goals and let regulated industries find their own least expensive and burdensome ways to meet them (but see Wagner (2000) for a counterargument) is also based on some inconsistencies. Performance standards, which would tend to be less disruptive on market structure, are hard to enforce (Coglianese and Lazer 2003)—but more tellingly, in some cases, businesses clamor for "flexibility" only to later rebuke government agencies for *not* providing technological specifications that give them assurances of how to comply (Finkel and Sullivan 2011). But perhaps the weakest argument against allowing the democratic process (through participatory regulatory governance) to identify and support "winning" technologies that reduce risks is the fact that we have long allowed government to do this anyway, in many accepted though opaque ways. In the USA, the federal government has provided the coal industry with more than $70 billion in subsidies since 1950 (Taxpayers for Common Sense 2009); the effects on newer energy sources of this sort of market distortion are difficult to estimate but could well be monumental in size. In the pharmaceutical industry, the policy of allowing unlimited off-label use for drugs once they have been approved for a specific use amounts to a "leg up" over both established and innovative therapies for the same diseases (Comanor and Needleman 2016)—this amounts not only to "picking winners" but giving these favored technologies the kind of head start over competitors that could last for generations. The most pervasive arena in which government already picks winners and losers is probably that of international trade. Anecdotally, the USA and EU negotiated a pair of reciprocal tariffs several decades ago, with the EU disfavoring American cars and the USA placing a heavy tariff on European light trucks—this, of course, has had the effect of helping our domestic truck manufacturers "win," to the detriment of the domestic passenger car sector. So while promoting industrial policy for SynBio raises hackles, it should not be the very idea of favoring some industries over others that causes us to turn our backs on such policies.

OR

(d) (2) *Choose a mix of solutions implemented together in quantities sufficient to fulfill the need or want but parceled out in such a way that the sum of all net risks is even lower than the net risk of the relatively most favorable single solution.* It is possible that the optimal policy would involve a portfolio of solutions, with each one governed by policies that would accentuate its benefits while keeping its downside risks relatively low (and especially keeping them below any sharp nonlinearities in the technology's exposure-risk function). Such an approach would require vigilance and planning but might blunt some of the concern about brighter-line policies that would elevate one solution to "winner" status while greatly or completely curtailing others' roles in the economy.

(e) *Consider those governance tools—qualitative regulation (bans), quantitative regulation (exposure limits or controls of a given exposure reduction efficiency), or any of a variety of "soft law" mechanisms—that best produce the desired optimal net risk profile.* The gap between seeking net risk reduction and fulfilling that desire must fall to one or more tools of regulatory governance. Finkel, Deubert et al. (2018b) elaborates in order of stringency on various subtypes of "nudges" (information dissemination, guidance documents, and the like), public-private partnerships (Marchant and Finkel 2012), enforcement of general norms, and enforcement of newly written regulations, as each might apply to the problem of repeated head trauma and brain disease in professional football. However, there are several useful kinds of governance tools not mentioned in that article, including using civil liability as a powerful incentive to reduce downside risks (McCubbins et al. 2013; De Jong 2013) or requiring developers of new technologies to post bonded warranties against unforeseen harms (Baker 2009). On the other hand, Finkel, Deubert et al. emphasize one innovative governance idea that is not often included among the portfolio of "soft law" ideas commonly recommended (Mandel and Marchant 2014): an "enforceable partnership" in which a regulated industry develops its own code of practice and/or exposure controls but explicitly agrees to agency citations and penalties for violating that code. Such an arrangement might be especially appropriate for SynBio applications, since the developers generally can revise their views about which controls are most effective much faster than the public rulemaking process ever could. One other way to array the various governance options, as seen in Fig. 1, is to deemphasize the specific tools and instead portray the range of orientations from most supportive of emerging technologies to least supportive. In any event, the literature makes various recurring points about the nuances of emerging technology governance, particularly (1) that it is most "artful" when it seeks "effective compromise" such that while not all participants will be satisfied, all will agree that their views were heard and that the regulator's logic was transparent and reasonable (Zhang et al. 2011; Coglianese 2015); (2) that the choice of instrument and the stringency of control should vary depending on the stage at which the technology currently exists (e.g., laboratory work vs. field trials vs. first full-scale releases vs. routine releases) and that government should establish "checkpoints" to appraise the most sensible controls at each stage

(Bedau et al. 2009); and (3) that agencies sometimes can make good use of "soft law" mechanisms early in the lifespan of an emerging technology but should be ready to eventually "harden" those tools into traditional regulatory forms lest the regulated industries correctly perceive that the agency is using "soft law" as a crutch (Cortez 2014).

(f) *Consider structural change in government to better organize itself to adminis-ter and enforce the tools chosen.* Most of the sparse literature that considers improving the capacity of government to regulate emerging technologies focuses on "small gaps" where no agency has authority to solve a particular problem or where duplicative authorities foster controversy and delay (Paradise and Fitzpatrick 2012). For example, Taylor (2006) pointed out that the US Food and Drug Administration has jurisdiction over the safety of cosmetics but lacks statutory authority to oversee, prior to their marketing, cosmetics made with nanotechnology components. Similarly, Mandel and Marchant (2014) recom-mend that EPA seek authority to require a pre-manufacture notice from develop-ers of new microorganisms, not just for those that combine genetic material from two or more organisms from different genera but those that combine genetic material from species within the same genus. Most scholars construe these problems as solvable with minor statutory changes (see, e.g., Carter et al. 2014) or with interagency coordination provided by a White House office (see, e.g., PCSBI 2010). However, at least one investigator (Davies 2009) has gone further to recommend the reorganization of several current agencies (EPA, OSHA, NIOSH, the Consumer Product Safety Commission, the National Oceanic and Atmospheric Administration, and the US Geological Survey) to create a Cabinet-level "Department of Environmental and Consumer Protection" to regulate existing and emerging technologies that affect human health, safety, and the environment.

Although this process of conceiving of, comparing, and choosing among solu-tions can be intricate and can demand creative and bold thinking about "tragic choices" (Calabresi and Bobbitt 1978), its core tenet can be simply described: *in contemplating whether to encourage or discourage an emerging technology solution to a human need, society should tolerate more potential downside risk when the solu-tion has greatly improved potential for unprecedented risk reduction.* For "wants," the logic would be the related statement that "society should tolerate more potential downside risk when the solution can fulfill the want in unprecedented new ways or to a new extent." And the fundamental corollary to each of these principles would be that "society should be especially wary of courting new downside risks when the risk reductions they offer are negligible or when the consumer benefits are marginal."

For example, a SynBio (or, for that matter, a conventional) product that makes clothes whiter is arguably less worth taking risky chances on than one that could substitute for gasoline in cars; and further, if the SynBio product only makes clothes marginally more white than the next-best conventional alternative, it may be even less worth risking harm for.

How "radical" (Løkke 2006) is this precept? We are already comfortable declar-ing that some larger risks are more acceptable than related smaller risks, when we can explain this as a consequence of voluntary choice versus involuntary imposition (Starr 1969). But here I am arguing that we should consider certain risks less or more acceptable not because of qualities of the harms but qualities of the *solutions* that may make risks more or less worth bearing. Setting an ambient air quality standard only requires the decision-maker to consider the likely costs of achieving it against the benefits of doing so; requiring automobiles, on average, to achieve a given higher level of fuel efficiency goes a bit further toward favoring certain technologies over others but implicitly considers *any* technology with a positive risk-risk profile as acceptable. So it *may* be unprecedented to take the next logical but large step and compare risk profiles in order to favor technologies with significant new net benefits over marginal ones.

To the contrary, I suggest that placing hurdles in the way of products with small marginal benefits and worrisome new risks is in fact very similar to proposals made beginning several decades ago (Nussbaum 2002) that the FDA should treat truly novel pharmaceuticals more permissively than it treats "me-too" drugs that only offer slight variations on existing substances, because the former have novel benefits that may be more likely to justify their new risks. As Angell (2004) pointed out, the FDA currently treats both novel and derivative drugs equally, approving them if they are both safe and are *more effective than a placebo*: "the ['me-too' drug] needn't be better than an older drug already on the market to treat the same condition; in fact, it may be worse. There is no way of knowing, since companies generally do not test their new drugs against older ones for the same conditions at equivalent doses." Angell and others (Gagne and Choudhry 2011) have repeatedly called for FDA to make "approval of new drugs contingent on their being better in some important way than older drugs already on the market."[11]

A solution-focused approach, applicable to the other end of the marginal benefit spectrum, is also being suggested with respect to the FDA and the drug approval process. A major part of the "21st Century Cures Act," signed into law in 2016, provides for expedited approval for new medical devices that may benefit patients with "unmet medical needs for life-threatening or irreversibly debilitating conditions" (Avorn and Kesselheim 2015). Similarly, FDA has issued several regulations streamlining the drug approval process to treat certain very serious conditions that have no effective current therapies, stating that "these procedures reflect the recognition that physicians and patients are generally willing to accept greater risks or side effects from products that treat life-threatening and severely-debilitating illnesses, than they would accept from products that treat less serious illnesses. These procedures also reflect the recognition that the benefits of the drug need to be evaluated in light of the severity of the disease being treated" (FDA 2014).

---

[11] There are of course important counterarguments to a policy of disfavoring "me-too" products. Miller (2014) points out that derivative drugs may differ from the prior compound only in that they have fewer adverse side effects or in that they are effective in a different patient subpopulation. In either case, society might benefit from access to both products, although this would be consistent with Angell's criterion of "better in some important way."

These kinds of benefit-aware risk comparisons, of course, are precisely what a comparative risk-risk (solution-focused) analysis of SynBio versus conventional approaches to solve a problem would do—allow, and encourage, those approaches that are "better in some important way" than the status quo, either because of the paucity of effective solutions at present or because of a truly groundbreaking advance over approaches that are satisfactory but not ideal.

Overt and Hidden Values in Risk Assessment and Management

Both in assessing the net risks of any technology (SynBio or otherwise) and in deciding whether and how to manage any risks identified, we need more than methodological improvements in risk estimation and in decision-making under uncertainty; we need a much more transparent mode of analysis such that the large number of *hidden influential value judgments* that pervade the analysis can be brought to light. In Finkel 2018b, I identified more than 70 steps within a typical cost-benefit analysis where influential value judgments are made and generally kept implicit or are disclosed but misleadingly labeled as objective or purely scientific choices. These judgments range in scope from narrow and quantitative choices that influence key numerical quantities in one portion of a risk assessment or a CBA (e.g., the use of a particular single discount rate to render future consequences less salient than present ones) to fundamental definitional choices that influence the entire direction of the analysis (e.g., whether the "optimal" decision is tacitly defined as the one that maximizes total net benefit, one that achieves an arbitrarily "sufficient level" of benefit at the bare minimum cost, or as some other legitimate resting place). The main problem with embedding one value-laden choice out of many at multiple places in an analysis is, of course, that affected citizens may not realize that they profoundly disagree with the particular value chosen and would welcome the (possibly quite different) results of an analysis that substitutes one or more values they do agree with.

Some of the dozens of hidden value-laden assumptions I and others have identified would arise only infrequently in the kind of net-risk-versus-net-risk comparisons advocated here for making policy about SynBio applications—either because they affect portions of the analysis (particularly the estimation of the economic costs of regulatory control) that are not crucial to the comparison or because they involve aspects of the policy process (e.g., post hoc evaluation of the results of regulatory or other interventions) that do not affect the comparisons themselves. In comparing the risk profiles of SynBio and conventional applications, some of the more important recurring value judgments include:

- Should harms to non-human species be included among the "risks that matter" (assuming said harm does not indirectly affect people at all)?
- Should the non-utilitarian concerns of some citizens, particularly the aversion to "tinkering with the natural order" for good or ill, be given weight apart from the consequences themselves?

- Should analysis take account of risk reduction benefits or new harms to citizens outside the USA when making choices about domestic policy?
- Should harms that would affect subsequent generations be discounted at the same rate as intra-generational harms or at a lower rate so they don't effectively vanish from the equation?
- Should we treat risks from naturally occurring substances or organisms as equivalent to equal risks from synthetic ones?
- Should a risk profile with a lower expected value but longer right-hand tail than another one be treated as preferable (on the basis of expectation) or the opposite (on the basis of a worst-case comparison)? (see Finkel, Trump et al. 2018a for the claim that the risk profile of the Oxitec SynBio mosquito, compared to pesticides and other conventional approaches to controlling dengue fever, may have a favorable expectation but a longer right tail)?[12]

I advocate for substantial efforts to reveal all of the value judgments permeating evidence-based policy analysis—not through a laborious process of highlighting them each time but rather by the publication (as a single document affecting all health, safety, and environmental agencies or perhaps by agency-specific documents) of a free-standing "value statement" that would flag them all, explain which judgments the agency would generally make by default in the absence of specific reasons to the contrary (and *why* this value judgment was chosen), and offer one or more alternative value judgments that could be made instead if sufficient reason was provided in a specific assessment. This may be a daunting task, but there is a much more practical and imminent first step; even if analyses of SynBio and other

---

[12] When this chapter was in proof, a research group (Evans et al. 2019) made headlines by publishing a set of findings that pointed to a higher downside risk and a lower efficacy for the Oxitec SynBio mosquito than previously believed. The group studied the aftermath of the release of roughly 50 million transgenic mosquitoes in the city of Jacobina, Brazil, and found that from 10 % to 60 % of a (small; roughly 10–20 mosquitoes per group) number of insects they genotyped showed a mixed genome—with one or more genes from the Oxitec mosquito having been introduced into the wild-type genome. They assumed that this resulted from a small percentage of the progeny being able to survive to adulthood, contrary to the intent of the lethal gene Oxitec inserted as a fail-safe mechanism, and they also claimed that the total insect population rebounded during the experiment to nearly the levels pre-release. Although Evans et al. found that the mixed-genome mosquitoes were no more infective with respect to dengue or Zika than the wild type, they speculated that the surviving insects could have acquired other unfortunate characteristics, such as insecticide resistance. If confirmed, this finding would complement our conclusion in Finkel, Trump et al. (2018a) that the net risk profile of the Oxitec application may be favorable on average but have a negative reasonable-worst-case profile, although it might also change the expected value if the efficacy was overestimated by Oxitec. However, Oxitec has issued a preliminary response to Evans et al. (Oxitec 2019), claiming that the survival of a few percent of progeny was anticipated and widely disclosed before the field trials, that there is no evidence that any introduced genes conferred any untoward characteristics such as insecticide resistance, and that the eventual rebound of the mosquito population was completely expected because the releases were of limited duration. I emphasize that ongoing controversy over the risk profile of a SynBio application is instructive, and does not affect whether a solution-focused net risk assessment paradigm is the best way to structure governance decisions.

technologies cannot be made fully transparent to "the outside world" as to their embedded value judgments, the *analysts themselves* must recognize them and *ensure that in each case, the same judgment is used on both sides of the comparison.* If this is not done, the comparison will be worse than misleading, as it will foster the impression that the "safer" alternative was chosen rationally. Imagine, for example, a comparison of the risk profiles of a compound like artemisinin produced via natural sources (the wormwood plant) versus one produced by genetically engineered yeast, with the former profile tacitly considering the economic harm to industries anywhere in the world if the competing application was supported, while the latter profile tacitly only considered economic harm to US industries. In this hypothetical, the major unemployment effects would not be counted for the one alternative (drying up the market for wormwood) where they were substantial.

Cheerleading and Poor Risk (and Benefits) Communication in SynBio

Careful risk assessments, whether performed in the classical or the solution-focused paradigm, can be undone by tone-deaf risk communication. Among the many recurring deficiencies in efforts to communicate risks, lapses that "can create threats larger than those posed by the risks that they describe" (Morgan et al. 2002) are the deliberate or unintentional trivialization of risk, the overuse of jargon, the reliance on misleading or inappropriate comparisons to unrelated risks, and the tendency to provide only population-wide average risks and mask substantial interindividual differences (Finkel 2016). Many experts (Sandman 1993; NAS 1996) stress that intentional attempts to persuade people via risk communication sometimes work but eventually often backfire. I've read several of the leading general-interest books on SynBio (along with many peer-reviewed articles), which arguably give a good cross-section of how experts communicate to laypeople about these new applications. In my reading, I found some troubling signs not only in how risks are described but how benefits are (Kahn 2011):

- At the least important end of the spectrum, developers of SynBio technology sometimes take verbal shortcuts describing their advances. For example, Oxitec, developers of a hybrid male mosquito, often referred to them as "sterile," when the central advance Oxitec made is that the males *are* fertile but produce offspring that die before they are mature enough to bite humans (Finkel et al. 2018a). This distinction is largely semantic, and, as Oxitec has said, "there is no layman's term for 'passes on an autocidal gene that kills offspring'" (Specter 2012). However, the ambiguity (or discrepancy, depending on one's point of view) gave a critic from Friends of the Earth an opening to say that Oxitec has been "less than forthcoming" with its public statements that allow the more reassuring interpretation that the mosquitoes cannot produce offspring (Specter 2012).

- Of greater concern is a tendency of SynBio developers to condescend to the public by suggesting that it is irrational to fixate on the downside risks. For example, scientific giant Craig Venter has made the sweeping statement that "few of the questions raised by synthetic genomics are truly new" (Venter 2013, at 152), which of course sidesteps the question of whether "old" risks created anew can be unacceptably high. Similarly, Brassington (2011) used an odd phrase to "quantify" SynBio risks: "However, while these risks are not *vanishingly small*, they can be met not by forbidding SynBio research, but by pursuing it wisely" (emphasis added). Something "large" is also not "vanishingly small," but the phraseology here strongly implies that we know SynBio risks to be "small"—perhaps non-zero, but arguably so small as to be impalpable and hence unworthy of concern.
- A similar tendency involves a kind of acknowledgment that public fears are not unfounded but one that sequesters this concern and ultimately steps away from it. Consider this quote by Venter (2013, at 155): "For me, a concern is 'bioerror': the fallout that could occur as the result of DNA manipulation by a non-scientifically trained biohacker or 'biopunk.'" This is essentially a "safe if used as directed" warning, which is not a warning at all but a denial of the inherent danger(s) in favor of dangers brought on by insufficient policing of human actors. Without taking a position on the merits, this does seem reminiscent of the "guns don't kill people; people do" argument that seeks to channel concern away from "the right users."
- Most generally, pioneers in synthetic biology sometimes invoke their own expertise, or that of the cadre of developers more broadly, as a kind of talisman that can turn estimated risks into irrelevancies (Rampton and Stauber 2002). When a *New York Times* reporter (Rich 2014) brought up various potential risks of de-extinction technology, the lead scientist at "Revive and Restore" simply asserted that "We have answers for every question… We've been thinking about this for a long time." Perhaps more tone-deaf still is this assertion from Venter, who invoked Isaac Asimov's "three laws of robotics" to reassure readers that nothing can go seriously awry: "One can apply these principles equally to our efforts to alter the basic machinery of life by substituting 'synthetic life form' for 'robot'" (Venter 2013, at 153). Here citizens concerned about untoward risks of SynBio are met with *fictional* solutions to a problem —in Asimov's created world, robots could be hardwired to always obey and never to harm, but of course hybrid organisms do not have programmable brains, and wishing for a fail-safe mechanism is quite different from building one.

If at the same time that SynBio advocates were understating risks or hyping untested ways to eliminate any risks that remain, they were also overstating the benefits of their innovations, citizens might be doubly disadvantaged as they try to make sense of the trade-offs. However, my sense is that the potential benefits of SynBio are *not being stressed enough* and that categories of benefit that are less impactful are emphasized:

- In particular, developers and advocates often emphasize the "elegant" features of SynBio advances—and not just in applications such as "glowing fish" that may have no tangible benefits other than their novelty. For example, Lee Silver (2007) quoted MIT professor Tom Knight as stating that "the genetic code is 3.6 billion years old. It's time for a rewrite"—without linking that intellectually compelling prospect to any specific (or even hypothetical) advantages it might confer. This wide-eyed enthusiasm for the "can," rather than the "should," may also serve to heighten concern about the possible downside risks that are not mentioned.
- Even some medical applications of SynBio are praised for their ability to move the human organism closer to "perfection," which again mentions an inchoate benefit, and here one that reasonable people may actually consider a disbenefit (Hurlbut 2013).
- There are, by contrast, examples where supporters of SynBio emphasize the tangible and pragmatic benefits of applications, such as this observation from Rooke (2013). I suggest that more successful risk-benefit communication ought to look more like this example than the previous ones:

> Technological advances in the field of health continually bring us closer to a world where a healthy life is a real option for every individual on the planet, regardless of geography, culture, or socioeconomic status. However, these benefits tend to accrue disproportionately to the developed world; the need is still great for solutions that can diagnose illness, protect against infection, and treat disease in a broad array of low-cost settings with developing-world healthcare systems and limited infrastructure.

Broad/Tentative Observations About Comparative Risk Profiles of SynBio Categories

In other work performed with Sloan Foundation support, my colleagues and I published a detailed case study of the Oxitec SynBio mosquito (Finkel et al. 2018a) but also investigated in broad terms the kinds of incremental benefits and risks that various types of different SynBio applications might pose. Table 2 presents some tentative observations, using exemplar applications from each of ten categories where SynBio developers are working, suggesting that in some kinds of applications (e.g., biological pesticides), the SynBio alternative *may* have large incremental risks that do not justify the small incremental benefits they offer over conventional solutions to this problem. We also suggest that in other categories (e.g., specialty chemicals), the new downside risks would likely be small, but so would the incremental benefits. In contrast, we see the general categories of disease vector controls and medical treatments as ones where the new risks from SynBio may be comparable to or smaller than the risks we currently tolerate from conventional approaches and where the efficacy of a new approach may make the SynBio application a win/win for fulfilling a human need.

Table 2 General observations about risk-risk aspects of synthetic biology solutions, by category

Type of solution	Examples (selected)	Major potential benefits	Major risks	Efficacy of best conventional alternative(s)	Risks of best conventional alternative(s)	Tentative risk-risk appraisal
Medical treatments	Pulmonary arterial hypertension (Machado 2012); semisynthetic artemisinin; bacteria that ingest cholesterol	Novel or more efficient treatment of debilitating disease or medical condition	Adverse effects on patients; transmission to healthcare workers; job loss to traditional harvesters	Moderate to limited	Toxic side effects; potential for development of resistance with infectious organism; imprecise application of treatment	SynBio-enabled treatment may be the only option available (raising "right to try" issues) or meet a significant need for medicine and public health
Disease vector controls	Oxitec mosquito for dengue/Zika control (Finkel et al. 2018a)	Rapid reduction in transmission of certain infectious diseases; cost savings relative to conventional vector control options	Possible jump of autocidal gene to a different organism; possible ecological repercussions of decimating a (invasive, in this case) species	Limited (vaccines, pesticides, environmental cleanup)	Toxicity of pesticides; cost and labor of environmental cleanup; secondary toxicity concerns; development of target species' resistance to chemical control	SynBio offers a possible step change toward smaller net risk profile
Biofuels	Diesel or jet fuel (Amyris); algal ethanol; hydrogen	Lower cost (?), less reliance on fossil fuels	Accidental release of hybrid organisms (outcompete wild types); demands on water, land, and energy use (NAS 2012)	Moderate to limited (sugarcane/corn ethanol)	Land use; price/ dislocation effects in corn market; nitrogen pollution downstream (Donner and Kucharik 2008)	Not a clear marginal net benefit—after all, just substituting one source of greenhouse gas emissions for a similar one made a different way

Type of solution	Examples (selected)	Major potential benefits	Major risks	Efficacy of best conventional alternative(s)	Risks of best conventional alternative(s)	Tentative risk-risk appraisal
Specialty chemicals	Isoprene (Goodyear) in *E. coli*	Substitution of renewable source of the chemical for a fossil fuel source	Accidental release; job loss to traditional harvesters	Moderate to high (conventional or synthetic rubber)	Contribution to greenhouse-gas emissions; job loss to traditional harvesters	SynBio appears to duplicate existing products, with similar toxicity and similar or greater use of land and other resources (Morais et al. 2015)
Environmental sensors	Arsenic detection (French et al. 2011)	Rapid identification of environmental contaminants	Uncontrolled release of genetically modified material that persists within the environment	Limited (Naujokas et al. 2013)	Labor-intensive; worker risks	SynBio-engineered sensors have significant upside within toxic/contaminated environments yet have significant uncertainty regarding their ability to be contained and controlled in the environment
Environmental cleanup	In situ biodegradation of toxic waste (de Lorenzo et al. 2018)	Rapid remediation of environmental contaminants	Uncontrolled release of genetically modified material that persists within the environment; mutation or unintended phenotypic change of organism's consumption of resources in the environment	Limited (chemical dispersants or soil washing)	Toxicity of dispersants; worker risks; high labor/resource cost-to-cleanup (Bates et al. 2015)	SynBio-engineered remediation options have significant upside within toxic/contaminated environments and have considerable cost-saving potential yet have significant uncertainty regarding their ability to be contained and controlled in the environment

(continued)

Table 2 (continued)

Type of solution	Examples (selected)	Major potential benefits	Major risks	Efficacy of best conventional alternative(s)	Risks of best conventional alternative(s)	Tentative risk-risk appraisal
Medical diagnosis	Bacteria that show leishmaniasis infection; bacteria that can sense incipient cholera infections (Perkel 2013)	Rapid, accurate, and targeted detection of infection or foreign organism in body	Misdiagnosis (false-positive or false-negative); absorption and uptake of engineered genetic material; harmful interaction with host immune system/gut flora	Moderate (conventional testing with chemicals or sensors)	High cost and labor of testing; inaccessibility of equipment to at-risk populations; potential for misdiagnosis	SynBio-engineered organisms can be substantially cheaper, faster, and more easily distributed than testing via conventional equipment. However, they also raise the potential for negative interactions of engineered genetic material within the human body, including the uptake of engineered material and/or the unintended interaction or disruption of host biological processes
Agriculture	Plants that make their own fertilizer	More efficiency in cost, labor, and resource use to fertilize crops; greater geospatial range of agricultural production in semi-arable land	Loss of biodiversity; secondary impacts of fertilizer contamination into the environment at certain intervals; conversion of sensitive land area into agricultural production; unintentional hybridization of engineered and non-engineered crops	Moderate (conventional fertilizers)	High cost of conventional fertilizers; environmental runoff; waste	SynBio-engineered crops are substantially more resource-efficient yet have the potential to disrupt local biodiversity or hybridize and spread beyond their initial containment area

Type of solution	Examples (selected)	Major potential benefits	Major risks	Efficacy of best conventional alternative(s)	Risks of best conventional alternative(s)	Tentative risk-risk appraisal
De-extinction	Reviving a woolly mammoth species from extinction (Rich 2014)	Novelty; "making amends" for prior human incursions	"Playing God"; unexpected gene transfer; quality of life of "new" organisms; stability of modified genetic constructs; incapability of revived organism to persist within new environment	N/a	N/a	Inchoate risks and moral objections, for limited new benefit
Biological pesticide	Mousepox calamity in Australia (Jackson et al. 2001)	Disease vector control	Spread to non-target organisms or non-target species; eventual mutation and resistance of surviving population	Moderate to low (chemicals, barriers, pheromones, introduction of natural predators to local area)	Toxicity; chemical resistance; secondary toxicity events	SynBio pesticide options are more targeted and fast-acting than existing options but are likely irreversible and have a strong likelihood of escaping containment or spreading to non-target species

Solution Generation as the Complement to Alternatives Assessment

The discussion to this point has not exhausted the potential for solution-focused thinking, as the various proposals (including full-blown marginal risk profile analysis of solutions) have all presupposed that a development of a new application will then prompt discussion about the new risks it poses and new risk reductions it offers, in context of other ways to meet the same need or fulfill the same want. But what if instead of a problem that has already attracted multiple solutions, *we are faced with a problem desperately in need of even one good solution?* The complement to regulators seeing competing solutions available and asking "why?" (or "which"?) would be someone "dreaming things that never were and saying 'why not'?" (Shaw 1949).

One way to organize creative thought around "solutions we need" is to extrapolate from existing lines of SynBio research to instances where similar technology might be able to do vastly more good. For example, various researchers are trying to engineer microbes that would have salutary effects on human health and quality of life if introduced into the human gut microbiome. It is also the case, though, that collectively the digestive systems of domesticated ruminant animals worldwide (primarily cows, sheep, and bison) add enormous amounts of methane to the atmosphere—roughly 20% of all anthropogenic methane (Lassey 2007), a potent greenhouse gas (Friedman et al. 2018). Investigators have been attempting to reduce methane generation by changing the animals' diets, and by selective breeding, but have not succeeded in making a dent in the total. But very recently, students at the University of Nebraska began, but "ran out of time," experimenting with introducing a gene from a red alga (*C. pilulifera*), one that codes for the enzyme bromoperoxidase, into *E. coli* for introduction into the digestive systems of cattle (University of Nebraska-Lincoln 2017). Interestingly, cattle can be fed bromoperoxidase directly by feeding them large amounts of seaweed, but the bromoform produced in seaweed farming is a potent depletor of stratospheric ozone, which the students described aptly as "fix[ing] one environmental issue by creating another." Currently there appears to be little experimental or commercial interest in using SynBio to attack the problem of methanogenesis in ruminants and its role in exacerbating global climate change, surely a problem in need of a breakthrough.

Similar "if only…" thinking can also be applied to existing products that satisfy consumer demand, *simply by looking for products with the largest environmental or human health "footprints."* Surely high on such a list would be palm kernel oil, whose worldwide production converts several million hectares each year from tropical forest to monoculture, with implications for endangered species like the orangutan and with widespread use of child labor for harvesting (Rosner 2018). But while industrial feedstocks like isoprene have attracted much interest from SynBio developers, there appears to be only one company actively trying to engineer organisms to produce synthetic palm oil (that company, Solazyme, has been criticized for choosing algae as its host organism, in a system that requires large amounts of sugarcane to be harvested to feed the algae; SynBioWatch 2016). As with the

Nebraska team, a group of students at the University of Manchester also participated in an iGEM competition (Univ. of Manchester 2013) and explored the possibility of producing synthetic palm oil in *E. coli* instead but apparently lacked the resources to bring this idea past the conceptual stage.

So given the economic realities that set developers' sights based on market potential rather than on reducing environmental or other externalities (whether caused by the paucity of solutions or by the footprints of existing products), how can governments focus on "solution generation" to complement solution appraisal, and how can they attract entrepreneurs to fill the vacuums they identify?

Here the useful ideas are conceptually simple though politically fraught and are ones the US and other nations have grappled with already (for a prime example, see the Orphan Drug Act of 1983, which provided tax incentives and extended patent protection to developers of drugs that are intended to treat a disease affecting fewer than 200,000 Americans). Government or private philanthropies could identify areas where a novel solution would be immensely beneficial and then offer a "grand challenge" prize for its development (Adler 2011; also see Table 1 in Rooke 2013) or directly subsidize the early stages of research and development. Manzi (2014) summarizes the salutary results from subsidies, concluding that "the Breakthrough Institute has produced excellent evidence that government subsidies for speculative technologies and research over at least 35 years have played a role in the development of the energy boom's key technology enablers: 3D seismology, diamond drill bits, horizontal drilling, and others." He recommends that our "existing civilian infrastructure … can be repurposed, including most prominently the Department of Energy's national laboratories, the National Institutes of Health, and NASA. Each of these entities is to some extent adrift the way Bell Labs was in the 1980s and should be given bold, audacious goals. They should be focused on solving technical problems that offer enormous social benefit, but are too long-term, too speculative, or have benefits too diffuse to be funded by private companies." In other words, these giant agencies could devote some of their resources to identifying "orphan problems" that we have learned to live with but where innovation could conceivably reveal that this acquiescence wasn't necessary.

A related idea has been championed by Outterson (2014), who has suggested in Congressional testimony that the federal government could offer a guaranteed payment stream to the successful developer of a needed antibiotic or other drug (one with large social benefits but marginal profitability for the developer) in exchange for the right to market the product.

Of the two parts to the "solution generation" puzzle—identifying situations where society needs a new solution and providing the "activation energy" so that developers will have the incentives and resources to explore such a solution—the former is clearly already occurring, as evidenced by the fact that I began by idly speculating about the benefits of a SynBio approach to methanogenesis in ruminant animals or a SynBio alternative to palm oil monoculture, only to find that various university groups were already working on the broad outlines of these very breakthroughs. But the other side of the coin—that to my knowledge neither idea has left the university environment to be brought forward to bench-scale fruition—suggests that new policies and organizational arrangements are needed to move good ideas forward in the absence of clear short-term profitability.

Conclusions

It should not be controversial that we are better off *knowing* whether a new technology has net risk reduction benefits that would outperform the status quo, whether or not we have the will to act on that knowledge (especially the will to act in ways that would cause the less efficient solutions to make way for more efficient ones). Of course, the probability and severity of risky scenarios are always uncertain, and risk comparisons are more uncertain than risk estimates (Finkel 1995), so judgment will always be needed to weigh the differential costs of error (boosting a new technology on the basis of its likely superiority, but one that will turn out to have risk-increasing consequences, versus impeding a new technology such that risk-increasing solutions will be allowed to persist).

The controversy comes when we contemplate intervening in the market to promote risk-reducing solutions over risk-increasing ones. Most ideologies other than pure libertarianism welcome the idea of government choosing *policies* that promote social welfare, whereas many liberals and conservatives bristle at the idea of government promoting individual *companies* over others (despite how often we tolerate government doing so via earmarks, subsidies, and the like). In between the extremes of "picking winning policies" and "picking winning firms" lies the notion of picking technologies or industries that solve problems with fewer untoward harms. Here I agree with much prior scholarship, particularly that of Rycroft and Kash (1992), that we need to repudiate the idea that "the politicians and bureaucrats who make these critical decisions would have neither the incentives nor the ability to pick winners as well as the private market place now does" (quoting a 1983 speech by Martin Feldstein, then chairman of the White House Council of Economic Advisers). Markets do reasonably well at allocating resources based on consumer preferences (as influenced by those doing the marketing), but much less well at allocating resources to minimize externalities. Comparative net risk analysis of solutions to a human need (or of ways to satisfy a want) provides the evidence that government needs to consider the non-market benefits and harms of technologies, which will allow government to *consider strengthening the barriers to entry for innovations that tend to increase net risk while attenuating those barriers for innovations that tend to decrease net risk.*

In other words, SFRA can tee up governance decisions that reject "permissionless innovation" when the SynBio or other new application is duplicative, ineffective, or harmful but that equally reject laissez-faire market primacy when the innovation is what we truly need to solve pressing health, safety, environmental, or other problems.

Acknowledgments I gratefully acknowledge financial and intellectual support for this chapter from the Alfred P. Sloan Foundation and thank Ben Trump for many helpful suggestions on Table 2.

References

Ackerman, F., & Heinzerling, L. (2005). *Priceless: On knowing the price of everything and the value of nothing*. New York: The New Press.

Adler, J. H. (2011). Eyes on a climate prize: Rewarding energy innovation to achieve climate stabilization. *Harvard Environmental Law Review, 35*(1), 1–45.

Aldrich, S. C. (2018). Integrating scenario planning and cost-benefit methods. In *Governance of emerging technologies: Aligning policy analysis with the public's values,* Gregory E. Kaebnick and Michael K. Gusmano, eds., *Hastings Center Report,* 48(S1), (pp.S65-S69). Garrison: The Hastings Center.

Andrews, C. J. (2002). *Humble analysis: The practice of joint fact-finding*. Westport Connecticut: Praeger.

Angell, M. (2004). The truth about the drug companies. *New York Review of Books,* July 15. Available at https://www.nybooks.com/articles/2004/07/15/the-truth-about-the-drug-companies/

Ashford, N. (2000). An innovation-based strategy for a sustainable environment. In J. Hemmelskamp, K. Rennings, & F. Leone (Eds.), *Innovation-oriented environmental regulation: Theoretical approach and empirical analysis* (pp. 67–107). Heidelberg: Springer Verlag.

Assmuth, T., & Finkel, A. M. (2018). Principles and ideals behind the 'rationality' of choices in response to risks. Ms. in review, Joint Research Centre, European Commission.

Avorn, J., & Kesselheim, A. S. (2015). The 21st century cures act: Will it take us back in time? *New England Journal of Medicine, 372*(26), 2473–2475.

Baker, T. (2009). Bonded import safety warranties. In C. Coglianese, A. Finkel, & D. Zaring (Eds.), *Import safety: Regulatory governance in the global economy*. Philadelphia: University of Pennsylvania Press.

Bates, M. E., Grieger, K. D., Trump, B. D., Keisler, J. M., Plourde, K. J., & Linkov, I. (2015). Emerging technologies for environmental remediation: Integrating data and judgment. *Environmental Science & Technology, 50*(1), 349–358.

Bedau, M. A., Parke, E. C., Tangen, U., & Hantsche-Tangen, B. (2009). Social and ethical checkpoints for bottom-up synthetic biology, or protocells. *Systems and Synthetic Biology, 3*, 65–75.

Brassington, I. (2011). Synthetic biology and public health. *Theoretical & Applied Ethics, 1*(2), 34–39.

Bunting, M. (2007). Scientists have a new way to reshape nature, but none can predict the cost. *The Guardian,* Oct. 21, available at https://www.theguardian.com/commentisfree/2007/oct/22/comment.comment

Calabresi, G., & Bobbitt, P. (1978). Tragic choices: The conflicts society confronts in the allocation of tragically scarce resources. In *Fels lectures on public policy analysis*. W.W. Norton & Co.

Cameron, D. E., Bashor, C. J., & Collins, J. J. (2014). A brief history of synthetic biology. *Nature Reviews (Microbiology), 12*, 381–390.

Carter, S.R., Rodemeyer, M., Garfinkel, M.S., & Friedman, R.M. (2014). *Synthetic biology and the U.S. biotechnology regulatory system: Challenges and options*. J. Craig Venter Institute report, available at https://www.jcvi.org/sites/default/files/assets/projects/synthetic-biology-and-the-us-regulatory-system/full-report.pdf

Caruso, D. (2008). *Synthetic biology: An overview and recommendations for anticipating and addressing emerging risks*. Washington, DC: Center for American Progress.

Coglianese, C. (2012). *Regulatory breakdown: The crisis of confidence in U.S. regulation* (p. 304). University of Pennsylvania Press. Philadelphia, PA USA.

Coglianese, C. (2015). *Listening· learning· leading: A framework for regulatory excellence.* Report to the Alberta Energy Regulator, Penn Program on Regulation, available at https://www.law. upenn.edu/live/files/4946-pprfinalconvenersreport.pdf

Coglianese, C., & Lazer, D. (2003). Management-based regulation: Prescribing private management to achieve public goals. *Law and Society Review, 37*, 691–730.

Comanor, W. S., & Needleman, J. (2016). The law, economics, and medicine of off-label prescribing. *Washington Law Review, 91*, 119–146.

Cooke, R. M., Wilson, A. M., Tuomisto, J. T., Morales, O., Tainio, M., & Evans, J. S. (2007). A probabilistic characterization of the relationship between fine particulate matter and mortality: Elicitation of European Experts. *Environmental Science & Technology, 41*(18), 6598–6605.

Cortez, N. (2014). Regulating disruptive innovation. *Berkeley Technology Law Journal, 29*(1), 175–228.

Cox, L. A. (2008). What's wrong with risk matrices? *Risk Analysis, 28*(2), 497–512.

Cox, L. A. (2012). Confronting deep uncertainties in risk analysis. *Risk Analysis, 32*(10), 1607–1629.

Cox, L. A., & Popken, D. A. (2014). Quantitative assessment of human MRSA risks from swine. *Risk Analysis, 34*(9), 1639–1650.

Dana, G. V., Kuiken, T., Rejeski, D., & Snow, A. A. (2012). Four steps to avoid a synthetic-biology disaster. *Nature, 483*, 29.

Davies, J. C. (2009). *Oversight of next generation nanotechnology.* Woodrow Wilson International Center for Scholars, PEN #18, April 2009, 39 pp. Available at https://www.nanotechproject. org/process/assets/files/7316/pen-18.pdf

De Jong, E. R. (2013). *Regulating Uncertain Risks in an Innovative Society: A liability law perspective.* In E. Hilgendorf & J.-P. Günther (Eds.), *Robotik und Recht Band I* (pp. 163–183). Baden-Baden: Nomos Verlag.

de Lorenzo, V., Prather, K. I., Chen, G. Q., O'Day, E., von Kameke C., et al. (2018). The power of synthetic biology for bioproduction, remediation, and pollution control: The UN's Sustainable Development Goals will inevitably require the application of molecular biology and biotechnology on a global scale. *EMBO Reports, 19*(4): e45658 ff.

Donner, S. D., & Kucharik, C. J. (2008). Corn-based ethanol production compromises goal of reducing nitrogen export by the Mississippi River. *Proceedings of the National Academy of Sciences, 105*(11), 4513–4518.

Evans, B. R., Kotsakiozi, P., Costa-da-Silva, A. L., Ioshino, R. S., Garziera, L., et al. (2019). Transgenic Aedes aegypti mosquitoes transfer genes into a natural population. *Nature Scientific Reports, 9*, 13047.

Finkel, A. M. (1995). Towards less misleading comparisons of uncertain risks: The example of aflatoxin and alar. *Environmental Health Perspectives, 103*(4), 376–385.

Finkel, A. M. (2007). *Distinguishing legitimate risk-risk tradeoffs from straw men.* Presentation at the Annual Meeting of the Society for Risk Analysis, San Antonio, TX, Dec. 11.

Finkel, A. M. (2008). Protecting people in spite of—or thanks to—the 'veil of ignorance', Chapter 17. In R. R. Sharp, G. E. Marchant, & J. A. Grodsky (Eds.), *Genomics and environmental regulation: Science, ethics, and law* (pp. 290–342). Baltimore: Johns Hopkins Univ. Press.

Finkel, A. M. (2011). Solution-focused risk assessment: A proposal for the fusion of environmental analysis and action. *Human and Ecological Risk Assessment, 17*(4), 754–787. (and 5 invited responses/commentaries, pp. 788–812).

Finkel, A. M. (2012). Harvesting the ripe fruit: Why is it so hard to be well-informed at the moment of decision?, Chapter 3C. In R. Laxminarayan & M. K. Macauley (Eds.), *The value of information: Methodological frontiers and new applications in environment and health* (pp. 57–66). Dordrecht: Springer Science & Business Media.

Finkel, A. M. (2016). *Risksplaining: A counter-productive cottage industry.* Invited presentation to Dow Chemical Co. March 14. Slide presentation available at https://tinyurl.com/ finkel-dow-risk-slides

Finkel, A. M. (2018a). *I thought you'd never ask: Structuring regulatory decisions to stimulate demand for better science and better economics.* Ms. in review, available from author.

Finkel, A. M. (2018b). Demystifying evidence-based policy analysis by revealing hidden value-laden constraints. In G. E. Kaebnick & M. K. Gusmano (Eds.), *Governance of emerging technologies: Aligning policy analysis with the public's values* (*Hastings Center Report*) (Vol. 48(S1), pp. S21–S49). Garrison: The Hastings Center. Available at https://onlinelibrary.wiley.com/doi/epdf/10.1002/hast.818.

Finkel, A. M., & Gray, G. M. (2018). Taking the reins: How decision-makers can stop being hijacked by uncertainty. *Environment Systems and Decisions, 38*(2), 230–238. https://doi.org/10.1007/s10669-018-9681-x.

Finkel, A. M., & Sullivan, J. W. (2011). A cost-benefit interpretation of the 'substantially similar' hurdle in the Congressional Review Act: Can OSHA ever utter the E-word (ergonomics) again? *Administrative Law Review, 63*(4), 707–784.

Finkel, A. M., Trump, B. D., Bowman, D., & Maynard, A. (2018a). A 'solution-focused' comparative risk assessment of conventional and synthetic biology approaches to control mosquitoes carrying the dengue fever virus. *Environment Systems and Decisions, 38*(2), 177–197.

Finkel, A. M., Deubert, C. R., Lobel, O., Cohen, I. G., & Lynch, H. F. (2018b). The NFL as a workplace: The prospect of applying occupational health and safety law to protect NFL workers. *Arizona Law Review, 60*, 291–368.

Food and Drug Administration, U.S. (2014). *Drugs intended to treat life-threatening and severely-debilitating illnesses.* Code of Federal Regulations, Title 21, Part 312.80, revised as of April 1, 2014.

French, C. E., de Mora, K., Joshi, N., Elfick, A., Haseloff, J., & Ajioka, J. (2011). Synthetic biology and the art of biosensor design, Appendix A5. In E. R. Choffnes, D. A. Relman, L. Pray, & Rapporteurs (Eds.), *The science and applications of synthetic and systems biology: Workshop summary* (pp. 178–201). Washington, DC: Institute of Medicine, National Academy Press.

Friedman, L., Pierre-Louis, K., & Sengupta, S. (2018). The meat question, by the numbers. *New York Times,* Jan 25.

Friends of the Earth, International Center for Technology Assessment, and ETC Group. (2012). *The principles for the oversight of synthetic biology,* 20 pp.

Gagne, J. J., & Choudhry, N. K. (2011). How many 'me-too' drugs is too many? *JAMA, 305*(7), 711–712.

Gigerenzer, G. (2006). Out of the frying pan into the fire: Behavioral reactions to terrorist attacks. *Risk Analysis, 26*, 347–351.

Goldstein, B. D. (2018). Solution-focused risk assessment. *Current Opinion in Toxicology, 9*, 35–39.

Graham, J. D., & Wiener, J. B. (Eds.). (1997). *Risk vs. risk: Tradeoffs in protecting health and the environment.* Cambridge, MA: Harvard University Press.

Hammitt, J. K. (1990). Subjective-probability-based scenarios for uncertain input parameters: Stratospheric ozone depletion. *Risk Analysis, 10*(1), 93–102.

Hayes, K. R., Barry, S., Beebe, N., Dambacher, J. M., De Barro, P., Ferson, S., et al. (2015). *Risk assessment for controlling mosquito vectors with engineered nucleases: Sterile male construct: Final report.* Hobart: CSIRO Biosecurity Flagship. Available at https://publications.csiro.au/rpr/pub?pid=csiro:EP153254.

Hurlbut, W. B. (2013). St. Francis, Christian love, and the biotechnological future. *The New Atlantis: A Journal of Technology and Society,* Winter/Spring, 93–100.

Hylton, W. S. (2012). Craig Venter's bugs might save the world. *New York Times,* May 30.

Jackson, R. J., et al. (2001). Expression of mouse interleukin-4 by a recombinant ectromelia virus suppresses cytolytic lymphocyte responses and overcomes genetic resistance to mousepox. *Journal of Virology, 75*: 1205–1210.

Kahn, J. (2011). Synthetic hype: A skeptical view of the promise of synthetic biology. *Valparaiso University Law Review, 45*(4), 29–46.

Kaplan, S., & Garrick, B. J. (1981). On the quantitative definition of risk. *Risk Analysis, 1*(1), 11–27.

Kousky, C., Pratt, J., & Zeckhauser, R. J. (2010). Virgin versus experienced risks. In E. Michel-Kerjan & P. Slovic (Eds.), *The irrational economist: Making decisions in a dangerous world* (pp. 99–106). New York: Public Affairs Press.

Lassey, K. R. (2007). Livestock methane emission: From the individual grazing animal through national inventories to the global methane cycle. *Agricultural and Forest Meteorology, 142,* 120–132.

Lempert, R. J., & Collins, M. T. (2007). Managing the risk of uncertain threshold responses: Comparison of robust, optimum, and precautionary approaches. *Risk Analysis, 27*(4), 1009–1026.

Løkke, S. (2006). *Chemicals regulation: REACH and innovation.* Conference proceedings, 16 pp. Available at: http://www.norlca.man.dtu.dk/-/media/Sites/Norlca_Nordic_Life_Cycle_ Association/symposium2006/proceedings/loekke.ashx?la=da

Machado, R. D. (2012). Seeking the right targets: Gene therapy advances in pulmonary arterial hypertension. *European Respiratory Journal, 39*(2), 235–237.

Mandel, G. N., & Marchant, G. E. (2014). The living regulatory challenges of synthetic biology. *Iowa Law Review, 100,* 155–200.

Manzi, J. (2014). The new American system. *National Affairs, Spring 2014,* 3–24.

Marchant, G. E., & Finkel, A. M. (2012). *Attempts to forge government-industry partnerships that are neither unduly coercive nor unduly meaningless.* Invited presentation at "Soft Law Governance workshop," Center for Law, Science, and Innovation, Arizona State University, Tempe, AZ, March 5, 2012.

Maynard, A. D., & Finkel, A. M. (2018). *Solution-focused risk assessment and emerging technologies.* Video (6′49″) available at https://www.youtube.com/watch?v=n7XkdbpYWHk

McCubbins, J. S. N., Endres, A. B., Quinn, L., & Barney, J. N. (2013). Frayed seams in the 'patchwork quilt' of American federalism: An empirical analysis of invasive plant species regulation. *Environmental Law, 43,* 35–81.

Miller, H. I. (2014). Critics of 'me-too' drugs need to take a chill pill. *Wall Street Journal,* January 1.

Moe-Behrens, G. H. G., Davis, R., & Haynes, K. A. (2013). Preparing synthetic biology for the world. *Frontiers in Microbiology, 4,* 1–10.

Mokhtari, A., Moore, C. M., Yang, H., Jaykus, L.-A., Morales, R., Cates, S. C., & Cowen, P. (2006). Consumer-phase *Salmonella enterica* serovar enteritidis risk assessment for egg-containing food products. *Risk Analysis, 26*(3), 753–768.

Montague, P., & Finkel, A. (2007). Two friends debate risk assessment and precaution. *Rachel's Democracy and Health News,* No. 920. Available at http://www.rachel.org/?q=en/newsletters/ rachels_news/920#Two-Friends-Debate-Risk-Assessment-and-Precaution

Morais, A. R. C., et al. (2015). Chemical and biological-based isoprene production: Green metrics. *Catalysis Today, 239,* 38–43.

Morgan, M. G., Fischhoff, B., Bostrom, A., & Altman, C. J. (2002). *Risk communication: A mental models approach.* New York: Cambridge University Press.

National Academy of Sciences. (1983). *Risk assessment in the Federal Government: Managing the process.* Washington, DC: National Academy Press.

National Academy of Sciences. (1994). *Science and judgment in risk assessment.* Washington, DC: National Academy Press.

National Academy of Sciences. (1996). *Understanding risk: Informing decisions in a democratic society.* Washington, DC: National Academy Press.

National Academy of Sciences. (2006). *Preventing the forward contamination of Mars.* Washington, DC: National Academy Press.

National Academy of Sciences. (2009). *Science and decisions: Advancing risk assessment.* Washington, DC: National Academy Press.

National Academy of Sciences. (2012). *Sustainable development of algal biofuels in the United States.* Washington, DC: National Academy Press.

Naujokas, M. F., et al. (2013). The broad scope of health effects from chronic arsenic exposure: Update on a worldwide public health problem. *Environmental Health Perspectives, 121*(3), 295–302.

Nelson, K. C., & Banker, M. J. (2007). *Problem formulation and options assessment handbook: A guide to the PFOA process and how to integrate it into environmental risk assessment of genetically modified organisms.* 252 pp., available at https://gmoera.umn.edu/sites/gmoera.umn.edu/files/pfoa_handbook_bw.pdf

Nussbaum, N. J. (2002). Making 'me-too' drugs benefit the public. *American Journal of Medical Quality, 17*(6), 215–217.

Outterson, K. (2014). *Testimony to the House Energy and Commerce Committee, Sept 19.* Available at https://docs.house.gov/meetings/IF/IF14/20140919/102692/HHRG-113-IF14-Wstate-OuttersonK-20140919.pdf

Oxitec Inc. (2019). *Oxitec responds to article entitled "Transgenic Aedes Aegypti Mosquitoes Transfer Genes into a Natural Population."* Website dated September 18, 2019, https://www.oxitec.com/news/oxitec-response-scientific-reports-article. Last accessed 11 Oct 2019.

Paradise, J., & Fitzpatrick, E. (2012). Synthetic biology: Does re-writing nature require re-writing regulation? *Penn State Law Review, 117*, 53–87.

Perkel, J. M. (2013). Streamlined engineering for synthetic biology. *Nature Methods, 10*(1), 39–42.

Powers, C. M., Dana, G., Gillespie, P., Gwinn, M. R., Hendren, C. O., Long, T. C., Wang, A., & Davis, J. M. (2012). Comprehensive environmental assessment: A meta-assessment approach. *Environmental Science & Technology, 46*, 9202–9208.

Presidential Commission for the Study of Bioethical Issues. (2010). *New directions: The ethics of synthetic biology and emerging technologies.* Washington, DC: PCSBI. Available at https://bioethicsarchive.georgetown.edu/pcsbi/sites/default/files/PCSBI-Synthetic-Biology-Report-12.16.10_0.pdf.

Rampton, S., & Stauber, J. (2002). *Trust us, we're experts: How industry manipulates science and gambles with your future.* New York: TarcherPerigee.

Rascoff, S. J., & Revesz, R. L. (2002). The biases of risk tradeoff analysis: Towards parity in environmental and health-and-safety regulation. *University of Chicago Law Review, 69*, 1763–1836.

Rich, N. (2014). The mammoth cometh, *New York Times,* March 2 (Sunday magazine, p. MM24).

Rodemeyer, M. (2009). *New life, old bottles: Regulating first-generation products of synthetic biology,* Woodrow Wilson International Center for Scholars, March 2009, 57 pp. Available at http://www.synbioproject.org/publications/synbio2/

Rooke, J. (2013). Synthetic biology as a source of global health innovation. *Systems and Synthetic Biology, 7*, 67–72.

Rosner, H. (2018). Palm oil is unavoidable: Can it be sustainable? *National Geographic,* December 2018, available at https://www.nationalgeographic.com/magazine/2018/12/palm-oil-products-borneo-africa-environment-impact/

Rycroft, R. W., & Kash, D. E. (1992). Technology policy requires picking winners. *Economic Development Quarterly, 6*(3), 227–240.

Sandman, P. M. (1993). *Responding to community outrage: Strategies for effective risk communication.* American Industrial Hygiene Association. Available at http://petersandman.com/media/RespondingtoCommunityOutrage.pdf

Schmidt, M., & de Lorenzo, V. (2012). Synthetic constructs in/for the environment: Managing the interplay between natural and engineered biology. *FEBS Letters, 586*, 2199–2206.

Schwermer, H., De Koeijer, A., Brulisauer, F., & Heim, D. (2007). Comparison of the historic recycling risk for BSE in three European countries by calculating the basic reproduction ratio R_0. *Risk Analysis, 27*(5), 1169–1178.

Shaw, G. B. (1949). From "Back to Methuselah" (Act I; The Serpent says these words to Eve).

Shlyakhter, A. I. (1994). Improved framework for uncertainty analysis: Accounting for unsuspected errors. *Risk Analysis, 14*, 441–447.

Silver, L. (2007). Scientists push the boundaries of human life. *Newsweek,* June 3. Available at http://www.newsweek.com/scientists-push-boundaries-human-life-101723

Specter M. (2012). The mosquito solution. *The New Yorker*, July 9.

Starr, C. (1969). Social benefit versus technological risk. *Science, 165*(3899), 1232–1238.

Suskind, R. (2007). *The one percent doctrine: Deep inside America's pursuit of its enemies since 9/11.* New York: Simon and Schuster.

SynBioWatch. (2016). *Solazyme: Synthetic biology company claimed to be capable of replacing palm oil struggles to stay afloat.* Available at http://www.synbiowatch.org/2016/02/solazyme-investigation/

Taxpayers for Common Sense. (2009). *Coal: A Long history of subsidies.* June 11, available at https://www.taxpayer.net/energy-natural-resources/coal-a-long-history-of-subsidies/

Taylor, M. R. (2006). *Regulating the products of nanotechnology: Does FDA have the tools it needs?* Woodrow Wilson International Center for Scholars, Project on Emerging Nanotechnology #5, October, 66 pp.

Thierer, A. (2016). *Permissionless innovation: The continuing case for comprehensive technological freedom*, revised and expanded ed. (Arlington, Virginia, George Mason University). Available at https://www.mercatus.org/system/files/Thierer-Permissionless-revised.pdf

University of Manchester. (2013). An impact analysis of a synthetic palm oil: Outlining a new approach to ethical considerations in the production of high-value chemicals. Available at http://2013.igem.org/wiki/images/9/9c/MANCHESTERIGEMimpactanalysisofsyntheticpalmoil.pdf

University of Nebraska-Lincoln. (2017). Helping reduce methane emissions from livestock. Available at http://2017.igem.org/Team:UNebraska-Lincoln/Description

Venter, J. C. (2013). *Life at the speed of light: From the double helix to the dawn of digital life.* New York: Penguin Books.

Wagner, W. (2000). The triumph of technology-based standards. *University of Illinois Law Review,* 83–113.

Wareham, C., & Nardini, C. (2015). Policy on synthetic biology: Deliberation, probability, and the precautionary paradox. *Bioethics, 29*(2), 118–125.

Willis, H. H., & Florig, H. K. (2002). Potential exposures and risks from beryllium-containing products. *Risk Analysis, 22*(5), 1019–1033.

Wright, O., Stan, G.-B., & Ellis, T. (2013). Building-in biosafety for synthetic biology. *Microbiology, 159*, 1221–1235.

Zhang, J. Y., Marris, C., & Rose, N. (2011). *The transnational governance of synthetic biology: Scientific uncertainty, cross-borderness and the 'Art' of governance.* BIOS working paper no. 4, London School of Economics and Political Science, 37 pp. Available at http://openaccess.city.ac.uk/id/eprint/16098/

A Solution-Focused Comparative Risk Assessment of Conventional and Emerging Synthetic Biology Technologies for Fuel Ethanol

Emily Wells, Benjamin D. Trump, Adam M. Finkel, and Igor Linkov

Introduction

Global energy demand is increasing due to global development initiatives and steady population growth. The US Energy Information Administration's International Energy Outlook 2017 (U.S. EIA 2017c) projects that the world energy consumption will raise from approximately 575 quadrillion Btu in 2015 to 736 quadrillion Btu by 2040—an increase of 28% (U.S. EIA 2017c). Fossil fuels, such as petroleum and natural gas, serve as the leading energy sources for various sectors, such as transportation. However, the International Energy Agency (IEA) forecasts that biofuel production will increase by 15% over the next 5 years to reach approximately 42.6 billion gallons (IEA 2018). Various types of renewable fuels or fossil fuel additives are being researched and developed as complements or supplements to fossil fuels. Ethanol, or ethyl alcohol, is one such additive, particularly for motor fuel in the United States and Brazil. Fuel ethanol has been proposed to offset dependence on petroleum, thereby reducing greenhouse gas emissions by up to 43% relative to gasoline (Flugge et al. 2017). Additionally, as advanced ethanol production processes are less sensitive to the vagaries of geography, as will be discussed later in

E. Wells (✉) · I. Linkov
Carnegie Mellon University, Pittsburgh, PA, USA

US Army Corps of Engineers, Washington, DC, USA
e-mail: emwells@andrew.cmu.edu

B. D. Trump
US Army Corps of Engineers, Washington, DC, USA

University of Michigan, Ann Arbor, MI, USA

A. M. Finkel
University of Michigan, Ann Arbor, MI, USA

© Springer Nature Switzerland AG 2020
B. D. Trump et al. (eds.), *Synthetic Biology 2020: Frontiers in Risk Analysis and Governance*, Risk, Systems and Decisions, https://doi.org/10.1007/978-3-030-27264-7_10

this chapter, countries can produce it domestically rather than having to rely on the geopolitics associated with the world petroleum market.

Ethanol is directly blended with petroleum to comprise approximately 10% the volume of each gallon of gasoline consumed through US gas stations (US EIA 2016). In 2017, approximately 27 billion gallons of ethanol was produced internationally, with a projected growth rate of 2% each year. It is predicted that fuel ethanol will account for approximately two-thirds of overall biofuel production growth and that by 2023, the annual output of ethanol will be 31 billion gallons internationally (IEA 2018).

Ethanol is currently produced by converting a variety of feedstock sources into useful sugars. While existing ethanol production has been derived primarily from corn and sugarcane feedstocks, advanced production methods have the potential to use various species of algae to produce an algal oil substitute. Strains of naturally occurring algae are capable of yielding such algal oil in limited quantities, but innovative technologies utilizing synthetic biology are being considered to improve the production process. Synthetic biologists are interested in developing strains of engineered algae in controlled environments to produce ethanol with more efficient and renewable ethanol yield rates.

Yet, engineered algal ethanol imposes unique risks, benefits, and other implications. As an emerging technology, synthetic biology processes introduce issues of uncertainty and complexity that derive from the novelty of the technology. Further, there are limited data pertaining to these emerging technologies, which makes it difficult to precisely quantify the risks and to subsequently improve best practices. However, these same emerging technologies can provide significant benefits to human and environmental health, such as improved air quality relative to fossil fuels. Understanding that current fossil fuel and conventional ethanol production entail risks of their own, it is crucial to compare energy sources based on both risk and potential benefits. Rather than asking whether engineered algal ethanol is efficacious and "safe enough" for deployment in its own right, a comparative approach is critical to assess various attributes of the emerging energy source against the risks and benefits of the best conventional solutions to meet national and international energy demand. Conventional quantitative risk assessments (QRA) measure quantitative data pertaining to an alternative's risks; as synthetic biology is emerging and field use is limited, the critical quantitative data are limited, and a modified approach to emerging technology risk assessment is necessary (Malloy et al. 2016; Linkov et al. 2018).

A solution-focused risk assessment (SFRA) (Finkel 2011; Finkel et al. 2018) is one such approach that can qualitatively and quantitatively evaluate synthetically engineered algal ethanol relative to conventional competitors. In general, SFRA tries to transcend traditional risk assessment questions ("is it safe enough?" or "what level of exposure yields an acceptably low risk?" to instead require risk assessors and decision-makers to collaborate from the earliest point and address broader questions of which of several competing technologies best fulfills a given human need (considering both risk reduction benefits and new downside risks) (see Finkel et al. 2018). Specifically, this chapter introduces an SFRA that assesses economic and social implications, sustainability, environmental implications, and risk considerations. These considerations will be compared across corn, sugarcane, and algal ethanol (natural and engineered) production processes. The trade-offs between risks and benefits are evaluated. The benefits of the various ethanol sources are weighed against their potential risks in order to conceptualize the net risk reduction for each ethanol source, relative to the others. Because the environmental and human health benefits of ethanol fuel, once it is produced, do not depend on the means by which it was generated, we only need to compare the downside risks of the various technologies (in many other cases, the products derived via synthetic biology approaches differ from the conventional product it seeks to displace, and so the risk reduction benefits may differ and need to be accounted for). Therefore, this SFRA approach compares whether the risks of conventional ethanol production (e.g., land use requirements for conventional ethanol sources) outweigh the novel risks of emerging ethanol production methods.

In this chapter, a general framework for an SFRA is laid out with recommendations for how to interpret input and outcome measures and for how future research could build from this framework. The SFRA presented here provides a framework to consider the risks of each feedstock option by using ranges of measures found in existing literature. The SFRA approach puts problem *decisions* at the forefront of risk reduction; in this case, what bioethanol feedstock options minimize adverse economic and environmental implications and risks. Rather than focusing on estimating an acceptable level of risk, SFRA aims to identify which decision or alternative has the greater net risk reduction. Ideally, this comparative approach allows for the benefits of certain, perhaps advanced or novel, alternatives to be realized in comparison to status quo technologies (e.g., petroleum production and consumption) should the advanced alternatives provide net risk reduction. The net risk reduction of alternatives is compared across four factors: sustainability, environmental implications, social and economic implications, novel risks. Future sensitivity analyses performed on these metrics could assess the decision thresholds across the four factors while fine-tuning the choice among technologies within specific locations and economies and across uncertain parameters. This is particularly crucial for synthetically engineered algal ethanol, for which limited public empirical data exist. The less predictable, novel risks associated with synthetically engineered algae are discussed with guidance on how to overcome the ambiguity associated with incorporating and comparing "known" and "unknown" risks.

Background: Development of Fuel Ethanol

Before discussing the current sources of ethanol feedstock and their production processes, it is necessary to review the history of ethanol development and eventual commercialization. The production of the various types of ethanol dates to the Neolithic Period (4500–2000 BC) when sugar was fermented into ethanol for alcohol production (Roach 2005). Early ethanol production centered on the distillation of wine and spirits for alcoholic beverages, where these ethanol precursors were derived from grapes, rice, and other agricultural plants. Ethanol production for fuel use took off in the early nineteenth century, when Swiss chemist Nicolas-Théodore de Saussure determined ethanol's chemical formula in 1807 (de Saussure 1807). This formula served as the basis for early synthetic ethanol production from ethylene or coal gas. The early modern use of ethanol centered on lamp fuel in the mid-nineteenth century, although various tariffs and taxes on ethanol use prohibited large-scale commercialization in the United States (Solomon et al. 2007; Tyner 2008; Campbell et al. 2008; Segall and Artz 2007). These efforts were driven by the belief that ethanol fuel could serve as a more efficient and cleaner burning alternative to traditional oils or coal, which had been widely used throughout the early Industrial Revolution.

The first modern and widespread commercial application of ethanol as automobile fuel for an internal combustion engine dates to early vehicles in the 1910s and 1920s (DiPardo 2000). These vehicles established the framework for future gasoline-ethanol blends, where Ford's early automobiles were able to operate on either gasoline or ethanol (DiPardo 2000). Today, virtually all of the commercial fuel ethanol production worldwide is produced by private companies in the United States, including Valero, Poet, Flint Hills Resources LP, Green Plains Renewable Energy, and ADM, by the state-run Brazilian company Petrobras, or external companies such as Raizen (Lovins 2005; Renewable Fuels Association 2016a, b). By 2011, companies (state-run or fully private) were responsible for approximately 87% of worldwide fuel ethanol production, or over 19 billion gallons of ethanol (Renewable Fuels Association 2011; Renewable Fuels Association 2012).

Similar to their American counterparts at Ford, Brazil's conversion of sugarcane into ethanol began in the late 1920s with the introduction of automobiles to the country (Valdes 2011). Ethanol production from sugarcane grew dramatically during World War II as oil shortages arose, which led the Brazilian government to mandate 50% ethanol fuel blends (Kovarik 2008). While sugarcane ethanol production declined post-war in the midst of cheap gasoline, it increased again during the oil crises in the 1970s and 1980s. Due to these oil crises, the Brazilian government has since directly funded private and state-run ethanol companies in an effort to phase out dependency on foreign fossil fuels in favor of domestic biofuels like sugarcane ethanol (Bastos 2007). The Brazilian national government formalized their efforts to promote sugarcane ethanol production in Programa Nacional do Álcool, or the National Alcohol Program, started in 1975 (Bastos 2007).

Conventional Ethanol Production Processes

Conventional ethanol production requires a crop or biomaterial to be transformed and manipulated from its native state into a liquid. Specifically, this occurs in different physical and chemical processes, including biomaterial growth, collection, dehydration, and fermentation. The dehydration and fermentation stages are used to convert the raw biomass into ethanol by removing excess water from the biomass and chemically converting plant sugars into energy.

The conversion of biomass to ethanol is a multiphase process that involves significant fuel expenditure (Pimental 2005). These steps needed to convert a crop to ethanol or biodiesel may differ based on the particular crop or biomatter used for fuel conversion yet generally follow the sequence of growth, collection, dehydration, and fermentation to yield ethanol. Each stage of the generic life cycle is further described below (Von Blottnitz and Curran 2007) (Fig. 1).

The first stage in the generic ethanol production process includes the growth of the feedstock for eventual conversion into ethanol fuel. This crop growth does not substantially differ from how the crops are grown for food. Additionally, crop growth can be multipurpose, where ripened crops may be used for ethanol or food, depending on the stakeholder's interests. With respect to corn, approximately 40% of all corn grown in the United States, or roughly 130 million tons, will be used for corn ethanol (Mumm et al. 2014). The timeframe for growth will differ based upon the crop grown and seasonality, with corn being cold-intolerant and planted in the

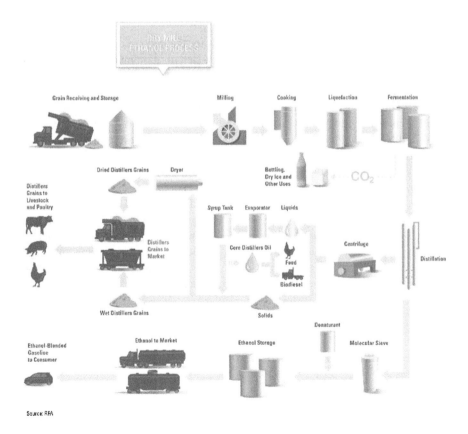

Fig. 1 Generic ethanol fuel life cycle. (Source: Renewable Fuels Association (2016c))

summer months (Pollack 2011). Sugarcane is generally only grown in warm temperate to tropical regions in South America and South Asia, with 75 tons of raw sugarcane produced annually in Brazil per hectare of cultivation (Da Rosa 2012). This makes sugarcane production in Brazil an economically important industry, with benefits for both improved energy efficiency and a significant source of employment for locals.

Once grown, the crops are harvested and organized based upon their intended purpose (ethanol, food, etc.). For corn and sugarcane, each individual ear is harvested by hand or by a mechanical picker and is stored in bins that are designed to keep moisture levels low via "grain dryers" (Van Devender 2011). For sugarcane, each plant is capable of multiple harvests, so collection methods are careful not to damage the sugar-producing plant. According to Rakkiyappan et al. (2009), mechanical methods of collecting sugarcane are capable of collecting approximately 100 tons per hour, while a seasoned sugarcane harvester can cut roughly 500 kilograms per hour, where by-hand harvesting accounts for more than half of sugarcane collection annually, ensuring a steady demand for physical labor. Regardless of the method used, the collected sugarcane must be processed quickly once harvested, as it almost immediately begins to lose its sugar content once harvested (Rakkiyappan et al. 2009).

After harvesting, crops intended as biomass for ethanol are dehydrated and distilled to prepare them for eventual fermentation. Dehydration involves the drying of crops and is generally conducted using one of three processes, including azeotropic distillation, extractive distillation, and molecular sieves (Kumar et al. 2010; Rouquerol et al. 1994). Overall, the general purpose of each of these methods is to quickly remove any retained liquid from the feedstock. This prevents the material from spoiling during the ethanol production process and prepares the feedstock for its eventual fermentation.

The last step in ethanol creation is fermentation, through which sugars such as fructose, sucrose, and glucose are converted into energy (Stryer 1975). More specifically, the conversion of sugars into energy produces ethanol and carbon dioxide as waste material, where the ethanol may be sequestered for eventual use as fuel (Stryer 1975). Once produced, ethanol is then blended with gasoline and burned— normally for an internal combustion engine. While not directly covering any stage of ethanol production, the "burning" phase is reviewed in order to determine the environmental impact associated with burning ethanol and releasing toxic material into the environment.

Advanced Ethanol Production Processes

Within the United States and abroad, conventional research within the subject of ethanol production has focused on two general strains of inquiry. The first includes the refinement of existing ethanol production such as with corn and sugarcane, where researchers in private companies and US government agencies like the EPA and USDA have sought to improve the energy yield while reducing environmental pollution throughout the ethanol life cycle. The second focuses upon novel methods of ethanol production, including non-genetically modified algae, and the process of cellulosic ethanol production. While ethanol production has continually grown since World War II, significant research and investment into new ethanol production strategies blossomed in the early 2000s, where world ethanol production tripled between 2000 and 2007.

Conventional ethanol research is motivated by a mixture of economic and social drivers. Socially, the rising food versus fuel debate (discussed further in the Implications section) has raised questions about the impact of ethanol fuel production on global food prices, where organizations such as the World Bank have asserted that the rising land use of foodstuffs for ethanol production directly contributed to rising global food costs that have significant economic impacts in sub-Saharan Africa (US EPA 2007). Ethanol research is also driven by economic factors, where government agencies and private companies in the United States continue to seek an alternative to corn ethanol, which has a relatively low energy balance score of 2.3. The net energy balance approximates the amount of energy produced given the amount of energy consumed. The net energy balance for each ethanol source will be comparatively assessed later in this chapter. The rapid growth of worldwide ethanol production coupled with these social and economic factors has driven the field's conventional research in order to find an alternative that has a minimal impact on global foodstuffs while improving energy balance ratios and reducing reliance upon fossil fuels.

Experimentation with cellulosic ethanol has occurred since the first cellulosic ethanol plant opening in South Carolina in 1910. However, high production costs have hindered consistent and widespread commercialization (Wang 2009). Using a mixture of wood, grasses, or other inedible plant pieces, cellulosic ethanol is produced via biochemical or thermochemical processing (Pimentel and Patzek 2005). A general production cycle is illustrated in Fig. 2.

Cellulolysis is the process which makes use of lignocellulosic material (or the inedible and structural parts of plants) to create ethanol. Specifically, hydrolysis is used to cleave chemical bonds of the lignocellulosic material using water, where the resulting sugars are eventually fermented and distilled into ethanol (Fujita et al. 2002). The process of cellulolysis is generally subdivided into five stages, including pretreatment, hydrolysis and sugar separation, fermentation, distillation, and dehydration (Lynd et al. 1991; Zhu et al. 2009). The pretreatment phase of cellulolysis is used in order to refashion the biomaterial prior to hydrolysis. Specifically, the lignocellulose within the available biomaterial is treated with chemicals to break its rigid structure, where the chemical method used differs based upon the biomaterial chosen for ethanol conversion. Next, the treated lignocellulosic material is converted via hydrolysis in order to break down the material's sugar molecules in order to isolate those sugar molecules for further fermentation. This generally occurs using one of two forms of hydrolysis, including an acidic chemical reaction or an enzymatic reaction. Chemical hydrolysis has been around since the nineteenth century and involves introducing an acid to the cellulose to separate its sugar molecules.

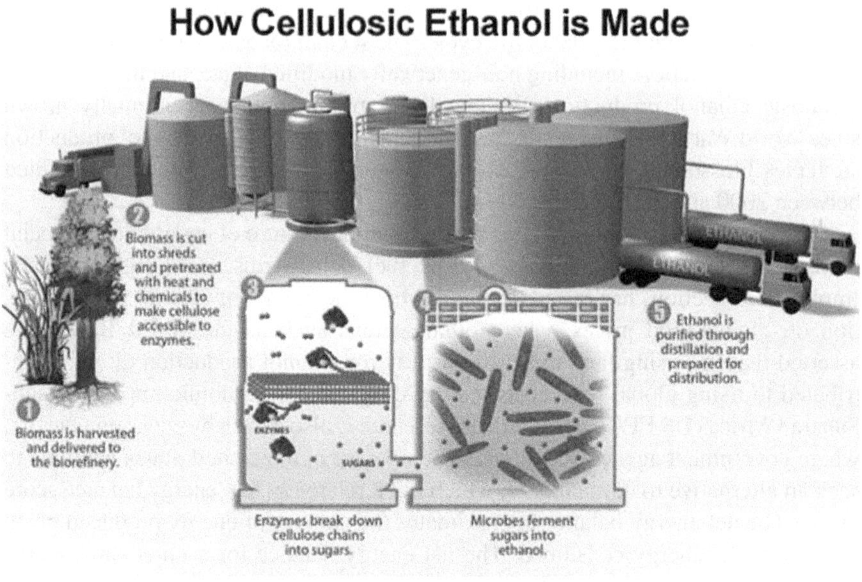

How Cellulosic Ethanol is Made

Fig. 2 Cellulosic ethanol production process. (Image source: US DOE (2007))

The enzymatic process uses enzymes to break down cellulose sugar chains to allow for collection of cellulose sugars.

After hydrolysis, the sugars acquired from hydrolysis are fermented through the use of yeast. These sugars (glucose, sucrose, and fructose) are converted into energy that will be eventually converted into ethanol. After fermentation, distillation of the converted sugars is used to produce 95% alcohol, which allows for eventual conversion into ethanol to be combined with gasoline. Distillation is carried out similarly as with general corn or sugarcane ethanol production. Lastly, dehydration converts the 95% alcohol into an alcohol liquid with a 99.5% ethanol concentration, which makes the ethanol ready for public consumption as vehicle fuel or other gasoline-driven purposes.

The second method of producing cellulosic ethanol includes gasification, or the chemical approach toward producing ethanol from cellulosic material. Rather than using chemical decomposition via cellulolysis, carbon in the cellulosic material is converted into gas, which fuels combustion, and then fermented. This generally occurs in three steps. In the first stage, carbon molecules are broken apart to make carbon monoxide, carbon dioxide, and hydrogen. These molecules are eventually used in fermentation to be converted into energy. Unlike the yeast used for fermentation in the cellulosic approach noted above, gasification uses the *Clostridium ljungdahlii* for fermentation. The bacteria consume carbon dioxide, carbon monoxide, and hydrogen and produce an output of ethanol and water. Lastly, the ethanol and water mixture produced from the *Clostridium ljungdahlii* bacteria are separated via distillation, leaving only the ethanol for commercial consumption.

Cellulosic ethanol is estimated to have an energy balance ranging from 2 to 36, where the large range in energy balance scores reflects different types of biofuels and energy conversion processes used to generate the ethanol (Schmer et al. 2008). This indicates that the method has the potential to produce significantly more energy than corn ethanol (which has an energy balance ratio of approximately 1.3) and potentially even higher than sugarcane (which has an energy balance ratio of 7–8). The technology takes advantage of abundant raw materials, where over 300 million tons of cellulose-containing materials that could create cellulosic ethanol is thrown away each year in the United States. However, the technology remains economically unviable, where cellulosic ethanol has lower energy content than traditional fossil fuels and would cost an estimated $120 per barrel of oil. Research on this technology continues to attempt improvements in energy efficiency and reduction in cost, along with further diversification of feedstock to be used in cellulolysis. For example, kudzu has been suggested as a potential source of cellulosic biomass.

Other than cellulosic ethanol production, an additional alternative method for ethanol production currently under research includes the use of algae. First discussed as a potential fuel source in 1942, German scientists Harder and von Witsch argued that microalgae could be cultured and grown in a controlled setting as a source of lipids for fuel or even food (Harder and von Witsch 1942). In the immediate aftermath of World War II, research regarding the conversion of algae into biodiesel fuel further spread to the United States, Israel, Japan, and England, where motivation for an alternative fuel source remains strong due to fuel limitations throughout the 1940s (Burlew 1953). However, the declining cost of fossil fuels reduced the need for an alternative energy source, although algae fermentation continued to be researched for applications of food and wastewater treatment (Borowitzka 2013).

The international oil embargo in the late 1970s rekindled interest in the development of algal biofuel (DOE). This interest was particularly strong in the United States, which invested $25 million into the Aquatic Species Program over an 18-year period with the intent of promoting a commercialized algal biofuel. However, scientists within this program came to find that natural algae (or those algal organisms lacking any genetic modification via synthetic biology) had several limitations that could hinder large-scale commercialized production, particularly limitations of economically feasible growth in a controlled environment (Sheehan et al. 1998). The final report issued by the Aquatic Species Program suggested that genetic engineering was necessary in order to overcome these natural hindrances and limitations, where a genetically engineered algae would grow and populate faster in a variety of environmental conditions (Sheehan et al. 1998). The Aquatic Species Program was disbanded in 1996, and it was not until a sharp increase in oil prices in the 2000s that funding for such biofuels increased, particularly in the United States, Australia, and the European Union (Pienkos and Darzins 2009). Along with providing domestic energy security, the Australian government has stated that biodiesel production from algal lipids may provide economic opportunities and jobs to various under-served or rural areas (SARDI Aquatic Sciences). By March 2013, the American energy company Sapphire Energy initiated the first commercialized sale of algal biofuels (SARDI Aquatic Sciences).

Today, algae can be used to generate a variety of fuels, where the lipid portion of the algae is converted into biodiesel with the potential for future conversion to etha-nol (Ellis et al. 2012). Algae are cultivated and harvested in 1–10 days and can be grown in areas that are unsuited for agricultural production or exposed to untreated wastewater (Chisti 2007). Currently, most research and production of algal biofuel takes place in photobioreactors (a series of glass tubes which are exposed to water) or open ponds, where ponds are less costly than photobioreactors but more vulner-able to contamination (Mata et al. 2010).

Current State of Fuel Ethanol

Ranging from the conversion of corn to fuel in the United States to sugar to fuel in Brazil, the current state of ethanol research and development is driven in an attempt to foster a sustainable fuel source that reduces or eliminates domestic reliance on nonrenewable fossil fuels. A number of different types of feedstock products may be used to generate ethanol, including barley, hemp, sweet potatoes, and cellulose. Yet, production is dominated by corn in the United States and sugarcane in Brazil, with smaller production levels in Europe, China, and elsewhere (Table 1) (Renewable Fuels Association 2016c). Overall, the United States and Brazil accounted for approximately 83% of global ethanol production in 2015.

Global ethanol production has increased on a yearly basis since 2005, and the United States has seen the greatest production rate increase (British Petroleum 2016). Between 2005 and 2015, total ethanol production in the United States increased from 3.9 to 14.9 billion gallons (Renewable Fuels Association 2016c). The majority of US ethanol production occurred in the Mid-West region, where corn optimally grows. Additionally, South America and Central America nearly doubled ethanol production between 2005 and 2015 (British Petroleum 2016).

Table 1 Global ethanol fuel production in millions of gallons produced in 2015 based on Renewable Fuels Association data

2015 World ethanol fuel production (billion gallons)					
Producer	United States	Brazil	Europe	China	Canada	Rest of world
Gallons	14.8	7.1	1.4	0.8	0.4	1.1
Percentage of global production	56%	27%	5%	3%	2%	7%

Energy Efficiency of Conventional and Advanced Ethanol Feedstock

Throughout the ethanol fuel life cycle, one of the fundamental concepts governing the efficiency and viability of turning a specific feedstock into ethanol is energy balance. Specifically, this includes the total amount of energy input into the process of converting biomass against the energy released by burning the ethanol, represented as (Shapouri et al. 2002):

$$Net \text{ energy balance} = \frac{energy\ produced}{energy\ consumed.}$$

The numerator contains the potential energy that may be used upon burning the created ethanol, while the denominator contains all of the energy invested into producing the ethanol (including field preparation and crop cultivation). An "energy-positive" ethanol is one where the energy produced is greater than energy consumed, while an "energy-negative" ethanol is one where the energy production is lower than the energy consumed. With regard to investments, all energy expenditures in the growth, collecting, drying, and fermentation of biomass are included in the energy balance computation (Agler et al. 2008; Hill et al. 2006; Murphy and Power 2008). Generally, fossil fuel energy is utilized on the investment side of the energy balance equation, where coal, oil, or natural gas is used to convert biomaterial into ethanol (Hill et al. 2006; Murphy and Power 2008). Energy-negative ethanol production methods cost more energy via fossil fuels to create 1 liter of ethanol than would be produced, while energy-positive ethanol production methods offer a net energy gain by the end of the production process. Overall, energy balance is a critical element in determining the efficacy of ethanol fuel production, where if a particular method or feedstock generates a net negative energy balance, it would unlikely be commercialized for the long term. For any potential algal feedstock, the product would eventually have to foster not only a net positive energy balance score but may also need to offer similar or improved energy balance scores to conventional biomaterials should the risks associated with algal ethanol outweigh the risks associated with conventional biomaterials.

Isaias Macedo (1998) conducted studies regarding the energy balance values of corn and sugarcane ethanol, respectively, indicate that sugarcane ethanol has a net positive energy balance number yet corn ethanol is not substantially positive and may even be negative in certain conditions of crop spoilage or improper conversion to ethanol fuel (De Oliviera et al. 2005; Macedo 1998). For corn ethanol, 1 unit of fossil-fuel energy is required to create 1.3 energy units of ethanol (Macedo 1998). This figure was calculated by Macedo in his review of corn ethanol, but more recent analyses suggest that the corn ethanol production processes are becoming more efficient and reach net energy balances ranging from 2.6 to 2.8 (Renewable Fuels Association 2016a, b). While corn ethanol does contribute to a net positive energy balance, the energy improvement is quite limited and may not warrant the environmental degradation caused by harvesting corn and the pollution accrued by converting corn feedstock into ethanol fuel. Sugarcane ethanol is substantially more efficacious, where 1 unit of fossil fuel energy is required to create approximately 8 to 9 energy units of ethanol (Bourne and Clark 2007). This net energy balance indicates that sugarcane is significantly more energy efficient than corn, as it requires significantly less feedstock to produce a greater amount of liquid ethanol to be mixed with various gasoline blends. However, sugarcane may only be grown productively in tropical climates, whereas corn is more flexibly grown across a wider range of climates.

Advanced ethanol production processes may offer higher net energy balances than conventional approaches. By converting cellulosic biomass into ethanol, a wide range of net energy balance values have been presented across the literature. Reported net energy balances of cellulose range from 1.42 to 36, with 36 being a massive net energy producer. The range of values derives from the variance in perennial herbaceous plants that can be harvested for ethanol (Schmer et al. 2008). Naturally occurring cyanobacteria and microalgae that are grown agriculturally can yield net energy balances ranging from 0.7 to 7.8. The range of values here reflects differences in growth environments, where algae with higher net energy outputs may be grown in more suitable environments (Shen and Luo 2011; Brentner et al. 2011) (Table 2).

Overall Observations of Conventional and Newer Ethanol Production Processes

Ethanol fuel additives offer a mechanism to offset petroleum consumption through a variety of feedstock alternatives and production processes. Conventional feedstocks, including corn and sugarcane, have experienced widespread commercialization in the United States and Brazil. More advanced processes, such as those using cellulose and algae, may be more energy efficient but have experienced less commercialization, largely due to their high principal and R&D costs.

Table 2 Net energy balance values for each ethanol source. Higher values indicate greater energy efficiency. The ranges of values reported here reflect the approximate minimum and maximum net energy balance scores presented across prior research and agency reports

Ethanol source	Net energy balance	Source
Corn	1.3–4	Macedo (1998); Renewable Fuels Association, (2016a, b)
Sugarcane	8–9	Bourne and Clark (2007)
Cellulose	1.42–36	Shahrukh et al. (2016); McLaughlin et al. (2011); Schmer et al. (2008)
Cyanobacteria/microalgae (natural)	0.7–7.8	Shen and Luo (2011); Brentner et al. (2011)

A Synthetic Biology Solution to Biofuel Production

Synthetic biology serves as a possible mechanism for improving upon current conventional and advanced ethanol feedstock options. The use of synthetic biology to improve ethanol production is similar to existing conventional research in that its motivation is to find an economically feasible feedstock source that is energy efficient. While synthetic biology applications mirror conventional research in the cultivation of algae as a biomass for fuel production, the technology differs in that algal blooms are specifically engineered to enhance fermentation and photosynthesis processes, increase lipid content, increase pathogen resistance, produce higher-value co-products, and/or diminish unwanted cellular regulation (Georgianna and Mayfield 2012; Gimpel et al. 2013). Overall, the primary goal of synthetic biology's algal ethanol option is to dramatically reduce the energy needed to convert biomass into fuel ethanol such that engineered algae could produce significant amounts of algal oil (an immediate precursor to ethanol fuel) without the significant fossil fuel and manpower resources needed to produce ethanol from conventional biomass. With these R&D aims, algal ethanol is anticipated to be more energy efficient than corn or sugarcane, where engineered algae would only require initial start-up energy costs to produce several substantial harvests of various ethanol fuels. Synthetic biology ethanol technologies aim to improve the existing limitations of corn ethanol (i.e., increasing the net energy balance), sugarcane ethanol (i.e., desensitize feedstock to grow in diverse environments), cellulosic ethanol (i.e., reduce downstream production costs), and naturally occurring algal ethanol (i.e., increase net energy balance).

The synthetically engineered algal ethanol process would be accomplished by converting the algae's lipids into biodiesel, which is identical to the process noted above for non-engineered algal oils. Subsequently, the algal cells' carbohydrates can be fermented into bioethanol in a process very similar to existing conventional practices in corn or sugarcane.

Synthetic biology was proposed by Craig Venter in 2011 as a tool to make algal cells more economically viable and technologically feasible as an ethanol production source while improving the capabilities of algal ethanol production in terms of energy requirements, environmental impact, and economic potential. By fine-tuning the genome of specific algae using synthetic biology techniques, it is possible to create a modified species of algae that is a highly cost-effective alternative to other forms of biomass while being compatible with existing bioethanol manufacture and supply infrastructures. For example, where many existing bioethanol products have low energy density and are incompatible with existing fuel infrastructure (Stephanopoulos 2007; Atsumi et al. 2008), Craig Venter of Synthetic Genomics Inc. claims that engineered algae could be engineered and developed to produce 5–10 times more fuel per acre than contemporary feedstock. Likewise, where biodiesel is plagued by issues such as high cost and limited availability of necessary biomass, engineered algae are sustainable in that algae can be manipulated to continually produce ethanol via sunlight without killing the algae cell in general

(Demirbaş 2002). In 2009, ExxonMobil funded and began a collaborative effort with Synthetic Genomics Inc. In 2017, the pair announced a breakthrough in advanced biofuel production—they increased algae's oil content from 20% to 40% (Ajjawi et al. 2017).

Genetically manufactured algae can serve as a renewable, economically viable, and energy-efficient method of replacing limited fossil fuels (Georgianna and Mayfield 2012). Additionally, the ability to engineer such algae to have similar properties to petroleum-based fuels allows for its use in existing transportation infrastructure, which can limit indirect costs involved in switching fuel sources (Peralta-Yahya et al. 2012). Such algae would be required to exhibit certain characteristics, including (Alper and Stephanopoulos 2009):

(a) High substrate utilization and processing capacities
(b) Fast pathways for sugar transport
(c) Good tolerance to inhibitors
(d) High metabolic fluxes, and
(e) Producing a single fermentation product

As such, while the potential for algae to serve as the next wave of ethanol biofuels is apparent, it is still uncertain how much biosynthesis and genetic manipulation is required to produce an "ideal" product. Additionally, tens of thousands of algae species exist, further complicating the identification of an ideal candidate for further research and use.

Proposed Role of Synthetic Biology in State of Fuel Ethanol Production

Synthetic biology has emerged as a technical approach to enhance algal ethanol production, aiming to make algal ethanol more energy efficient, regenerative, and less costly. However, because of the novelty and uncertainty surrounding synthetic biology, it is unclear whether synthetically engineered algae are a viable bioethanol alternative. To determine optimal ethanol feedstock sources and processes, it is necessary to comparatively review the alternatives using risk assessment. The risk assessment should include traditional quantitative assessments but further be informed by a "solution-focused" orientation to risk. By using SFRA, the scope of assessment expands beyond the cost considerations emphasized in the EPA's latest Renewable Fuels Standards Program, and the unique benefits of each feedstock alternative are compared. Existing assessment protocols are not fully able to capture the complexity of ethanol production processes, and traditional risk assessments also have difficulty dealing with the uncertainty of synthetically engineered organisms (Trump et al. 2019). A solution-focused risk assessment provides a lens to think comparatively and holistically about the impact of synthetically engineered biofuels. As the problem is complex in nature, assessments need to be comprehensive and consider a variety of factors. The intent of this assessment is not to provide a conclusion on the viability or ethics of synthetically engineered algal ethanol, but rather to pave the way for thinking about risks as the technology continues to develop.

Method: Solution-Focused Risk Assessment

Thinking about complex operations and risks involved in the energy sector is difficult due to its multi-faceted implications for the economy, the environment, and human health. Because data on emerging technologies are limited due to their novelty, it is challenging to derive accurate estimates of the environmental and social impacts of emerging energy technologies. However, uncertainty analysis can help evaluate the hazard and exposure scenarios associated with emerging energy technologies, in this case synthetic biology-enabled products. As a first step in exploring the potential risks imposed by emerging algal ethanol production, relative and comparative assessment characterizes the potential benefits and unique risks posed by engineered algae relative to the risks of conventional sources. SFRA is one platform to review various implications that a synthetic biology option for biofuels might have, including comparative consideration of technological risks, costs, and benefits. These implications can be compared across ethanol feedstock options to determine which option optimally satisfies the goal of attaining a cheap, renewable, efficient energy source with minimal downside risks to human health and the environment. A synthesis of qualitative and quantitative information will be included to compare conventional and synthetic biology options for fuel ethanol. The information will be divided into four factors:

1. Sustainability
2. Economic implications
3. Environmental implications
4. Novel risk potential of synthetic biology

Traditional risk assessment quantifies the safety of a product or process according to hazard, exposure, and effect data (EPA 2017). This risk assessment approach helps identify scenarios in which products or processes are generally considered safe enough for commercial use. However, traditional quantitative risk assessment does not fully consider and weigh the costs and benefits of technological alternatives, especially those that are just emerging. While the method can deem whether a current technology is safe or unsafe, it bypasses the opportunity to solve problems by considering unique benefits that may result from an emerging technology—particularly as data availability on a newer approach to an old problem may be limited. SFRA utilizes concepts of traditional quantitative risk assessments and considers whether there are potential emerging technologies that can be developed and commercialized for a more optimal outcome. Thus, SFRA includes both existing

quantitative data and potential qualitative information in order to compare technological alternatives.

In this chapter, the development and use of synthetically engineered algal ethanol is considered through the SFRA approach. The possible benefits and impacts of
using synthetic biology to enhance algal ethanol are compared against conventional
and advanced ethanol production processes. A solution-focused assessment has
been applied to multiple factors that will determine the viability and efficacy of
ethanol production alternatives. The analysis presented in this chapter primarily
serves as an initial framework for which to compare ethanol production sources;
future sensitivity and uncertainty analyses should be conducted to refine the risk and
benefit parameters presented here.

A literature search for relevant data was conducted using peer-reviewed articles
and government agency and private sector reports to inform each assessment. The
assessment constructs and measures each ethanol feedstock alternative by the four
factors, defined as:

1. *Sustainability:* determined by land use and resource availability by geography
   and climate
2. *Environmental implications:* determined by greenhouse gas emissions and water
   requirements
3. *Economic implications*: determined by cost per gallon produced and direct and
   indirect employment rates
4. *Novel risk potential of synthetic biology:* includes qualitative information related
   to environmental, legal, and technological risks unique to synthetically engineered algae

Results of Solution-Focused Risk Assessment

Based on data compiled through the literature review, conventional feedstock
options and synthetically engineered algae were holistically compared.

Sustainability Metrics and Data

To determine the sustainability of ethanol feedstock, the available supply of ethanol
resources and land use requirements were assessed. Available supply was determined by where ethanol feedstock resources are geographically located and to what
magnitude. Land use requirements were assessed by the volume of ethanol produced for each feedstock alternative according to liters (L) produced per hectare
(ha) per year (yr). This is a common metric used across the literature to assess land
use; any resources that reported land use data with different units were converted to
liters per hectare per year. High land use requirements contribute to the ecological

footprint of an energy source, resulting in the potential loss of biodiversity and increased erosion (Dias de Oliveria et al. 2005).

Renewable fuel sources are defined by the US Department of Energy as "... combustible liquids derived from grain starch, oil seed, animal fats, or other bio-mass; or produced from a biogas source, including any non-fossilized, decaying, organic matter capable of powering spark ignition machinery" (Alternative Fuels Data Center 2017). These fuel sources are regenerative, unlike depleting sources of coal and petroleum. Sustainability is necessary to consider for emerging energy technologies, as fossil fuels will eventually be scarce and humans will increasingly need to utilize alternative fuel sources. Bioethanol is one possible option.

In the United States, corn feedstock drives the majority of biofuel production. In 2015, corn feedstock led to the production of 14,659 billion gallons of biofuel—a vast majority relative to the 450 billion gallons produced from wheat feedstock and the 1.5 billion gallons produced from sugarcane (Bergtold et al. 2016). However, this production does not come without costs. In the United States in 2016, almost ten million acres of land was used to grow corn. One-third of US domestic corn is used for alcohol for fuel use (Fig. 3), suggesting that over three million acres of US land is used for corn ethanol production (USDA Economic Research Service 2017). Thus, corn ethanol is a land-intensive production process that utilizes land that may

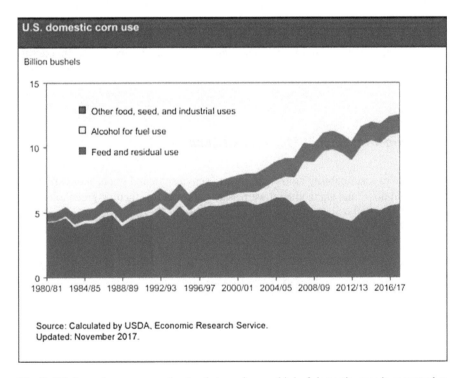

Fig. 3 US domestic corn use estimates that nearly one-third of domestic corn is converted to alcohol for fuel use

otherwise be used for food production. Sugarcane ethanol is primarily produced in Brazil; in 2015, Brazil's sugarcane ethanol industry contributed to 28% of global ethanol production (Renewable Fuels Association 2016a, b). Despite Brazil's high sugarcane ethanol output, only 4.6 million hectares of Brazil's total 851 million hectares of land area is utilized for ethanol production. Thus, only 0.5% of Brazilian land area is needed for ethanol production (UNICA 2016). Additionally, Brazil has undertaken agro-ecological zoning regulations that ensure sugarcane expansion is sensitive to biodiversity and native vegetation (UNICA 2016). Relative to conventional production's land use, it is estimated that only 4% of US land would be needed for algae to replace the energy supply of *all* domestic and imported petroleum used in the United States (Georgianna and Mayfield 2012). The potential benefits of algal ethanol production are further pronounced considering that even if all US corn and soybean (another conventional ethanol feedstock) were dedicated to biofuel production, this would only meet 12% of US gasoline demand and 5% of diesel demand (Hill et al. 2006).

To directly measure land use requirements for each feedstock alternative, a measure of liters of ethanol produced per hectare in a year was included in this comparative review (Table 3). Higher values indicate that greater volumes of ethanol can be produced from a hectare-sized area of land. Corn has the lowest volume output of ethanol per hectare per year at 4600 L produced (Georgianna and Mayfield 2012). Sugarcane has the second lowest output, at 9000 L per hectare (Goldemberg 2008). The advanced production feedstock yields significantly greater ethanol output than conventional feedstock, and synthetically engineered algal ethanol is estimated to produce the highest volume of ethanol per hectare in a year at 93000 to 112,000 liters produced (Waltz 2009).

Environmental Impact Metrics and Data

In addition to sustainability considerations, the environmental impacts of corn, sugarcane, cellulose, and algal ethanol were assessed using two factors: greenhouse gas emission rates and water requirements for production. These metrics were included as they are commonly used in environmental life cycle assessments (Georgianna

Table 3 Liters produced of ethanol per hectare per year for each ethanol feedstock alternative

Ethanol source	Land use (liters per hectare)	Source
Corn	4600	Georgianna and Mayfield (2012)
Sugarcane	9000	Goldemberg (2008)
Cellulosic biomass	1000–2000	Robertson et al. (2017); Sanford et al. (2017)
Microalgae (natural)	36,000–115,000	Georgianna and Mayfield (2012); US Bioenergy Technologies Office (2016)
Cyanobacteria/microalgae (synthetic)	93,000–112,000	Waltz (2009)

and Mayfield 2012). Greenhouse gas (GHG) emissions are necessary to consider because their release into the atmosphere adversely traps heat and subsequently increases the global average temperature. Between 2005 and 2015, ethanol production in the United States increased from 3.9 to 14.9 billion gallons (Renewable Fuels Association 2016c). Given the increasing ethanol production rate, it is critical to asses which ethanol production processes optimally reduce carbon dioxide and other GHG emissions. Additionally, the fuel ethanol industry could have significant impacts on global GHG emissions, as transportation contributes to approximately 29% of total GHG emissions in the United States and 14% of total emissions worldwide (EPA 2018). Greenhouse gas emissions are a critical consideration, as the rate of global GHG emissions increased by approximately 2.7% from 2017 to 2018, reaching 37 billion tons (Global Carbon Project 2018).

In this analysis, GHG emissions for each ethanol source are presented as a rate change relative to petroleum GHG emissions. For each source, a variety of percent changes were presented across the literature. Therefore, when applicable, this data is presented as a range from the lowest reported percent reduction in GHG emissions to the highest reported reduction in GHG emissions.

Table 4 represents the percent reduction rate of GHG emissions relative to petroleum for each ethanol feedstock source. All conventional and advanced feedstock options reduce GHG emissions relative to petroleum. Conventional feedstock options are reported to reduce GHG emissions by a lesser percent than advanced options; specifically, some corn ethanol estimates yield a 21.8% reduction rate, whereas estimates of cellulosic biomass reach 89–94% reduction rates. Microalgae reduce GHGs by about 70%, thereby making them on par or less emission-intensive than other ethanol sources. Data on synthetically engineered algal ethanol is still in development, but assuming synthetic biologists fulfill their aims of engineering more efficient algae that consume CO_2 as a primary food source, synthetic algal ethanol production has the potential to serve as a greenhouse gas mitigation technique.

Additionally, water requirements are assessed for comparison. Water requirements are presented in terms of gallons of water required for each gallon of ethanol produced. The resulting values reflect aggregated water input required for each stage of the production cycle, including harvesting, hydrolysis/liquefaction, fermentation, distillation, and transportation. Like the reported GHG emission rates, a range of gal water/gal ethanol is presented for each feedstock alternative.

Table 4 Percent reduction in greenhouse gas emissions relative to the emission rate of petroleum

Ethanol source	Percent reduction in GHG emissions (relative to petroleum)	Source
Corn	22–76%	Renewable Fuels Association (2016c); EPA (2007)
Sugarcane	56–80%	EPA (2007); Junqueira (2017)
Cellulosic biomass	89 –94%	Schmer et al. (2008); Wang et al. (2011)
Microalgae	69%	Algenol (2017)

On average, 3–15 gallons of water is required to produce 1 gallon of ethanol (Wu et al. 2009). The estimated number of gallons required to production 1 gallon of ethanol for each feedstock option is shown in Table 5. The ranges are quite spread for many of the feedstock options, largely because climatic and environmental conditions influence the amount of water needed for feedstock harvesting and cultivation. For instance, under optimal environmental conditions, corn growth would require less water than it would under suboptimal environmental conditions (i.e., high temperatures). Based on data found through the literature review, conventional ethanol production has higher water requirements than advanced ethanol production methods. Naturally occurring microalgae potentially have the lowest water requirements, as only 0.6 gallons of water are needed to produce a gallon of ethanol (Martín and Grossmann 2013). However, an upper bound suggests that microalgae could require up to 964 gallons of water to produce 1 gallon of ethanol. Like conventional feedstock, the estimate is dependent on growth conditions and specific production processes. Should the goals for synthetically engineered algae be achieved, the water requirements of microalgae could be further reduced, particularly through the closed-feedback growth cycles of photobioreactors. Additionally, algae can be engineered to use and recycle non-potable water, such as saltwater and brackish water. While conventional feedstock requires freshwater for cultivation, algae reduce dependence on freshwater consumption. In the future, the range of water use for each ethanol source should be further analyzed in terms of probability distributions over the range, and Monte Carlo simulations could be used to derive the average and the reasonable ranges of performance for each ethanol source.

Costs and Social Well-Being Implication Metrics

While sustainability and environmental impact assessments are critical to include in the SFRA, socioeconomic implications were assessed to develop the holistic approach to the feedstock comparison. Costs of each feedstock were determined by the cost per gallon of ethanol produced, and social implications focused on job creation and loss. Specifically, costs per gallon produced are calculated using a metric called the gasoline gallon equivalent. The energy density of ethanol is about 60–66%

Table 5 Water requirements for feedstock type based on gallons of water needed to produce 1 gallon of ethanol

Ethanol source	Water use (gal water/gal ethanol)	Source
Corn	1–324 gal	Wu et al. (2009); National Academy of Sciences and National Research Council (2012)
Sugarcane	927–1391 gal	Wu et al. (2009)
Cellulosic biomass (switchgrass)	1.9–9.8 gal	National Academy of Sciences and National Research Council (2012)
Microalgae (natural)	0.6–964 gal	Martín and Grossmann (2013)

that of gasoline, as gasoline yields approximately 34 MJ/L and ethanol yields approximately 18–23 MJ/L (Jolly 2001). Thus, researchers often use the gasoline gallon equivalent to compare the cost of different energy resources, which controls for energy output by volume (EIA 2017a, b). The gasoline gallon equivalent determines the cost per gallon by including feedstock cost, equipment costs, and final product yields. Therefore, it accounts for facility and equipment costs that may impose capital cost restraints for a feedstock option. All prices were adjusted to the 2016 US dollar. It is important to note that these cost estimates do not include any potential subsidies or government-imposed financial incentives.

Table 6 presents the cost per gasoline gallon equivalent for each feedstock type. These prices capture current production costs given energy density relative to gasoline. These costs are not, however, would not necessarily be consumer-facing, as they do not account for regulation or subsidies. The average cost of a gallon of gasoline in 2016 was $2.43 (EIA 2016). The cost per gasoline gallon equivalent of ethanol in 2016, averaging across all ethanol feedstock options, was estimated to be between approximately $2 and $2.50 in the United States (EIA 2017a, b; AFDC 2017). Conventional ethanol feedstock cost estimates are lower than advanced ethanol feedstock, with corn's GGE cost estimated to be less than gasoline itself (USDA 2006). Cellulosic biomass feedstock stands as the least costly advanced ethanol production process, at $2.20 to $5.50 GGE. Microalgae, whether through hydrothermal liquefaction production processes or the current industrial "state-of-the-art" technology, are the most expensive ethanol feedstock. As these cost estimations are accounting for facility and equipment costs, these high costs are likely driven by research and development equipment investments. The industrial state-of-the-art synthetic algal ethanol currently costs about $13–17 to produce per gallon, which is significantly more expensive than conventional feedstock options. This expense may drive consumers away, as they would purchase cheaper ethanol derived from different feedstocks; however, should the other risks (e.g., land use) of algal ethanol outweigh those imposed by conventional feedstock options, government subsidies on algal ethanol could be imposed. Additionally, it has been suggested that algae

Table 6 The cost per gasoline gallon equivalent for each ethanol feedstock type

Energy source	Cost per gasoline gallon equivalent (GGE)	Source
Gasoline	$2.43	EIA (2016)
Ethanol (general)	$1.96–$2.53	EIA, (2017a, b), AFDC, Clean Cities Price Report (2017)
Corn	$1.21–$1.23	USDA (2006)
Sugarcane	$0.95–$2.76	USDA (2006)
Cellulosic biomass	$2.20–$5.49	U.S. DOE (2015); Adusumilli et al. (2013)
Microalgae (hydrothermal liquefaction process)	$2.11–$7.23	Zhu et al. (2013)
Microalgae (industrial state of the art)	$13.35–$17.00	U.S. DOE (2017)

would not be killed during the collection of ethanol, allowing for a continual use of the organism to produce fuel (Georgianna and Mayfield 2012). While it is perhaps too early to be certain, it is likely this ability to continuously reutilize engineered algal cells would contribute to a further decline in cost per gallon yield, as the same cells would produce several harvests of ethanol fuel with only site maintenance and the provision of algal food to keep production going.

In addition to this production cost comparison, economic considerations such as direct and indirect employment rates are valuable assessments to gauge how conventional and advanced ethanol production markets might respond to a disruptive emerging ethanol production technology. Conventional feedstock production processes actively employ thousands of individuals in the United States and over a million in Brazil. Therefore, should an emerging technology, such as synthetically engineered algal ethanol, erupt, these markets could be significantly disrupted. Ethanol production in 2015 led to the employment of nearly 86,000 direct jobs across the United States and added $44 billion to the US gross domestic product (GDP) and $24 billion in household income (Renewable Fuels Association 2016a, b). In addition to direct employment and profit, corn ethanol market further entails indirect economic impacts and employment opportunities. When direct, indirect, induced, construction, agriculture, and R&D jobs supported by ethanol production are included, the number of employment opportunities in the United States was estimated at more than 357,400 jobs in in 2015 (Urbanchuk 2017). It is important to note that this estimate may not capture all the jobs that were already displaced in the shift from corn for food to corn for ethanol.

Brazil produces the majority of global sugarcane ethanol and remains the second leading producer of ethanol worldwide (Renewable Fuels Association 2016a, b). The Brazilian sugarcane industry is comprised of three main sectors: sugarcane cultivation, sugar production, and ethanol production (Moraes et al. 2015). The sugarcane industry as a whole employed approximately 1.2 million workers in 2015 and generated $36 billion USD in gross annual revenue (UNICA 2016). In Brazil, the number of new and closed sugarcane ethanol mills has been steadily decreasing, likely reducing the number of employees hired by the industry. For instance, 430 mills were running in 2010, while only 383 mills existed by 2016. For each year between 2005 and 2011, there were more net sugarcane ethanol mill openings than closures; however, since 2012, there have been more net mill closures than openings (Renewable Fuels Association 2016c). The industry's declination in active sugarcane production mills may be attributed to changes in the Brazilian political and socioeconomic climate (Granco et al. 2015).

Should synthetically engineered algae ethanol technology become more widespread and commercialized, it could swing the fuel economy in both beneficial and disruptive ways. As a net positive, the introduction of synthetic algal biofuels into several nations which currently do not produce significant corn, sugar, or cellulosic ethanol would enable such states to produce their own domestic renewable fuel. This would be particularly advantageous to those states with limited arable land or few crops with a significant positive energy balance score, as algal blooms are able to produce ethanol in a variety of terrains as long as they have access to sunlight, water, and CO_2 (Georgianna and Mayfield 2012; Darzins et al. 2010).

However, countries whose GDP is significantly bolstered by their current ethanol industry may be negatively impacted by the economic disruption of algal biofuel on their existing ethanol production. In Brazil, ethanol production has declined already in part because of the expansion of corn ethanol production in the United States. If synthetically engineered algal ethanol expands in such a robust way that it can be grown on non-arable land, the United States and Brazil may face declining rates of ethanol exportation. Further, the synthetic biology approach will likely limit the number of employees required, as the algae will need less maintenance than corn or sugarcane. Therefore, the number of direct ethanol production jobs will potentially decrease, causing employment rates to drop particularly in the US Midwest and Brazil.

Novel Risk Considerations of Synthetic Biology Approaches

Each of the conventional, advanced, and emerging ethanol feedstock options poses some degree of unique drawbacks. An observed drawback of increased conventional ethanol production includes a corresponding increase in prices of crops used for fuel, which can lead to a rise in food prices locally and globally and diminished food production (Babcock 2012; Inderwildi and King 2009). Additionally, while studies indicate a reduction in CO_2 emissions by corn ethanol in comparison to unleaded gasoline and reductions in CO_2 emissions by sugarcane ethanol, the conversion of fields for crop harvesting contributes to a significant one-time spike in CO_2 that may take decades to balance out with the fuel's reduced CO_2 (Bourne and Clark 2007; Rosenthal 2008).

Synthetically engineered algae may offer distinct benefits over conventional and advanced production processes, such as decreased land and water requirements, increased energy efficiency, and the ability to grow on non-arable land. However, there are unique risks potentially imposed by this emerging technology that are not relevant to conventional ethanol and advanced ethanol feedstock conversion. These novel risks may be present in the production cycle itself or during subsequent interaction with the natural environment. Considerations include how synthetically engineered algae may yield biosecurity and biosafety risks. Biosafety risks largely apply to the concept of horizontal gene transfer, which is defined as the transfer of genes between organisms in a manner other than traditional reproduction. Synthetic biology technologies in particular face this risk as horizontal gene transfer is a common and "somewhat uncontrolled" trait in the microbial biosphere (Cardinale and Arkin 2012). If engineered algae cells transfer synthetic information into the natural world, unanticipated and potentially adverse consequences could result (Cardinale and Arkin 2012; Michalak et al. 2013). Therefore, horizontal gene transfer may instigate risks to the biodiversity of the natural environment that are not yet well characterized. Proper containment of engineered organisms is critical yet difficult as research efforts have focused on mutations at the micro-organismic level. Photobioreactors offer greater containment security than open pond systems but also entail higher capital costs of instalment.

Biosecurity concerns present risks of nefarious agents or bioterrorists harnessing synthetic biology mechanisms and technologies to create biological weapons with devastating consequences (Schmidt et al. 2009; National Research Council 2004). Biosecurity entails concerns of "dual use"—where synthetic biology technologies designed to benefit humans and the environment are deliberately misused for human or environmental harm. As synthetically engineered algae present an opportunity as an energy resource, risks to domestic energy security may be imposed should the technology be misused.

Additionally, as engineered algae are in its research and development phase, it is not yet possible to ensure that researchers will be able to engineer algae in an entirely predictable, consistent, and controlled manner. Off-target gene editing may occur, resulting in synthetically engineered algae that do not yield ethanol as desired. Substantial genetic modifications of cells may impose adverse consequences to humans and the natural environment (Mukunda et al. 2009; Moe-Behrens et al. 2014). To overcome similar research and development challenges associated with genetically modified algae used to produce algal ethanol, Henley et al. (2013) considered a range of impacts that genetically modified algae could have in the natural ecosystem. By listing conceivable risks associated with genetically modified algae as well as non-genetically modified algae, they were able to quantitatively and qualitatively compare natural and modified algae across a variety of hypothetical ecological, economic, and health-related risks. Henley et al. (2013) recommend that risk assessment protocols must first develop open mesocosm experiments for testing, prior to mass cultivation (Seager et al. 2017). Additionally, testing protocols should be adapted to the potential site of mass cultivation of genetically modified algae, which should be marked with detectable genetic markers. We recommend that synthetically engineered algal ethanol risk protocol uses similar testing protocol that is sensitive to local environments and ecosystems.Finally, the synthetic biology ethanol industry faces internal technical risks. Even if the algae are synthetically engineered to provide optimal benefits with minimal associated risks, commercial success is not guaranteed. For engineered algal ethanol to outsource conventional production processes, the technology will need to be massively scaled up. This will require large amounts of time and money for further research and infrastructure development (Connor and Atsumi 2010).

Discussion

Pursuing ethanol as a renewable alternative (or complement) to petroleum has demonstrated environmental benefits, such as reduced greenhouse gas emissions, and can lead countries with limited oil reserves toward oil independence. Synthetic biology offers opportunities to enhance ethanol production in such a way that it bypasses some of the current limitations facing conventional and advanced production processes. Synthetic biologists are engineering algae to achieve a more efficient, renewable fuel source as an alternative to diminishing fossil fuels. Specific to synthetically engineered algae, development and containment uncertainties may lead to biosecurity and biosafety concerns, such as unintended mutations, horizontal gene transfer, and negative human and environmental health consequences. Synthetically engineered algal ethanol also entails high capital costs of investment and may disrupt conventional ethanol production processes that US and Brazilian economies benefit from.

Beyond the hazard, exposure, and effect assessment set forth by traditional risk assessments, a solution-focused risk assessment was used to compare synthetically engineered algal ethanol to conventional and advanced ethanol feedstock options. SFRA methods provide a holistic and comparative assessment as to which ethanol feedstock pursuits offer the greatest benefits and reduced risks. Thus, a solution-focused risk assessment compared each feedstock option across four factors: sustainability, environmental implications, economic implications, and novel risk potential for synthetic biology. The SFRA method builds off traditional risk assessment in that it encompasses both quantitative data and qualitative information related to the safety and net benefits of multiple products or processes that each fulfills the same human need or want.

Based upon an SFRA of engineered algal ethanol against conventional and advanced ethanol production alternatives, there are potenital benefits of continuing research and development on engineering algae to increase the global renewable energy supply. Synthetically engineered algae are demonstrated to be less land intensive than other feedstock options (Table 3), and they allow for more net energy production in a vast array of environments and climates, as the algae harvesting and cultivation take place in controlled laboratories. The controlled growth process makes algae robust and capable of growing on non-arable land, which is beneficial for countries without domestic oil reserves or land capable of growing corn or sugarcane. Additionally, synthetically engineered algal ethanol yields higher volumes of ethanol per land area as opposed to conventional ethanol feedstock options, which are land intensive. Synthetically engineered algal ethanol is also being designed to have a higher net energy balance than other feedstock options, particularly in that algae are being engineered to produce ethanol without dying. Therefore, less energy will be expended into the harvesting and cultivation phases of the production cycle. According to quantitative estimations of GHG emissions and water requirements, synthetically engineered algal ethanol seems to outperform other conventional and advanced ethanol feedstock options. Thus, the associated benefits of engineered algal ethanol exemplify progress toward a sustainable, efficient renewable energy source.

The economic and environmental implications of these bioethanol feedstock options are sensitive to specific geographic locations of production and to the technologies used. To further refine the estimates presented here, mathematical analyses can be used to quantitatively compare bioethanol feedstock options and production technologies. In prior research, probabilistic analyses have been used to simultaneously compare multiple objectives associated with bioethanol production (Kostin 2013; Amigun et al. 2011). Kostin et al. (2012) assessed Argentina's sugarcane ethanol industry by developing a decision support tool for strategic supply chain management, taking into account both economic and environmental parameter constraints and uncertainties. Three mathematical models were used (deterministic, stochastic, multi-objective) for optimal industry planning and design. This sort of quantitative analysis could be applied to other countries and compare a variety of feedstock options, as was performed in this SFRA. An extension of this SFRA that incorporates stochastic models could handle levels of uncertainty in product demand, economic implications, and environmental implications that would better reflect the sensitivities and uncertainties of particular geographies and economies. Similarly, Amigun and Gorgens (2011) conducted a quantitative risk and cost assessment of advanced bioethanol production in South Africa using a stochastic Monte Carlo analysis. Monte Carlo analysis was used to quantify economic risk outcomes across three production technologies under a range of economic parameters. Both the mathematical programming and Monte Carlo approaches to sensitivity and uncertainty analyses of bioethanol feedstocks could include the economic and environmental benefits and risks presented here. For instance, the land use measures presented here are largely dependent on local environments and geography. To assess a similar problem, Tenerelli and Carver (2012) used multi-objective and uncertainty analyses for agro-energy spatial modeling to assess the land capabilities of various perennial crops used for energy. Their model served to assess the potential of different topographies and provided a range of these potentials for energy crop conversions. An uncertainty analysis was performed that simulated the influence of input data and model parameters (Tenerelli and Carver 2012). A similar method and simulation as applied to bioethanol feedstock options would aid in making more accurate risk reduction calculations than the general ranges provided here.

Future research that merges SFRA with quantitative sensitivity analyses will help identify decision thresholds specific to different geographies and economies for which particular feedstock options may have net risk reductions relative to other bioethanol feedstock options and fossil fuels. Prior research on bioethanol feedstock comparisons has shown that the net environmental impact of ethanol fuel depends on the structures of individual production processes, whose predicted outcomes are heavily influenced by the parameterized calculations used (Börjesson 2009). Therefore, to further develop this SFRA, a sensitivity analysis of the four factors (sustainability, environmental implications, economic implications, and novel risk potential for synthetic biology) will help determine optimal place- and economy-specific feedstock options.

While these sensitivity analyses are useful for known risks and benefits, the potential novel risks associated with synthetically engineered algae must also be considered. Emerging technologies bear the brunt of uncertainty and complexity, making it difficult for developers or risk analysts to quantify the risks associated with a technology that has not yet experienced commercialization. Synthetic biology involves various uncertainties regarding the likelihood and magnitude of adverse effects. Despite the potential benefits that synthetic biology products may offer relative to conventional technologies, the novel risks and uncertainties may slow regulation, thereby limiting development and market diffusion (Trump 2017). An adaptive approach to regulation can help governments adjust policies and regulations in an iterative manner as more information is acquired on genetically engineered algae (Greer and Trump 2019).

A more specific approach to quantifying specific risks associated with synthetic biology products is outlined in Trump et al. (2018). This approach could be applied to synthetically engineered algal ethanol and potentially serve to reduce some of the uncertainties and close some of the gaps in knowledge that currently exist. Under this framework, it is first necessary to identify each potential hazard associated with the engineered algal ethanol while understanding that some hazards may be unpredictable. Then, it would be necessary to pair each hazard with its individual risk characterization, which would be independently calculated. For the risk characterizations of each hazard, it is recommended that prior research is used to draw boundaries on plausible values of exposure effects; in this case, parameters might include the proliferation rate of the synthetically derived algae (relative to the proliferation rate of naturally occurring algae). Then, explicit experimental procedures can allow for measuring these parameters where the risk outcome (e.g., loss of containment) is sensitive to the parameter. The engineered algae used to produce ethanol can be tested in a freshwater source, such as a contained water source that is similar to the natural environment. The interaction of the algae with the natural environment will help estimate the magnitude and severity of the unique risks posed by the algae. The environmental risk can be further studied by sensitivity analyses that simulate the engineered algae breaking the contained testing area to potentially more sensitive, natural environments with greater biodiversity.

This approach to considering novel risks associated with algae has been similarly evaluated by Henley et al. (2013) in their consideration of the potential ecological, economic, and health-related implications of genetically modified algae. They focus particularly on the risk of horizontal gene transfer but predict that most traits introduced into genetically modified algae are not likely to hold a comparative advantage to naturally occurring algae, which would result in a low ecological risk. Henley et al. (2013) outline all possible risks associated with genetically modified algae—a very similar approach to that proposed by Trump et al. (2018). Henley et al. (2013) propose that coupling continual monitoring of genetic and mechanical containment strategies with novel cultivation techniques (e.g., matching genetically modified algal traits to unnatural conditions) will help reduce risks. Thus, through monitoring and mesocosm experimentation in contained areas, it is possible to get a sense of how genetically modified and synthetically engineered algae would interact with the natural environment. Despite the potential risks, continued and controlled experimentation is necessary to determine whether the benefits posited by synthetically engineered algae truly outweigh the expected and unknown associated risks.

While studies of this sort continue, especially within private corporations pursing synthetically engineered ethanol production mechanisms, future researchers and developers in this space should carefully consider how to prioritize and categorize hazards. For instance, if the algae had the potential for horizontal gene transfer with humans that would affect human body chemistry, this type of risk should be mitigated before synthetically derived algal ethanol is aggressively pursued. In assessing and managing potential risks such as horizontal gene transfer, a Bayesian approach to uncertain biogeography's and species distribution could be used (Landis et al. 2013). Under this approach, Markov chain Monte Carlo analyses are used over possible biogeographies, which allows the parameters of a biographic model to be estimated and compared (Landis et al. 2013). Specifically, this Bayes approach uses collected data to estimate the joint posterior probability of parameters to develop realistic biogeographic models (Landis et al. 2013). This approach could help estimate the proliferation and propagation of synthetically derived algae from data that have already been collected on the organism.

Based on the present SFRA approach and future integrations with more quantitative risk analyses, synthetically engineered algal ethanol may be a viable renewable energy resource that could offset fossil fuel consumption and make it possible for more countries to establish energy independence. Future research on ethanol production should continue to compare both the risks and benefits of the spectrum of different ethanol feedstock options. Solution-focused risk assessment offers a platform to make this comparison holistic and consider the impact that emerging synthetic biology technologies will have on conventional energy production and on the externalities accompanying it.

Acknowledgments Partial funding for this project was generously provided by a grant from the Alfred P. Sloan Foundation.

References

Adusumilli, N. C., Rister, M. E., Lacewell, R. D., Lee, T., & Blumenthal, J. (2013). Mitigating externalities related to land use change for biomass production for energy in the Tres-Palacios river watershed of Texas.

Agler, M. T., Garcia, M. L., Lee, E. S., Schlicher, M., & Angenent, L. T. (2008). Thermophilic anaerobic digestion to increase the net energy balance of corn grain ethanol. *Environmental Science & Technology, 42*(17), 6723–6729.

Ajjawi, I., Verruto, J., Aqui, M., Soriaga, L. B., Coppersmith, J., Kwok, K., et al. (2017). Lipid production in Nannochloropsis gaditana is doubled by decreasing expression of a single transcriptional regulator. *Nature Biotechnology, 35*(7), 647.

Algenol. (2017). Algenol integrated pilot-scale biorefinery: January 29,1010 – July 1, 2015 public version final report. Available online at: https://www.osti.gov/servlets/purl/1360777

Alper, H., & Stephanopoulos, G. (2009). Engineering for biofuels: Exploiting innate microbial capacity or importing biosynthetic potential? *Nature Reviews Microbiology, 7*(10), 715–723.

Alternative Fuels Data Center. (Retrieved 2017). Ethanol production and distribution. US Department of Energy. Retrieved from: https://www.afdc.energy.gov/fuels/ethanol_production.html

Amigun, B., Petrie, D., & Görgens, J. (2011). Economic risk assessment of advanced process technologies for bioethanol production in South Africa: Monte Carlo analysis. *Renewable Energy, 36*(11), 3178–3186.

Atsumi, S., Hanai, T., & Liao, J. C. (2008). Non-fermentative pathways for synthesis of branched-chain higher alcohols as biofuels. *Nature, 451*(7174), 86–89.

Babcock, B. A. (2012). The impact of US biofuel policies on agricultural price levels and volatility. *China Agricultural Economic Review, 4*(4), 407–426.

Bastos, M. B. (2007). Brazil's ethanol program-an insider's view. *Energy Tribune, 20.*

Bergtold, J. S., Sant'Anna, A. C., Miller, N., Ramsey, S., & Fewell, J. E. (2016). Water Scarcity and Conservation Along the Biofuel Supply Chain in the United States: From Farm to Refinery. In *Competition for Water Resources: Experiences and Management Approaches in the US and Europe* (p. 124). Amsterdam: Elsevier.

Bioenergy Technologies Office. (2016). National algal biofuels technology review. Department of Energy. Retrieved from www.energy.gov

Börjesson, P. (2009). Good or bad bioethanol from a greenhouse gas perspective–what determines this? *Applied Energy, 86*(5), 589–594.

Borowitzka, M. A. (2013). Energy from microalgae: A short history. *Algae for Biofuels and Energy, 5,* 1–15.

Bourne, J. K., & Clark, R. (2007). Biofuels: Boon or Boondoggle? Producing fuel from corn and other crops could be good for the planet-if only the process didn't take a significant environmental toll. New breakthroughs could make a difference. *National Geographic, 212*(4), 38.

Brentner, L. B., Eckelman, M. J., & Zimmerman, J. B. (2011). Combinatorial life cycle assessment to inform process design of industrial production of algal biodiesel. *Environmental science & technology, 45*(16), 7060–7067.

British Petroleum. (2016). BP statistical review for world energy June 2016. Available online at: http://oilproduction.net/files/especial-BP/bp-statistical-review-of-world-energy-2016-full-report.pdf

Burlew, J. S. (1953). *Algae culture: From laboratory to pilot plant* (pp. 1–357). Washington, DC: Carnegie Institution of Washington.

Campbell, J. E., et al. (2008). The global potential of bioenergy on abandoned agriculture lands. *Environmental Science & Technology, 42*(15), 5791–5794.

Cardinale, S., & Arkin, A. P. (2012). Contextualizing context for synthetic biology–identifying causes of failure of synthetic biological systems. *Biotechnology Journal, 7*(7), 856–866.

Chisti, Y. (2007). Biodiesel from microalgae. *Biotechnology Advances, 25*(3), 294–306.

Connor, M. R., & Atsumi, S. (2010). Synthetic biology guides biofuel production. *BioMed Research International, 2010.*

Da Rosa, A. V. (2012). Fundamentals of renewable energy processes. Academic Press.

de Saussure, T. (1807). Mémoire sur la composition de l'alcohol et de l'éther sulfurique. *Journal de physique, de chimie, d'histoire naturelle et des arts, 64,* 316–354.

Darzins, A., Pienkos, P., & Edye, L. (2010). Current status and potential for algal biofuels production. *A report to IEA Bioenergy Task, 39*(13), 403–412.

Demirbaş, A. (2002). Diesel fuel from vegetable oil via transesterification and soap pyrolysis. *Energy Sources, 24*(9), 835–841.

Dias De Oliveira, M. E., Vaughan, B. E., & Rykiel, E. J. (2005). Ethanol as fuel: energy, carbon dioxide balances, and ecological footprint. *AIBS Bulletin, 55*(7), 593–602.

DiPardo, J. (2000). Outlook for biomass ethanol production and demand. Available online at: http://www.ethanol-gec.org/information/briefing/6.pdf. Accessed July 2007.

Ellis, J. T., Hengge, N. N., Sims, R. C., & Miller, C. D. (2012). Acetone, butanol, and ethanol production from wastewater algae. *Bioresource technology, 111,* 491–495.

Environmental Protection Agency. (2007). *Greenhouse gas impacts of expanded renewable and alternative fuels use.* Washington, DC: United States Environmental Protection Agency. EPA420-F-07-035.

Finkel, A. M. (2011). Solution-focused risk assessment: A proposal for the fusion of environmental analysis and action. *Human and Ecological Risk Assessment, 17*(4), 754–787. (and 5 concurrent responses/commentaries, pp. 788–812).

Finkel, A. M., Trump, B. D., Bowman, D., & Maynard, A. (2018). A "solution-focused" comparative risk assessment of conventional and synthetic biology approaches to control mosquitoes carrying the dengue fever virus. *Environment Systems and Decisions, 38*(2), 177–197.

Flugge, M., J. Lewandrowski, J. Rosenfeld, C. Boland, T. Hendrickson, K. Jaglo, S. Kolansky, K. Moffroid, M. Riley-Gilbert, and D. Pape. (2017). A life-cycle analysis of the greenhouse gas emissions of corn-based ethanol. Report prepared by ICF under USDA Contract No. AG-3142-D-16-0243.

Fujita, Y., Takahashi, S., Ueda, M., Tanaka, A., Okada, H., Morikawa, Y., et al. (2002). Direct and efficient production of ethanol from cellulosic material with a yeast strain displaying cellulolytic enzymes. *Applied and Environmental Microbiology, 68*(10), 5136–5141.

Georgianna, D. R., & Mayfield, S. P. (2012). Exploiting diversity and synthetic biology for the production of algal biofuels. *Nature, 488*(7411), 329–335.

Gimpel, J. A., Specht, E. A., Georgianna, D. R., & Mayfield, S. P. (2013). Advances in microalgae engineering and synthetic biology applications for biofuel production. *Current Opinion in Chemical Biology, 17*(3), 489–495.

Global Carbon Project. (2018). Global Carbon Budget. Retrieved from https://www.globalcarbonproject.org/carbonbudget

Goldemberg, J. (2008). The Brazilian biofuels industry. *Biotechnology for Biofuels, 1*(6), 4096. https://doi.org/10.1186/1754-6834-1-6.

Granco, G., Sant'Anna, A. C., Bergtold, J. S., & Caldas, M. M. (2015). Ethanol plant location decision in the Brazilian cerrado. *State of the Art on Energy Developments, 11*, 31.

Greer, S. L., & Trump, B. (2019). Regulation and regime: the comparative politics of adaptive regulation in synthetic biology. *Policy Sciences*, 1–20.

Harder, R., & von Witsch, H. (1942). Bericht über Versuche zur Fettsynthese mittels autotropher Mikroorganismen. *Forschungsdienst Sonderheft, 16*, 270–275.

Henley, W. J., Litaker, R. W., Novoveská, L., Duke, C. S., Quemada, H. D., & Sayre, R. T. (2013). Initial risk assessment of genetically modified (GM) microalgae for commodity-scale biofuel cultivation. *Algal Research, 2*(1), 66–77.

Hill, J., Nelson, E., Tilman, D., Polasky, S., & Tiffany, D. (2006). Environmental, economic, and energetic costs and benefits of biodiesel and ethanol biofuels. *Proceedings of the National Academy of Sciences, 103*(30), 11206–11210.

Inderwildi, O. R., & King, D. A. (2009). Quo vadis biofuels? *Energy and Environmental Science, 2*(4), 343–346.

International Energy Agency. (2018). United States – 2018 update. Available online at: https://www.ieabioenergy.com/wp-content/uploads/2018/10/CountryReport2018_UnitedStates_final.pdf

Jolly, L. (2001). The commercial viability of fuel ethanol from sugar cane. *International Sugar Journal, 103*(1227), 117–143.

Junqueira, R. D. (2017). Ideologia de gênero": a gênese de uma categoria política reacionária–ou: a promoção dos direitos humanos se tornou uma "ameaça à família natural". Debates contemporâneos sobre educação para a sexualidade, 25-52.

Kostin, A. M., Guillén-Gosálbez, G., Mele, F. D., Bagajewicz, M. J., & Jiménez, L. (2012). Design and planning of infrastructures for bioethanol and sugar production under demand uncertainty. *Chemical Engineering Research and Design, 90*(3), 359–376.

Kovarik, W. (2008). *Ethanol's first century*. In XVI International Symposium on Alcohol Fuels: Radford University.

Kumar, S., Singh, N., & Prasad, R. (2010). Anhydrous ethanol: A renewable source of energy. *Renewable and Sustainable Energy Reviews, 14*(7), 1830–1844.

Landis, M. J., Matzke, N. J., Moore, B. R., & Huelsenbeck, J. P. (2013). Bayesian analysis of biogeography when the number of areas is large. *Systematic Biology, 62*(6), 789–804.

Linkov, I., Trump, B. D., Anklam, E., Berube, D., Boisseasu, P., Cummings, C., et al. (2018). Comparative, collaborative, and integrative risk governance for emerging technologies. *Environment Systems and Decisions, 38*(2), 170–176.

Lovins A. B. (2005). *Winning the oil endgame*, p. 105.

Lynd, L. R., Cushman, J. H., Nichols, R. J., & Wyman, C. E. (1991). Fuel ethanol from cellulosic biomass. *Science, 251*(4999), 1318–1323.

Macedo, I. D. C. (1998). Greenhouse gas emissions and energy balances in bio-ethanol production and utilization in Brazil (1996). *Biomass and Bioenergy, 14*(1), 77–82.

Malloy, T., Trump, B. D., & Linkov, I. (2016). Risk-based and prevention-based governance for emerging materials. *Environmental Science and Technology, 50*, 6822–6824.

Martín, M., & Grossmann, I. E. (2013). Optimal engineered algae composition for the integrated simultaneous production of bioethanol and biodiesel. *AIChE Journal, 59*(8), 2872–2883.

Mata, T. M., Martins, A. A., & Caetano, N. S. (2010). Microalgae for biodiesel production and other applications: A review. *Renewable and Sustainable Energy Reviews, 14*(1), 217–232.

McLaughlin, W., Conrad, A., Rister, M. E., Lacewell, R. D., Falconer, L. L., Blumenthal, J. M., et al. (2011). The economic and financial implications of supplying a bioenergy conversion facility with cellulosic biomass feedstocks (No. 1371-2016-108871).

Michalak, A. M., Anderson, E. J., Beletsky, D., Boland, S., Bosch, N. S., Bridgeman, T. B., et al. (2013). Record-setting algal bloom in Lake Erie caused by agricultural and meteorological trends consistent with expected future conditions. *Proceedings of the National Academy of Sciences, 110*, 6448–6452.

Moraes, M. A. F. D., Oliveira, F. C. R., & Diaz-Chavez, R. A. (2015). Socio-economic impacts of Brazilian sugarcane industry. *Environmental Development, 16*, 31–43.

Mukunda, G., Oye, K. A., & Mohr, S. C. (2009). What rough beast? Synthetic biology, uncertainty, and the future of biosecurity. *Politics and the Life Sciences, 28*(2), 2–26.

Mumm, R. H., Goldsmith, P. D., Rausch, K. D., & Stein, H. H. (2014). Land usage attributed to corn ethanol production in the United States: Sensitivity to technological advances in corn grain yield, ethanol conversion, and co-product utilization. *Biotechnology for Biofuels, 7*(1), 61.

Murphy, J. D., & Power, N. M. (2008). How can we improve the energy balance of ethanol production from wheat? *Fuel, 87*(10), 1799–1806.

National Academy of Sciences and National Research Council. (2012). *Biosecurity challenges of the global expansion of high-containment biological laboratories: Summary of a workshop.* Washington, DC: The National Academies Press. https://doi.org/10.17226/13315.

National Research Council. (2004). *Biotechnology Research in an Age of Terrorism*. Washington, DC: The National Academies Press. https://doi.org/10.17226/10827.

Peralta-Yahya, P. P., Zhang, F., Del Cardayre, S. B., & Keasling, J. D. (2012). Microbial engineering for the production of advanced biofuels. *Nature, 488*(7411), 320–328.

Pienkos, P. T., & Darzins, A. (2009). The promise and challenges of microalgal-derived biofuels. *Biofuels, Bioproducts and Biorefining, 3*(4), 431.

Pimentel, D., & Patzek, T. W. (2005). Ethanol production using corn, switchgrass, and wood; biodiesel production using soybean and sunflower. *Natural Resources Research, 14*(1), 65–76.

Pollack, A. (2011). *US approves corn modified for ethanol*. The New York Times (February 11, B1).

Rakkiyappan, P., Shekinah, D. E., Gopalasundaram, P., Mathew, M. D., & Asokan, S. (2009). Post-harvest deterioration of sugarcane with special reference to quality loss. *Sugar Tech, 11*(2), 167–170.

Renewable Fuels Association. (2011). Global ethanol production to reach 88.7 billion litres in 2011.

Renewable Fuels Association. (2012). *Accelerating Industry Innovation – 2012 Ethanol Industry Outlook*. Washington, DC: Renewable Fuels Association.

Renewable Fuels Association. (2016a). *Fueling a high octane future*. Retrieved from: https://www.ethanolrfa.org/wp-content/uploads/2016/02/Ethanol-Industry-Outlook-2016.pdf

Renewable Fuels Association. (2016b). Re-examining corn ethanol's energy balance ratio. Retrieved from: http://www.ethanolrfa.org/wp-content/uploads/2016/03/Re-examining-Corn-Ethanols-Energy-Balance.pdf

Renewable Fuels Association. (2016c). Global ethanol production. Available online at: https://afdc.energy.gov/data/10331

Roach, J. (July 18, 2005). 9,000-year-old beer re-created from Chinese recipe. National Geographic News.

Robertson, G. P., et al. (2017). Cellulosic biofuel contributions to a sustainable energy future: Choices and outcomes. *Science, 356*, 1349.

Rosenthal, E. (2008). Biofuels deemed a greenhouse threat. *New York Times, 8.*

Rouquerol, J., Avnir, D., Fairbridge, C. W., Everett, D. H., Haynes, J. M., Pernicone, N., et al. (1994). Recommendations for the characterization of porous solids (Technical Report). *Pure and Applied Chemistry, 66*(8), 1739–1758.

Sanford, G. R., Oates, L. G., Roley, S. S., Duncan, D. S., Jackson, R. D., Robertson, G. P., & Thelen, K. D. (2017). Biomass production a stronger driver of cellulosic ethanol yield than biomass quality. *Agronomy Journal, 109*(5), 1911-1922.

Schmer, M. R., Vogel, K. P., Mitchell, R. B., & Perrin, R. K. (2008). Net energy of cellulosic ethanol from switchgrass. *Proceedings of the National Academy of Sciences, 105*(2), 464–469.

Schmidt, M., Ganguli-Mitra, A., Torgersen, H., Kelle, A., Deplazes, A., & Biller-Andorno, N. (2009). A priority paper for the societal and ethical aspects of synthetic biology. *Systems and synthetic biology, 3*(1-4), 3.

Seager, T. P., Trump, B. D., Poinsatte-Jones, K., & Linkov, I. (2017). Why life cycle assessment does not work for synthetic biology. *Environmental Science & Technology, 51*, 5861–5862.

Segall, S. D., & Artz, W. E. (2007). The Brazilian experience with biofuels. *Lipid Technology, 19*(1), 12-15.

Shahrukh, H., Oyedun, A. O., Kumar, A., Ghiasi, B., Kumar, L., & Sokhansanj, S. (2016). Comparative net energy ratio analysis of pellet produced from steam pretreated biomass from agricultural residues and energy crops. *Biomass and Bioenergy, 90*, 50–59.

Shapouri, H., Duffield, J. A., & Wang, M. (2002). *The energy balance of corn ethanol: an update.* Washington, DC: EERE Publication and Product Library.

Sheehan, J., Dunahay, T., Benemann, J., & Roessler, P. (1998). *A look back at the U.S. Department of Energy's Aquatic Species Program – biodiesel from algae.* Golden: National Renewable Energy Laboratory.

Shen, J., & Luo, W. (2011). Effects of monosulfuron on growth, photosynthesis, and nitrogenase activity of three nitrogen-fixing cyanobacteria. *Archives of Environmental Contamination and Toxicology, 60*(1), 34-43.

Solomon, B. D., Barnes, J. R., & Halvorsen, K. E. (2007). Grain and cellulosic ethanol: History, economics, and energy policy. *Biomass and Bioenergy, 31*(6), 416–425.

Stephanopoulos, G. (2007). Challenges in engineering microbes for biofuels production. *Science, 315*(5813), 801–804.

Tenerelli, P., & Carver, S. (2012). Multi-criteria, multi-objective and uncertainty analysis for agro-energy spatial modelling. *Applied Geography, 32*(2), 724-736.

Trump, B. D. (2017). Synthetic biology regulation and governance: Lessons from TAPIC for the United States, European Union, and Singapore. *Health Policy, 121*(11), 1139–1146.

Trump, B. D., Foran, C., Rycroft, T., Wood, M. D., Bandolin, N., Cains, M., et al. (2018). Development of community of practice to support quantitative risk assessment for synthetic biology products: Contaminant bioremediation and invasive carp control as cases. *Environment Systems and Decisions, 38*(4), 517–527.

Trump, B. D., Cegan, J., Wells, E., Poinsatte-Jones, K., Rycroft, T., Warner, C., et al. (2019). Co-evolution of physical and social sciences in synthetic biology. *Critical Reviews in Biotechnology, 39*(3), 351–365.

Tyner, W. E. (2008). The US ethanol and biofuels boom: Its origins, current status, and future prospects. *Bioscience, 58*(7), 646–653.

Urbanchuk, J. (2017). Contribution of the ethanol industry to the economy of the United States in 2016. Available online at: https://ethanolrfa.org/wp-content/uploads/2017/02/Ethanol-Economic-Impact-for-2016.pdf

U.S. Department of Agriculture. (2006). *Economic feasibility of ethanol production from sugar in the United States.* Archived from the original on 2007-08-15.

U.S. Department of Agriculture. Economic Research Service. (2017). US bioenergy statistics. Retrieved from: https://www.ers.usda.gov/data-products/us-bioenergy-statistics/

U.S. Department of Energy. (2007). Biofuels: bringing biological solutions to energy challenges. Available online at: https://genomicscience.energy.gov/pubs/Biofuels_Flyer_2007-2.pdf

U.S. Department of Energy. (2015). 2015 renewable energy data book. Energy Efficiency and Renewable Energy. Available online at: https://www.nrel.gov/docs/fy17osti/66591.pdf

U.S. Department of Energy. (2017). *Clean cities alternative fuel price report*. Golden: Energy Efficiency and Renewable Energy. Retrieved from www.afcd.energy.gov.

U.S. Energy Information Administration. (2016). US fuel ethanol production continues to grow in 2017. Retrieved from: https://www.eia.gov/todayinenergy/detail.php?id=32152

U.S. Energy Information Administration. (2017a). EIA projects 28% increase in world energy use by 2040. US EIA, Today in Energy. Retrieved from: https://www.eia.gov/todayinenergy/detail.php?id=32912

U.S. Energy Information Administration. (2017b). "Ethanol". Alternative fuels data center. Retrieved from: https://www.afdc.energy.gov/fuels/ethanol.html

U.S. Energy Information Administration. (2017c). International energy outlook 2017. Available online at: https://www.eia.gov/outlooks/ieo/pdf/0484(2017).pdf

U.S. Environmental Protection Agency. (2007). EPA finalizes regulations for a renewable fuel standard (RFS) program for 2007 and beyond. Office of Transportation and Air Quality.

U.S. Environmental Protection Agency. (2017). Global greenhouse gas emissions data. Retrieved from: https://www.epa.gov/ghgemissions/global-greenhouse-gas-emissions-data

U.S. Environmental Protection Agency. (2018). Inventory of U.S. greenhouse gas emissions and sinks. Available online at: https://www.epa.gov/ghgemissions/inventory-us-greenhouse-gas-emissions-and-sinks

UNICA. (2016). *Sugarcane: One plant, many solutions. Sugar, ethanol, bioelectricity, and beyond*. Retrieved from: http://sugarcane.org/resource-library/books/Folder%20and%20Brochure.pdf

Valdes, C. (2011). Can Brazil meet the world's growing need for ethanol? Available online at: https://www.ers.usda.gov/amber-waves/2011/december/can-brazil-meet-the-world-s-growing-need-for-ethanol/

Van Devender, K. (2011). *Grain drying concepts and options*.

Von Blottnitz, H., & Curran, M. A. (2007). A review of assessments conducted on bio-ethanol as a transportation fuel from a net energy, greenhouse gas, and environmental life cycle perspective. *Journal of cleaner production, 15*(7), 607–619.

Waltz, E. (2009). Cellulosic ethanol stimulus. *Nature Biotechnology, 27*(4), 304–304.

Wang, G. S., Yu, M. H., & Zhu, J. Y. (2011). Sulfite Pretreatment (SPORL) for Robust Enzymatic Saccharification of Corn Stalks. *Advanced Materials Research, 236*, 173–177. Trans Tech Publications.

Wu, M., Mintz, M., Wang, M., & Arora, S. (2009). *Consumptive water use in the production of ethanol and petroleum gasoline (No. ANL/ESD/09–1)*. Argonne: National Laboratory (ANL).

Zhu, M., Lü, F., Hao, L. P., He, P. J., & Shao, L. M. (2009). Regulating the hydrolysis of organic wastes by micro-aeration and effluent recirculation. *Waste Management, 29*(7), 2042–2050.

Zhu, Y., Albrecht, K. O., Elliott, D. C., Hallen, R. T., & Jones, S. B. (2013). Development of hydrothermal liquefaction and upgrading technologies for lipid-extracted algae conversion to liquid fuels. *Algal Research, 2*(4), 455–464.

An Initial Framework
for the Environmental Risk Assessment
of Synthetic Biology-Derived Organisms
with a Focus on Gene Drives

Wayne G. Landis, Ethan A. Brown, and Steven Eikenbary

Introduction

The goal of this chapter is to present a path to estimate risk due to synthetic biology being released into the environment. Our examples are for organisms released with gene drives specifically designed to alter the fitness of specific populations that either transmit disease or are nonindigenous and pose a hazard to the ecosystem services of a specific ecological structure. We apply the structure of source-stressor-habitat-effect-impact pathway derived from the relative risk model (Landis and Wiegers 2005) and as was demonstrated to be applicable in the National Academy of Sciences, Engineering and Medicine (NASEM) 2016 report *Gene Drives on the Horizon*. This relative risk model is now calculated employing Bayesian networks and has been applied to forestry management (Ayre and Landis 2012), infectious disease (Ayre et al. 2014), invasive species (Herring et al. 2015), contaminated sites (Landis et al. 2017a; Johns et al. 2017), and watershed management (Hines and Landis 2014; Graham et al. 2019).

The following sections outline the process for developing a conceptual model for the estimation of risk due to a gene drive-carrying organism and the requirements for transitioning it to a Bayesian network. The network incorporates both deterministic and stochastic components into a computational model describing causality in order to estimate the probabilities of specific outcomes. Two case studies are used as examples, (1) release of gene drive mosquitos to reduce the population carrying disease and (2) the reduction of a mouse population altering the structure of an island ecosystem.

W. G. Landis (✉) · E. A. Brown · S. Eikenbary
Institute of Environmental Toxicology, Huxley College of the Environment,
Western Washington University, Bellingham, WA, USA
e-mail: Wayne.Landis@wwu.edu

© Springer Nature Switzerland AG 2020   257
B. D. Trump et al. (eds.), *Synthetic Biology 2020: Frontiers in Risk Analysis and Governance*, Risk, Systems and Decisions, https://doi.org/10.1007/978-3-030-27264-7_11

Background

Definition and Description of Risk and Uncertainty

The term *risk* has a number of definitions used in everyday language. Often "risk" is used to describe a hazard due to chemical or activity. Examples are "smoking as a risk of cancer" or "a risk of chemical x exposure is developing a rash." Risk also can be used to denote probability as in "the risk of rain this afternoon is 50 percent." Neither of these uses captures the use of the term in a risk assessment context.

In order to be precise about the use of "risk" in evaluating the existence of gene drives in the environment, the NASEM committee defined risk as:

> The *probability* of an effect on one or more specific endpoints due to a specific stressor or stressors. (NASEM 2016)

In other words, how often does (as a probability/frequency) a specific change or changes in the environment will affect something of value to society, the endpoint. Examples of endpoints are the human health, outdoor recreation, survival of an endangered species, and preservation of water quality among many others. Now to describe the calculation of risk in more detail.

The calculation of risk results in a probability distribution usually using a combination of data and a model that describes the causal interactions between the stressors, the environment, and the endpoints. Although a simple statement, the reality is challenging.

Usually, there is incomplete knowledge of the properties of a stressor, its fate in the environment, the exposure pathways to the endpoint, and the relationship of the change of the endpoint to the amount of the stressor. Even with well-studied interactions, there is natural variability or a certain lack of ability to measure the variables. The lack of knowledge, intrinsic variability, and inability to measure certain variables contributes to what is termed as *uncertainty*. In the case of risk assessment, uncertainty is not an emotion but a quantitative assessment of knowledge, variability, and measurement error that describes a probability distribution reflecting the predictive limitation of the assessment. Perhaps the best description of uncertainty is that it reflects what is known and how well it is known and how it affects the final assessment of risk. Uncertainty is not a feeling.

Advantages of Ecological Risk Assessment

There are a number of specific advantages to using risk assessment for informing a decision.

- Assess risk though a specific and quantitative process.
- Trace cause-and-effect pathways. The conceptual model, the computational backbone of the calculation, can describe what is known about the causal

interactions in the assessment. For example, the environmental factors controlling insect populations and their inheritance of the gene drive can be explicitly described in the model.

- Incorporate concerns of the relevant publics. Building the conceptual model requires that endpoints are determined not just from regulation but also from various groups of stakeholders from industry, government, non-governmental agencies, and the general public. This is especially critical in areas with a diversity of cultures, economies, and values.
- Identify sources of uncertainty. It will be clear what the sources of variability and error are likely to be in the final assessment. These assessments also aid to develop the next steps for research programs.
- Compare benefits and harms. Often there are trade-offs between managing different endpoints. The improvement of one desired outcome may adversely affect another. These interactions are difficult to evaluate without a clear framework.
- Compare alternative management strategies. Especially in an adaptive management framework, it will be possible to include different types of strategies to reduce risk and improve the efficacy of the approach.
- Inform research and public policy decisions. A clear conceptual model and the resultant output are a great tools for describing the interactions among the variables, the calculations provided, the probabilities of different outcomes, and the alternative management strategies, and finally the known unknowns can be listed.

Risk Assessment Adopted as a Tool for Decision-Making

Given the applicability of risk assessment as a decision-making approach for synthetic biology and specifically risk, the NASEM (2016) recommended its application. The report also had two clear conclusions.

The first was that there is enough knowledge to apply the methodologies of risk assessment to the problem of gene drives in the environment.

> There is currently sufficient knowledge to begin constructing ecological risk assessments for some potential gene-drive modified organisms, including mosquitos and mice. In some other cases it may be possible to extrapolate from research and risk analyses of other modified organisms and non-indigenous species.
>
> However, laboratory studies and confined field tests (or studies that mimic field tests) represent the best approaches to reduce uncertainty in an ecological risk assessment, and are likely to be of greatest use to risk assessors.

The second conclusion was that current guidance of risk assessment as decision-making approach is not sufficient to describe a robust process for gene drive technologies.

> In the United States, the relevant guidelines and technical documents are not yet sufficient on their own to guide ecological risk assessment of gene drive technologies, because they focus predominantly on evaluating the risks to populations or ecosystems posed by toxic

chemicals, and do not yet adequately address the assessment of multiple stressors and end-points or cumulative risk.

The lack of guidance from the U.S. federal government applicable to ecological risk assessment for the gene drive research community is a critical gap. Relevant U.S. guidelines and technical documents are not yet sufficient on their own to guide ecological risk assessment for gene drive technology.

The remainder of this chapter describes how to adapt risk assessment to the issues of risk assessment for gene drive approaches.

Adaptation of Risk Assessment to Gene Drives and Synbio

An outline of how to adapt risk assessment to informing decisions regarding the use of gene drive was introduced in Chapter 6 of the NASEM (2016) report. What follows is a brief summary with updates to include the demonstrated utility of the Bayesian network and the applicability of adaptive management. The final section of the chapter presents the two case studies.

Risk Assessment for Gene Drives

Ecological risk assessment has focused on chemical contaminants. In the case of gene drives, the agent replicates and has the potential to spread among the population or hybridize with closely related species, and its application is designed to alter the presence of a disease or pest species. As in NASEM (2016), we use the general cause-effect pathway as developed by Landis (2004 for nonindigenous species) (Fig. 1). There are five components in the framework. Each of these is in reference to a spatial component as determined by a digital map of the region under consideration.

Source(s)

This is the location of the various stressors and how they are released into the environment. For the introduction of the gene drive organism, there may be multiple release points at different times of the year or under certain conditions. Also included are the sources of stressors that may alter the survival of the gene drive organism or

Fig. 1 Illustration of the conceptual model framework for the relative risk model

affect the endpoints under management. Perhaps an insecticide is released into the environment to control the population that is also toxic to the gene drive carrier. Rodenticides applied to the environment to control a rat may also be toxic to the gene drive-carrying mouse.

Stressor(s)

In this formulation, the gene drive-carrying organism is the stressor, but not the only one in the environment that affects the endpoints. It is critical to include those confounding stressors that can affect disease prevalence or the size of the host population. If indirect effects to community structure are endpoints, then the key stressors determining the regional species composition also need to be incorporated.

Habitat/Location

Habitat/location describes the types of environment and their locations where the interactions between the stressors and the species or other types of environmental factors occur. These factors create a probability of exposure of the environmental stressors to the key environmental receptors. It is important to describe the key habitat types for the species or environmental characteristic that are considered as endpoints.

Effect(s)

Two broad categories of effects can be evaluated. The first is to the target population and includes an increase in the frequency of the gene drive, a change in population size, a change in population dynamics, a change in life history (including age structure), and an alteration in fitness of the individual and the population. The changes to the target population also can affect the prevalence of the disease. The second category of effects is non-target and off-target effects. These kinds of effects include the spread of the gene drive to non-target species, alterations in predator-prey dynamics, and changes in community structure including alterations to food webs and spatial interactions.

Impact(s)

The effects lead to alterations in the endpoints which represent value to society as determined by the stakeholders. In the case of gene drives released to the environment, there are a number of proposed endpoints. These endpoints can include reduction in human or animal disease, reduction in a nonindigenous species, reduction of a pest species, survivorship of threatened or endangered species, and ultimately ecosystem services. The endpoints represent the values of the stakeholders for that particular issue.

Bayesian Networks

The conceptual model is built to describe cause-effect relationships. The connections between the different components should describe interactions for which evidence is convincing. It is tempting to use associations, such as correlational statistics, and assume causation, but that adds model uncertainty to the estimate of risk.

Because of their flexibility, Bayesian networks (BNs) for estimating risk have a number of important advantages. BNs are acyclic graphs with the interactions between nodes determined by a conditional probability table. BNs are inherently probabilistic and can incorporate different types of data, and the sensitivity analysis is straightforward. The use of probability distributions and conditional probability tables to describe interactions makes uncertainty transparent. The structure of a conceptual model is similar in structure. However, in a classic BN, the interactions between nodes can be assumed to be either associative or causal in nature. In a BN used to describe a conceptual model, the interactions are designed to describe cause and effect as currently understood in the system. So, a BN employed in a risk assessment is built to describe causality.

The output from population models can be built into the BN. The various combinations of inputs to an age-structured population model and the outputs can be used to build conditional probability tables describing the probability of a reduction in population growth given a decrease in survivorship of immature organisms. Mitchell et al. (2018) demonstrated the feasibility of such an approach examining the effects of pesticides and water quality parameters on the future of Chinook salmon populations in the Yakima River, a tributary of the Columbia.

Graham et al. (2019) have used a BN to describe changes in community structure due to alternation in nutrient inputs. Community structure was measured by probing samples of estuaries for environmental DNA corresponding to specific types of organisms.

Adaptive Management

Landis et al. (2017b) has demonstrated that risk assessments incorporating BNs can complement a framework for adaptive management. The flexibility of the approach, the ability to easily update the model, and the transparency of the calculation are strong advantages. Different scenarios can be modeled, the probability of different outcomes compared, and monitoring programs designed incorporating the different ranges of outcomes.

Case Studies

Two case studies are used to illustrate conceptual models, the initial step in building a computational framework. Both are a part of ongoing projects to build Bayesian networks suitable for estimating risk. The descriptions follow the basic source-stressor-habitat-effect-impact structure.

Puerto Rico and Mosquito-Borne Disease

The case study on control of incidence of Zika and dengue fever in Puerto Rico includes the deployment of two different gene drive-modified mosquitos. There are two mosquito species in Puerto Rico that are vectors for the Zika virus and dengue fever: *Aedes aegypti* and *Aedes albopictus*. The *Aedes aegypti* mosquito is the primary vector for dengue fever, whereas both species of mosquito are responsible for transmission of Zika (CDC 2017, 2019; Matysiak and Roess 2017). The goal is to control and reduce mosquito populations in order to decrease Zika and dengue incidence, morbidity, and mortality. The partial cause-effect pathway shown in Fig. 2 describes causal linkages of factors affecting the use of gene drive-modified *Aedes aegypti* and *Aedes albopictus* to control disease in Puerto Rico. The model will be developed into a Bayesian network that will show probabilistic outcomes and is based heavily on the work done by Landis and Wiegers (2005), Ayre and Landis (2012), and Ayre et al. (2014) and is outlined in the National Academies of Sciences, Engineering, and Medicine (2016).

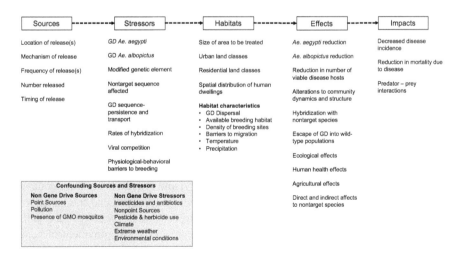

Fig. 2 Example of conceptual model for the introduction of gene drive mosquitos for the control of disease

Source

The source node will be composed of the spatial extent and conditions surrounding the initial release of the gene drive-modified organism. It is necessary for both the timing and location of release to be during mosquito breeding season and in areas where mosquito densities are high (NASEM 2016).

Stressor

The gene drive-modified *Aedes* spp. mosquitos are the stressors in this case study. The mosquitoes contribute the specific genetic sequence of the gene drive and the non-target sequences that may be affected. It is important to note how these changes may influence the life history of the gene drive-modified mosquitos. How the gene drive persists and is transported through the population must also be considered. The rates of hybridization of the gene drive-modified organisms with the wild-type population will be necessary to understand the persistence and spread of the gene drive within the target population. Viral competition within the host mosquito may influence specific rates of viral transmission (NASEM 2016).

Habitat

Habitat considers where the wild-type population mosquitos exist as well as where they breed and forage. The land classes and use within the study area will influence the amount of suitable habitat available (e.g., rural, urban, forested, etc.). The spatial distribution and overall patchiness of the environment will influence the density of breeding sites, the potential barriers to migration, and ultimately the dispersal of the gene drive. Other aspects that may relate to breeding habitat are temperature and precipitation. The socioeconomic status of the region may influence availability of suitable breeding habitat (NASEM 2016; Matysiak and Roess 2017).

Effects

The desired effects of the use and deployment of gene drive-modified *Aedes aegypti* and *Aedes albopictus* mosquitos would reduce populations of each. Reducing the number of mosquitos will reduce the number of viable female hosts for Zika and dengue fever. Hybridization with non-target species may occur, and the ability of the gene drive to escape into wild-type populations, causing further direct and indirect effects, would need to be accounted for. The goal of the deployment of gene drive-modified *Aedes* spp. is to affect human health endpoints, but other agricultural and ecological effects are likely (NASEM 2016).

Impacts

If successful, the deployment of gene drive-modified mosquitos as a vector for disease control would impact the incidence of Zika and dengue fever transmission and ultimately reduce human morbidity and mortality caused by these diseases. The reduction of mosquito populations will alter the ecological landscape in some form. This may be due to loss of a prey item for predators in the area (NASEM 2016).

Island Mouse Populations

Invasive rodent species represent a potential threat to island biodiversity. Traditionally, chemical rodenticides have been implemented to control and eliminate invasive rodent populations on islands; however, first-generation warfarin-based rodenticides are becoming less effective as mice and rats adapt to their smell, and second-generation anticoagulant rodenticides such as brodifacoum might be of concern to regulators due to their increased toxicity and persistence over their first-generation predecessors.

Utilizing gene drive technology to eliminate invasive island house mice (*Mus musculus*) is one of the case studies mentioned in *Gene Drives on the Horizon* (NASEM 2016). The proposed genetic element, called the "*t-sry* construct," is made up of two components: the *Sry* gene and the *t-haplotype*. The *Sry* gene causes female mice to develop as males and be sterile, and the *t-haplotype* is supposed to cause the *Sry* component to transfer to at least 90% of the offspring of male mice (Backus and Gross 2016). Hypothetically, gene drive-modified *Mus musculus* carrying the *t-sry* construct could be released into an island system and spread this gene through the population, eradicating themselves after a number of generations.

For this model (Fig. 3), the Farallon Islands were used as the risk region. The Farallones are a set of small islands west of the Golden Gate that serve as a national wildlife refuge and are home to various seabird, seal, bat, shark, and plant species. The islands have also become occupied with large quantities of nonindigenous house mice (*M. musculus*) (USFWS).

To model risks of the decision to release transgenic *Mus musculus* into the Farallones, a conceptual model (Fig. 3) was built showing the sources of stressors, stressors, habitats of relevant organisms, potential effects, and potential assessment endpoints.

Since the Farallones contain a national wildlife refuge and are closed to the public, the listed endpoints consisted solely of abundance of a select number of species that might be considered valuable from a conservation standpoint. The species consisted of the rhinoceros auklet, a seabird that has been extirpated from California and has only returned to the Farallones since rabbits were eradicated from the islands; Cassin's auklet, another seabird that is listed by the US Fish and Wildlife Service as a Bird of Special Concern; Brand't cormorant, a bird that is very dependent on the land mass of the Farallones due to their lack of waterproof feathers; and

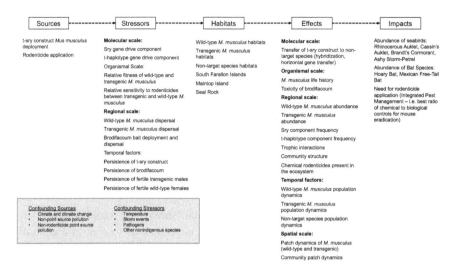

Fig. 3 Preliminary conceptual model for the introduction of gene drive-modified *M. musculus* to control mouse populations and protect native species

the ashy storm petrel, which is a candidate for listing under the Endangered Species Act (USFWS). Two other species were also listed as endpoints in the model: the hoary bat and the Mexican free-tailed bat.

Stressors in the model included the genetic element, the relative fitness and toxicant sensitivities of the wild-type and transgenic mice, the persistence of the element within the mice population, and the brodifacoum, which is one of the more commonly used second-generation rodenticides and has historically been used in island rodent eradication attempts (Thomas and Taylor 2002). Both stressors and effects were differentiated between spatial and temporal scales. Confounding sources and stressors were listed as assumptions.

Comparison to Other Approaches

Trump et al. (2018) have suggested an approach based on the classic approach as delineated by US EPA (1998). The framework included hazard identification, exposure assessment, a dose-response assessment, and a risk characterization step (Trump et al. 2018). In this regard, the proposed model is the same framework as outlined by US EPA but without the interaction with risk managers and decision-makers. In contrast to this approach, one of the conclusions of NASEM (2016) was that no conventional framework was sufficient as a method to estimate risk due to synthetic biology and introduced gene drives.

Instead NASEM (2016) proposed the relative risk methodology adopted in this report as a starting point. While certainly based on the risk assessment literature, it has taken a different approach being focused on establishing clear causal links, using ranks to combine multiple sources-stressors-habitats-effects and endpoints, and more recently using Bayesian networks as the computational framework.

As demonstrated in this chapter, we have created source-stressor-habitat-effect frameworks for two scenarios. The next step is to build the appropriate sets of causal pathways and transition the pathways into Bayesian networks to facilitate the computation.

Next Steps

An important next step is the development of a clearer understanding of the population genetics of gene drives for a variety of traits, in a variety of species, and in a variety of patch dynamics. Noble et al. (2018) suggest that the drives will be highly invasive. Conversely Unckless et al. (2017) have demonstrated that resistance to gene drives is likely to evolve. Given the variety of life history strategies, patch dynamics, species hybridization, and confounding factors such as the use of rodenticide and pesticides, further study is required.

Synopsis

This short chapter has outlined an approach to the conduct of an ecological risk assessment for gene drive organisms released to the environment. The approach inherently builds on causal interactions and is adaptable to a number of scenarios. Two case studies (rodent elimination and insect-transmitted disease) are summarized. A comparison to other approaches is presented as well as a short summary regarding next steps.

Bibliography

Ayre, K. K., & Landis, W. G. (2012). A Bayesian approach to landscape ecological risk assessment applied to the Upper Grande Ronde Watershed, Oregon. *Human and Ecological Risk Assessment., 18*, 946–970.

Ayre, K. K., Caldwell, C. A., Stinson, J., & Landis, W. G. (2014). Analysis of regional scale risk to whirling disease in populations of Colorado and Rio Grande cutthroat trout using a Bayesian belief network model. *Risk Analysis, 34*, 1589–1605.

Backus, G. A., & Gross, K. (2016). Genetic engineering to eradicate invasive mice on islands: Modeling the efficiency and ecological impacts. *Ecosphere, 7*(12), 1–14.

CDC. Centers for Disease Control and Prevention. (2017). About Zika. Retrieved from https://www.cdc.gov/zika/about/index.html

CDC. Centers for Disease Control and Prevention. (2019). Dengue. Retrieved from https://www.cdc.gov/dengue/entomologyecology/index.html

Graham, S. E., Chariton, A. A., & Landis, W. G. (2019). Using Bayesian networks to predict risk to estuary water quality and patterns of benthic environmental DNA in Queensland. *Integrated Environmental Assessment and Management, 15*, 93–111.

Herring, C. E., Stinson, J., & Landis, W. G. (2015). Evaluating non-indigenous species management in a Bayesian networks derived relative risk framework for Padilla Bay, Washington. *Integrated Environmental Assessment and Management, 11*, 640–652.

Hines, E. E., & Landis, W. G. (2014). Regional risk assessment of the Puyallup River watershed and the evaluation of low impact development in meeting management goals. *Integrated Environmental Assessment and Management, 10*, 269–278.

Johns, A. F., Graham, S. E., Harris, M. J., Markiewicz, A. J., Stinson, J. M., & Landis, W. G. (2017). Using the Bayesian network relative risk model risk assessment process to evaluate management alternatives for the South River and Upper Shenandoah River, Virginia. *Integrated Environmental Assessment and Management, 13*, 100–114.

Landis, W. G. (2004). Ecological risk assessment conceptual model formulation for nonindigenous species. *Risk Analysis, 24*, 847–858.

Landis, W. G., & Wiegers, J. (2005). Introduction to the regional risk assessment using the relative risk model. In W. G. Landis (Ed.), *Regional scale ecological risk assessment using the relative risk model* (pp. 11–36). Boca Raton: CRC Press.

Landis, W. G., Ayre, K. K., Johns, A. F., Summers, H. M., Stinson, J., Harris, M. J., Herring, C. E., & Markiewicz, A. J. (2017a). The multiple stressor ecological risk assessment for the mercury contaminated South River and Upper Shenandoah River using the Bayesian network-relative risk model. *Integrated Environmental Assessment and Management, 13*, 85–99.

Landis, W. G., Markiewicz, A. J., Ayre, K. K., Johns, A. F., Harris, M. J., Stinson, J. M., & Summers, H. M. (2017b). A general risk-based adaptive management scheme incorporating the Bayesian network Relative Risk Model with the South River, Virginia, as case study. *Integrated Environmental Assessment and Management, 13*, 115–126.

Matysiak, M., & Roess, A. (2017). Interrelationship between climatic, ecological, social, and cultural determinants affecting dengue emergence and transmission in Puerto Rico and their implications for zika response. *Journal of Tropical Medicine, 2017*. https://doi.org/10.1155/2017/8947067.

Mitchell, C., Chu, V. R., Harris, M. J., Landis, W. G., von Stackelberg, K. E., & Stark, J. D. (2018). Using metapopulation models to estimate the effects of pesticides and environmental stressors to Spring Chinook salmon in the Yakima River Basin, WA. https://cedar.wwu.edu/ssec/2018ssec/allsessions/146/. Accessed 11 Feb 2019.

National Academies of Sciences, Engineering, and Medicine (NASEM). (2016). *Gene drives on the horizon: Advancing science, navigating uncertainty, and aligning research with public values*. Washington, DC: National Academies Press.

Noble, C., Adlam, B., Church, G. M., Esvelt, K. M., & Nowak, M. A. (2018). Current CRISPR gene drive systems are likely to be highly invasive in wild populations. *eLife, 7*, e33423.

Thomas, B. W., & Taylor, R. H. (2002). A history of ground-based rodent eradication techniques developed in New Zealand, 1959-1993. In C. R. Veitch & M. N. Clout (Eds.), *Turning the tide: The eradication of invasive species* (pp. 301–310). Auckland: IUCN.

Trump, B. D., Foran, C., Rycroft, T., Wood, M. D., Bandolin, N., Cains, M., et al. (2018). Development of community of practice to support quantitative risk assessment for synthetic biology products: Contaminant bioremediation and invasive carp control as cases. *Environment Systems and Decisions, 38*(4), 517–527.

Unckless, R. L., Clark, A. G., & Messer, P. W. (2017). Evolution of resistance against CRISPR/Cas9 gene drive. *Genetics, 205*(2), 827–841.

USEPA. (1998). U.S. EPA guidelines for ecological risk assessment. EPA/630/R-95/002F. Published on May 14, 1998, Federal Register 63(93):26846–26924). U. S. Environmental Protection Agency, Washington, DC, USA.

USFWS. United States Fisheries and Wildlife Service. (2019). Retrieved from https://www.fws.gov/refuge/farallon_islands/. 24 Apr 2019.

Biology Without Borders:
Need for Collective Governance?

Todd Kuiken

"College students try to hack a gene – and set a science fair abuzz" (Swetlitz 2016); "Amateurs Are New Fear in Creating Mutant Virus"(Zimmer 2015); "DIY Gene Editing: Someone Is Going to Get Hurt" (Baumgaertner 2018); and "In Attics and Closets, Biohackers Discover Their Inner Frankenstein (Whalen 2009)"—these are the headlines the public reads in major publications like the *New York Times*, *Wall Street Journal*, and others about the increasing accessibility to biotechnologies. Read aloud; they sound like the opening trailers for horror movies. Have there been missteps? Stunts? Individuals that spark controversy? Of course. But pandemics? Environmental disasters? Of course not. What has occurred though, and the story that is rarely told, are the tens of thousands of students and everyday citizens that have been introduced to biology, biotechnology, and science more broadly, who might not otherwise have had the opportunity to explore it. As with any broad reaching loosely affiliated community, there will always be those pushing the boundaries and trying to steal the spotlight with hyperbole and stunts. And with the help of some in the press, have misbranded and misrepresented the entire community of citizens interested in biology. Unfortunately these stories overshadow the educational opportunities this community provides and dismisses the safety, security, ethical, and responsible innovation practices and programs they have established.

For nearly a decade, I have been involved with the International Genetically Engineered Machines (iGEM) competition, Do-It-Yourself Biology, and, more recently, the growing citizen health innovation movements. What I have discovered is that these sometimes separate and sometimes merged communities have been both proactive and adaptive in addressing safety, security, and ethical concerns.

By examining the safety, security, and human practices programs of iGEM (Part 1), the policies and practices the DIYbio community has established (Part 2), and a

T. Kuiken (✉)
North Carolina State University, Genetic Engineering & Society Center, Raleigh, NC, USA
e-mail: tkuiken@ncsu.edu

© Springer Nature Switzerland AG 2020   269
B. D. Trump et al. (eds.), *Synthetic Biology 2020: Frontiers in Risk Analysis and Governance*, Risk, Systems and Decisions, https://doi.org/10.1007/978-3-030-27264-7_12

strategy to enable citizen health innovators to conduct responsible research (Part 3), this chapter will present an argument for collective governance. If funded properly, collective governance could address the biosafety, biosecurity, and ethical concerns brought about by the rapid advances in biological and information technologies that have democratized biology and broken down the traditional mechanisms of governance.

The International Genetically Engineered Machines (iGEM) Competition

The iGEM competition is an annual synthetic biology event where undergraduates, graduate, high school students, and community biotech labs (DIYbio) compete to build genetically engineered systems using standard biological parts called BioBricks. According to the Registry of Standard Biological Parts, which is maintained by the iGEM Foundation, a BioBrick or a biological part "is a sequence of DNA that encodes for a biological function, for example a promoters or protein coding sequences. At its simplest, a basic part is a single functional unit that cannot be divided further into smaller functional units. Basic parts can be assembled together to make longer, more complex composite parts, which in turn can be assembled together to make devices that will operate in living cells" (iGEM 2017). Teams are provided with an initial kit that contains about 1700 parts, and throughout the competition, they create new parts and improve other parts contained in the Registry. All these parts are available for anyone to access, use, and share. There are over 20,000 documented genetic parts in the Registry, and "teams and other researchers are encouraged to submit their own biological parts to the Registry to help this resource stay current and grow year to year" (iGEM 2017).

iGEM began in January 2003 as an independent study course at the Massachusetts Institute of Technology (MIT) where students developed biological devices to make cells blink. This course became a summer competition with 5 teams in 2004 and continued to grow to 13 teams in 2005; it expanded to 340 teams in 2018, reaching 42 countries and over 5000 participants. Since 2004, over 40,000 students have participated in iGEM from across the globe (Fig. 1, iGEM map; Fig. 2, iGEM participation). Team projects have ranged from simple biological circuits to developing solutions to local and global environmental conservation issues.

iGEM places as high a priority on students learning the technical skills of synthetic biology as it does on them understanding and contextualizing how "human practices" (iGEM Competition 2018) will influence the impacts of their technology and how to best plan for potential consequences. Through the human practices component of iGEM, teams are required to study "how your work affects the world, and how the world affects your work" by imagining their projects in a social/environmental context and engaging with communities outside their lab to better understand issues that might influence the design and use of their technologies. Teams creatively engage with issues in ethics, sustainability, inclusion, security, and many other areas.

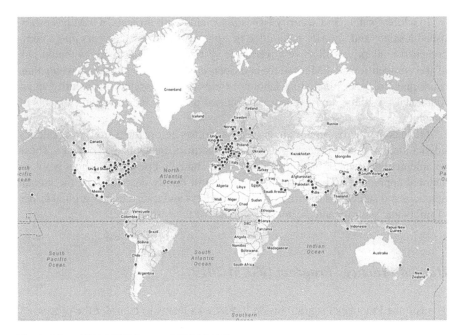

Fig. 1 Map of 2018 iGEM teams (iGEM 2018)

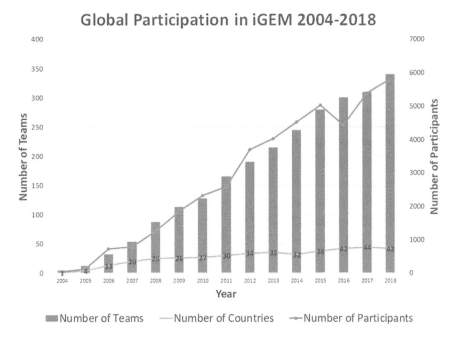

Fig. 2 Global participation in iGEM from 2004 to 2018 (iGEM 2018)

To address safety and security issues associated with projects, iGEM has established a safety and security committee that has evolved over the years into a comprehensive, adaptive collective governance system that manages the potential risk of the competition. iGEM's Safety and Security Program addresses a wider than usual range of issues including laboratory biosafety, laboratory biosecurity, environmental biosafety, dual-use research, animal use, and, increasingly, elements of bioethics.

iGEM comes across important issues, sometimes before there are formal rules or international regulations leading iGEM to sometimes create its own policies. These complement national or institutional rules and can help bridge differences between national approaches. The following sections will briefly describe how the human practices and safety and security programs address and train students around issues of safety and security and the societal impacts of their work.

Human Practices Program (iGEM 2019)

The Human Practices Program inside iGEM asks teams to consider the process of developing solutions to real-world problems in ways that are socially responsible, sustainable, safe, and inclusive. It recognizes that these issues are complex and do not have a single or simple answer. So human practices work requires teams to look beyond the lab. Inviting stakeholder input, building interdisciplinary collaborations, and understanding relevant regulations and codes of conduct in order to examine whether they are developing a responsible and impactful research project. Stakeholders can have different and sometimes conflicting values that can be equally valid. Human practices therefore require teams to think critically, be able to appreciate different views, and co-develop solutions that best serve the concerned communities. By reaching out to and learning from diverse communities, iGEM teams are creating opportunities for broader publics to help shape the practice of synthetic biology (iGEM 2019).

Teams are encouraged to explore whether their projects are both "good" and "responsible":

Responsible:

- How might your team's solution to one problem lead to other problems (e.g., social/political/ecological)? Could your project be misused?
- How can your team anticipate and minimize the impact of these concerns?
- What's your plan to inform and work with relevant authorities or stakeholders of potential risks related to your project?
- How might current policies and regulations apply to your project? Are they sufficient, and if not, how might they be changed?
- How does the iGEM community expect your team to be safe and responsible, both inside and outside of the lab?

Good:

- In what ways might your project benefit society?
- Which communities may be most interested or most affected by your project?
- Which communities may be left out or negatively impacted if your project succeeds?
- How might you get feedback on the viability and desirability of your approach? How will you adapt your project based on this feedback?
- How might your approach compare to alternative solutions to the same or similar problems (including approaches outside of biotechnology)?

*To examine the above questions, teams have (*iGEM 2019*):*

- Interviewed stakeholders who might make use of their work, like farmers, fashion designers, and factory workers
- Conducted environmental impact analyses
- Created museum exhibits and creative public engagement activities
- Written intellectual property guides
- Facilitated "white hat" biosecurity investigations
- Held forums with local legislators
- Spoken at the United Nations
- Developed tools to help other teams examine questions of ethics and responsibility

Through these activities, teams have engaged with topics and issues including ethics, safety, risk assessment, environmental impact, social justice, product design, scale-up and deployment, public policy, law and regulation, and much more. In each case, these activities have helped shaped the goals, execution, and communication of their projects (iGEM 2019).

Human practices work is a requirement of the competition in order for teams to qualify for awards and medals. To qualify for a bronze medal, teams must document how they came up with their idea and what inspired them. To qualify for a silver medal, teams must demonstrate how they have identified and investigated one or more human practices issues in the context of their project. To qualify for gold using their human practices work, teams must expand on their silver medal activities by demonstrating how their investigation of human practices issues has been integrated into the purpose, design, and/or execution of their project. Teams must demonstrate how they have responded to the conversations they have had with people outside the lab; how it influenced the goal, design, and execution of their project; and how they think about their work. Teams must demonstrate that their project (e.g., intended applications and their limits, potential users and stakeholders, experimental design, methods to deliver products and communicate results, etc.) has evolved based on their human practices work (iGEM 2019).

In addition to the medal requirements, teams can compete for two special prizes related to human practices. The Best Integrated Human Practices prize recognizes exceptional work based on the gold medal requirements for human practices.

To qualify for this award, teams must demonstrate how they have considered how their project affects society and how society influences the direction of their project. For example, how might ethical considerations and stakeholder input guide your project purpose and design and the experiments you conduct in the lab? How does this feedback enter into the process of your work all through the iGEM competition? Teams must document a thoughtful and creative approach to explore these questions and show how their project has evolved in the process to compete for this award (iGEM 2019).

Biosafety and Security

[The following section is taken in part from: (Millett et al. 2019)*]*

The Safety and Security Program expects teams to engage on these issues outside of their own community and even with non-specialists and the public. It does this through an approach that combines both incentives (such as through a Safety and Security Award for excellence) and penalties for non-compliance, up to and including disqualification (iGEM Foundation 2018a, 2018f).

The way iGEM addresses safety and security is an adaptive approach that builds on lessons learned each year in the competition. iGEM is a unique platform—offering both opportunities to innovate new tools and approaches but also to act as an international test bed for those developed elsewhere. iGEM believes biosafety and biosecurity are everyone's responsibility and need to be integrated throughout the competition's life cycle. The whole-of-life cycle approach iGEM currently employs requires teams to consider risk issues from the initial project design and continue to think about risks throughout their project, revisiting these issues as their plans change. Teams are also encouraged to think about any risks that might arise if their projects became final products. A separate, yet coordinated, biosafety system relates specifically to iGEM's Registry of Standard Biological Parts. iGEM believes safety and security is everyone's responsibility, from the team members to the instructors to the Safety and Security Program. The program is managed by the iGEM Director of Safety and Security, Piers Millett, and advised by the iGEM Safety Committee. The iGEM Safety Committee is a group of experts (all volunteers) in biosafety, biosecurity, and risk management. Its members come from diverse elements of industry, academia, and government. It includes members from North America, South America, Europe, and Asia. The committee is the ultimate arbiter of decisions on safety and security in iGEM.

iGEM requires teams to think about biosafety and biosecurity issues throughout the competition life cycle. These issues are included in project design and help shape what teams do in the lab and how they transfer the fruits of their work—both the tangible and intangible results.

As part of being responsible scientists and engineers, all iGEM teams are required to identify and manage risks associated with their project. This starts during the project design phase. All teams must share what risks they have identified and the procedures, practices, and other measures they have taken to mitigate them. When thinking about possible risks, teams need to consider potential harm to themselves, their colleagues, communities, and the environment. They are encouraged to think about both "What is being done" and "What is being used."

The competition makes use of a White List which details organisms and parts deemed safe to work with in a standard laboratory (iGEM Foundation 2018g). Teams are encouraged to reduce risks by using safer substitutes for more dangerous organisms/parts. iGEM recognizes that all biological lab work, even simple experiments, carries some risk. To manage these risks, iGEM teams must follow a set of safety and security rules:

- Teams must provide information on any safety and security risks from their project and steps taken to manage them.
- Teams must request permission before using parts and organisms not on the White List.
- The instructor or primary contact must sign off safety and security information provided by the team.
- All deadlines for providing safety and security information must be met.
- Teams must fully comply with the safety and security policies.
- Teams must work in the biosafety level appropriate for their project (and should not be using greater containment than necessary).
- Teams must follow shipment requirements when submitting samples.
- Teams must follow all biosafety and biosecurity rules of their institution and all biosafety and biosecurity laws of their country.
- Teams cannot conduct work with Risk Group 3 or 4 organisms.
- Teams cannot conduct research in a Safety Level 3 or 4 laboratory.
- Teams cannot conduct work with parts from a Risk Group 4 organism.
- Teams cannot release or deploy their project outside of the laboratory (including putting them in people) at any time during the competition.
- All experiments with human subjects (including noninvasive experiments, such as surveys) must comply with all relevant national and institutional rules.

The iGEM Safety and Security Committee has the authority to immediately disqualify any team found to be in non-compliance with these rules. If teams satisfy the Committee that they have modified their project to be in compliance, they may be re-qualified. As disqualification from the competition is the largest penalty iGEM can impose, we have found that this sends a clear message to the teams on the importance of thinking seriously about safety and security in their projects.

Working with Biological Parts

Because they are working with biological parts, teams need to consider the function of each part to determine whether, and how, it can be handled safely. When assessing the hazard posed by parts they want to use, teams need to think about the part's origin, its function, and how it may interact with other parts in their project. Teams are encouraged to avoid the use of dangerous parts and to seek safer alternatives. Even if the individual parts in a project are safe, they may have a dangerous function when combined with other parts or placed in specific systems. Teams are required to think about how their parts will work together. For example, could they imitate the function of a virulence factor? Could they be harmful to humans or the environment in some other way? In order to help teams understand any risks associated with parts developed in the past, iGEM puts "Red Flags" on any part in the Registry that might pose a risk when combined in certain systems with certain other part. iGEM does not accept dangerous parts (such as those that encode toxins). If a team wants to work with any part with a "Red Flag," they require special permission from the Safety and Security Committee.

On a regular basis, a commercial partner screens all parts in iGEM's registry for hazardous potential. The screening process looks at the likely origin of the part (by conducting blast searches against sequence databases) and approximate function using internal databases maintained by the partner firm. Any part that might pose a risk is identified and can result in the part receiving a "Red Flag."

Reviewing Biosafety and Biosecurity Information

All iGEM teams provide details of their risk assessment and how they are managing these risks, via a Safety and Security form. An initial draft of the form is required when most teams begin to move from the planning to laboratory phases of their projects. The form details what they plan to do in their project. They are expected to update their draft whenever their plans change. A final version becomes due as teams wrap up their lab work and begin to focus on how to communicate about their project (iGEM Foundation 2018d).

Whenever a team wants to use an organism or part not on the competition's White List, they have to seek approval from the Safety and Security Committee via a Check-In form. This provides additional details as to what they want to use, how they will obtain it, what they will do with it, what risks this might involve, and how they are managing these risks (iGEM Foundation 2018c).

If a team wants to use vertebrates (e.g., rats, mice, guinea pigs, hamsters), or higher-order invertebrates (e.g., cuttlefish, octopus, squid, lobster), they must seek approval from the Safety and Security Committee via an Animal Use form. This provides a thorough justification of why they want to use the animals based on the three Rs:

- *Replace*—whenever possible alternatives to animal models should be used. Teams must explain why no alternative approaches are possible.
- *Reduce*—if animals are to be used, the fewest possible needed to accomplish the goal of the research should be used. Teams must show they are using the appropriate number of animals to power their study.

- *Refine*—animal research must use methods that minimize or alleviate pain, suffering, or distress and enhance animal welfare. This includes appropriate housing, environment, stimulation, and feeding of animals (iGEM Foundation 2018b).

A second commercial partner screens all the forms provided by teams. They use a network of internationally certified biosafety and biosecurity professionals to review the details provided and highlight potential issues to the Safety and Security Committee.

Issues in Environmental Biosafety

iGEM has a strict no release policy. Projects have to stay inside the lab. Some projects, however, would envisage environmental release should they ever be sufficiently developed. Past examples have included the creation of engineered systems to clean up environmental contaminants or the use of biosensors to detect the presence of compounds of interest. Through their human practices work, teams working on these projects often explore what it might take to get regulatory approvals for such a product. Teams are also required to consider both immediate risks to the environment and potential risks should their project be fully realized.

In 2016, an iGEM team attempted to make a gene drive. They did not make a functional drive but did manage to get some of the components to work. As gene drives do not include any pathogens or parts connected with virulence or transmissibility, they do not appear on common control lists. None of their components was specifically captured by iGEM's safety and security rules and policies at that time. The Safety and Security Committee began working with team, noting that they were eloquent and engaged in considering broader implications of their project but had not anticipated the amount of scrutiny their project would receive (Minnesota 2016). iGEM has taken a number of gene drive-specific steps that have been shared with regulators around the world. They have been fed into a number of national policy development processes. In the months following the 2016 competition, where a team had attempted to develop a gene drive (Minnesota 2016), iGEM constructed the world's first policy on gene drives (iGEM Foundation 2018e). This project was reported in the wider press, noting that (a) international gene drive experts reported project was "not dangerous" and (b) the team had designed in specific safety precautions (Swetlitz 2016). iGEM's policy ensures robust review by requiring that any iGEM team's research on gene drives is dependent on special permission from the Safety and Security Committee. This requires a team to convince the Safety and Security Committee of the following:

- There will be no environmental release.
- The project is safe, based on host organism, parts, and containment measures.
- Best practices in containment developed by leading gene drive researchers are being implemented.

- The planned project has been discussed on a conference call with recognized international experts on gene drives and biosafety and biosecurity.
- Any commercially acquired parts are produced by companies that screen against regulated sequences (i.e., Australia Group List of Human and Animal Pathogens and Toxins for Export Control [The Australia Group 2019]).

Teams have to self-declare their intent to use gene drives—helping to address the challenge of identifying relevant work. A functional description of gene drives (rather than a list of specific parts) was developed to help teams describe the specific functions of the gene drive components they intend to use. Gene drive-specific language and examples were inserted into the White List to embed them into iGEM's routine safety and security activities. A ban on gene drives as parts in the competition Registry also helps to mitigate risks of accidental release.

iGEM has expanded the concept of safety and security by:

- Going beyond traditional agent-based risk assessment.
- Evaluating risk on "a case-by-case basis" as opposed to "in a broad and generic manner."
- Embracing a more whole-of-life cycle approach with the "aim to review the research before it begins and then periodically assess and evaluate the project concerning changes in the research that may present additional elements of importance for risk management."
- Utilizing multiple risk management approaches, including both biological tools, and human solutions.
- Embedding consideration of certain bioethics elements into biological risk assessments and management processes. For example, "What trade-off between the chance of benefit and the risk of harm is justifiable and acceptable and for whom?"
- Involving a wider set of stakeholders, including "scientists, biosafety officers, institutional leadership, and ethics consultants, with the aim of maximizing safety as well as scientific progress" (Lunshof and Birnbaum 2017).
- Human practices have been a core component of the competition, and successful teams universally consider "how their project affects the world and how the world affects their project" (iGEM 2019). More specifically, iGEM's belief that safety and security are the responsibility of all promotes the involvement of the widest possible group of stakeholders. This approach has proven successful in addressing a number of practical, real-world challenges for lab biosafety, environmental biosafety, biosecurity, and bioethics.

Do-It-Yourself Biology

Do-It-Yourself Biology, or DIYbio, is a global movement spreading the use of biotechnology and synthetic biology tools beyond traditional academic and industrial institutions and into other publics (Grushkin et al. 2013). Practitioners include a broad mix of citizen scientists, amateurs, enthusiasts, students, and trained scientists. Some of the practitioners focus their efforts on using the technology and gained knowledge to create art, explore biology, create new companies, or simply tinker. Others believe DIYbio can inspire a generation of bioengineers to discover new medicines, customize crops to feed the world's exploding population, harness microbes to sequester carbon, solve the energy crisis, or even grow our next building materials. The DIYbio movement now represents community labs, individual labs, and group-like incubator spaces spread across the globe (see Fig. 3).

The concept of amateur biotechnologists—what eventually became DIYbio—began to take shape around 2000 after a working draft of the human genome was completed by the Human Genome Project (Grushkin et al. 2013). People began setting up home labs (Carlson 2005), which evolved into dedicated labs in commercial spaces. The organizers pooled resources to buy, or take donations of, equipment and began what have become known as "community labs." The first opened in Brooklyn, NY, USA, in 2010. Community labs sustain themselves on volunteers, membership donations, and paid classes. DIYbio continues to grow rapidly. There are now community laboratories and other types of community incubator spaces spread across six continents (see Fig. 3). They participate in iGEM, provide educational and start-up opportunities and at a more basic level, and have exposed thousands of citizens to biology, biotechnology, and science more broadly who might not otherwise have had the opportunity.

The DIYbio community believes that wider access to the tools of biotechnology, particularly those related to the reading and writing of DNA, has the potential to spur global innovation and promote biology education and literacy that could have

Fig. 3 Map of community biotech labs and community incubator spaces as of 2018, based in part from http://sphere.diybio.org/ and personal communications

far-reaching impacts. These potential innovations raise valid questions about risk, ethics, and environmental release for all scientists, policymakers, and the public (Kuiken 2016). For instance, the Odin, a company that believes "the future is going to be dominated by genetic engineering and consumer genetic design," creates "kits and tools that allow anyone to make unique and usable organisms at home or in a lab or anywhere" (Odin 2018). Some of these kits raise serious environmental and ethical issues regarding animal welfare (Bloomberg 2018), along with societal questions about who should be able to access these technologies.

Efforts by the DIYbio Community to Address Safety, Security, and Ethics

"People overestimate our technological abilities and underestimate our ethics," Jason Bobe, one of the founders of DIYbio.org, told the *New York Times* in 2012 (Zimmer 2015). Safety, security, and ethics have been topics of discussions within the DIYbio community since its formation. In 2011, the Woodrow Wilson Center and DIYbio.org brought together the DIYbio leadership in Europe and United States to establish their own codes of ethics. Debated over the course of a few days, these codes came directly from the community at the time. Both codes are remarkably similar (Fig. 4). While the codes were never meant to be static or adopted by every member of the community, they help strengthen the culture of responsibility burgeoning in DIYbio. At the 2018 Global Community Biosummit (Biosummit 2018), a shared purpose statement was developed to complement these codes (Figs. 5 and 6).

Fig. 4 Graphic representation of the DIYbio codes of conduct workshop, London 2011

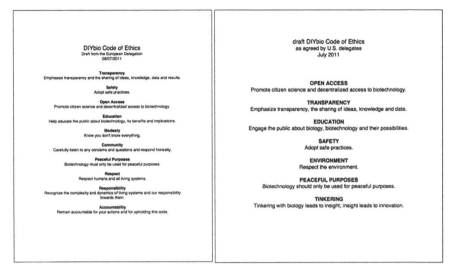

Fig. 5 DIYbio codes of conduct

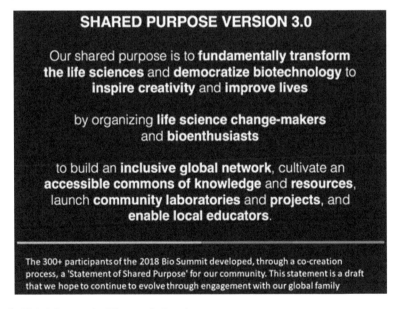

Fig. 6 Global Community Biosummit shared purpose

While codes of conduct can serve to provide a framework for responsible conduct, they are not a substitute for biosafety/security procedures. Over the years, the DIYbio community has developed collective governance mechanisms to address both safety and security. As part of the FBI's Biological Sciences Outreach Program, an agency effort designed to strengthen the relationship between the science and law enforcement communities, FBI representatives and some DIYbio leaders have begun a dialogue about safety and security. These dialogues inform the DIYbio community about the FBI's interests/concerns and inform the FBI agents about the types of work done at community labs, in particular what a DIYbio lab looks like (as opposed to a methamphetamine lab). Over the years, the program has built individual relationships between FBI agents and the DIYbio community. Because of these relationships, lab members have contacts within the FBI in the event of suspicious activity, and agents better understand the community and can respond appropriately to either false alarms or legitimate issues (Grushkin et al. 2013).

While each individual community lab has its own processes and procedures, many DIY community labs have strict rules about lab access and biosafety training programs and procedures in place. At Brooklyn's Genspace, for example, community lab directors evaluate each new member and their project for safety. In cases where the directors do not have the expertise to evaluate a project, they consult with the lab's safety advisory committee made up of university professors and biosafety officers. In the absence of such a committee, DIYbio.org provides the Ask a Biosafety Expert service (DIYbio 2013), where experts and members of the American Biological Safety Association answer safety questions (see Fig. 7). If the

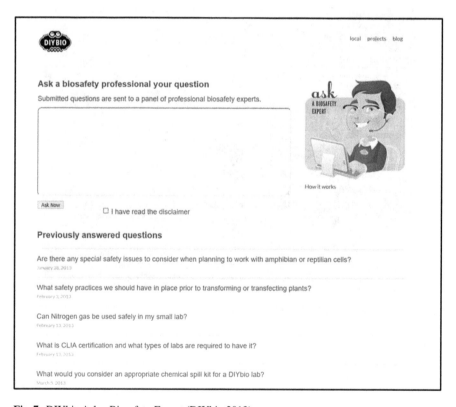

Fig. 7 DIYbio Ask a Biosafety Expert (DIYbio 2013)

potential member or project seems suspicious or unsafe, the project/person may not pass this screen. In addition, directors in most labs approve the reagents and biological materials that are purchased, brought in, and removed from the lab (Grushkin et al. 2013). With a grant from the Open Philanthropy Project (Open Philanthropy Project 2019), new hands-on training programs are being developed for 2019 but will need to be funded in the future to sustain them over the long term.

Taken collectively, these programs demonstrate that the DIYbio community has a responsible, proactive attitude that is well-suited for collective governance (Kuiken 2016).

The Bio-citizen

Taken in part from (Kuiken, Pauwels and Denton 2018; Pauwels and Denton 2018).

Stories of bio-citizens, people operating outside the traditional biomedical research community in order to address health-related issues, have astonished some and empowered others. Similar to the DIYbio movement, access to tools, technology, and information offers the lay public new opportunities to guide the direction of biomedical innovation and enables individuals to generate and mobilize new knowledge. The *Rise of the New Bio-Citizen* workshop (Kuiken et al. 2018) gathered key actors in citizen-driven biomedical innovation and advocacy, democratized biology (community bio-labs), and policy experts. Participants held an open discussion centered on the ethical, safety, and governance issues related to citizen-driven biomedical research. Collectively they discussed codes of conducts, guidelines, and policies that address governance issues identified in the Citizen Health Innovation Report (Pauwels and Denton 2018) and identified barriers and ways to enable increased participation among bio-citizens.

Under the designation "patient-led research" (PLR) or "citizen-driven biomedical research," citizens, patients, and families have increasingly become the leading force in the initiation or conduct of health research projects. Their activities may involve analyses of genomic data for diagnosing rare diseases, identification of potential therapeutic drugs, organization and crowdfunding of clinical trials' cohorts, and even self-surveillance or self-experimentation. Many of the participants in citizen-driven biomedical research are patients and families confronted with a condition that is the subject of their research, therefore facing new epistemic and governance challenges and often testing the ethical and regulatory limits within which health research has traditionally operated.

This new form of research where citizens and patients are the primary producers and mobilizers of knowledge promises to break new ground in underserved health domains. However, it suffers from a lack of legitimacy when it comes to assessing the quality of patients' experiential data. This endeavor also gradually transfers the responsibility of safety and ethics to lay experts, raising new ethical concerns—from blurring boundaries between treatments and self-experimentation, peer pressure to participate in trials, exploitation of vulnerable individuals or third parties (children), to a lack of regulation concerning quality control and risk of harm.

Patients often have in-depth experiential knowledge of their conditions along with a stake in making sure that a treatment or device will be effective, safe, and beneficial. Yet, facing regulatory uncertainty and potential stringency, they might not overcome the "chill factor"—a phenomenon described by citizen scientists and DIY inventors as the fear to confront regulators by sharing the recipe for a new invention.

Perspective from Regulators

Using patient experience data is not unprecedented in drug regulation, as the US Food and Drug Administration (FDA) approved Exondys 51 in September 2016 in part utilizing this type of information. Legislators describe "real-world evidence" (RWE) in Section 3022 of the twenty-first Century Cures Act as any drug performance data which does not come from randomized control trials. This information can originate from "ongoing safety surveillance, observational studies, registries, claims, and patient-centered outcomes research activities" (FDA 2018). Notable examples of RWE include electronic health records, personal health devices and/or apps, billing records, and social media. As defined by twenty-first Century Cures, RWE exclusively applies to drug regulation (potentially including regenerative therapies). This type of data would aim to enhance the generalizability of clinical trial findings (Sherman et al. 2016, p. 2293).

Twenty-First Century Cures directs FDA to create a trial framework for implementing the use of RWE by the end of 2018. This draft framework would use input from the public (e.g., industry, academia, patient groups) and apply only to drugs. FDA will then publish guidance on when RWE will be applicable and how best to collect this data. However, in July 2017, the FDA published draft guidance on utilizing RWE in medical device oversight (FDA 2017), suggesting RWE could become applicable across FDA regulation. RWE may help address issues with current clinical trial designs, which require large patient cohorts and high costs but still lack generalizability (Sherman et al. 2016). However, existing sources of RWE were not designed to aid regulatory decision-making and could present analytical challenges (Sherman et al. 2016). Patient experience data may be able to serve a similar role, but limited literature exists on the potential risks and benefits of using patient experience data in regulatory approval.

Interestingly, "patient experiences and perspectives," which the FDA has been tasked with measuring and analyzing, do not seem to align with citizen-driven biomedical research and patient-led health innovation. Since RWE applies to drug regulation, many of the case studies in this report would not fall under this classification of research because not all citizen-driven biomedical research aims to produce drugs that will require regulatory approval. At best, the definitions of these two terms—RWE and citizen-driven biomedical research—do not align; at worst, the FDA has been tasked with measuring and analyzing only a small subset of patient-led health innovations within the broader scope of citizen-driven health research. Even more recently, in November 2017, the FDA released information about the self-administration of gene therapy (FDA 2017; Smalley 2018). According to that statement:

> [the] FDA is aware that gene therapy products intended for self-administration and "do it yourself" kits to produce gene therapies for self-administration are being made available to the public. The sale of these products is against the law. FDA is concerned about safety risks involved. Consumers are cautioned to make sure that any gene therapy they are considering has either been approved by FDA or is being studied under appropriate regulatory oversight. (Ibid.)

These themes were present throughout the *Rise of the New Bio-Citizen* workshop discussions (Kuiken et al. 2018).

Breaking Barriers to Innovation

A recurring theme throughout the discussions at the workshop, "The Rise of the New Bio-Citizen: Ethics, Legitimacy, and Responsible Governance in Citizen-Driven Biomedical Research and Innovation" (Kuiken et al. 2018) was, broadly speaking, about regulations. What are the regulations that govern the bio-citizen? Should there be regulations that govern the bio-citizen? Are current regulations, or the threat of regulations, preventing or discouraging more bio-citizens from participating in biomedical research? How can the bio-citizen better understand the goals of regulation, and how can the regulator better understand the goals of the bio-citizen?

These questions around governance and regulatory systems require further discussion, but the overall sense from the participants is that regulations would not necessarily be a barrier to innovation. Providing resources for the bio-citizen to gain access to regulators in order to help reinterpret the regulations to fit their unique circumstances could help mitigate the potential for regulations to build barriers.

The diagram (Fig. 8. Context and constraints in bio-citizen spaces) below, which describes the innovation ecosystem, is one example of an accessible resource that may benefit bio-citizens, community bio-labs, and regulators. Community bio-labs have the potential to prototype and experiment in an environment with ongoing risk and safety oversight. In this way, community bio-labs could be a bridge between individual bio-citizens and regulators by serving as a safe space to experiment and test governance systems.

Fig. 8 Context and
constraints in bio-citizen
spaces. (Adapted from
Jeremy de Beer and Jain
(2018))

Informed Consent and Centralized Decentralization

Expanding the health innovation platform to include the bio-citizen raises the issue
of informed consent in a novel way. Participants in the workshop, *Rise of the New
Bio-Citizen* (Kuiken et al. 2018), wrestled with the concept asking questions like:

- Is bringing consent into the governance process too burdensome?
- What are the "right" levels of consent? Are there different levels of consent in
  different situations? If so, where does self-experimentation fall on this spectrum
  of consent?
- How much does one need to know to understand in order to give consent? How
  do we deal with known unknowns?
- How should we deal with incomplete information/knowledge transfer?
- Are the operating and rigid institutional framework of scientific and professional
  values problematic?
- Is the systematic institutionalization of ethical values problematic?
- Could you develop a citizen service provider for informed consent, a centralized
  institutional review board (IRB) that operates via decentralized community labs/
  IRBs to increase access?
- If you are filming and broadcasting everything that you are working on and/or
  doing, are you providing a resource and therefore a need for consent from those
  receiving that information?
- Where does the burden of consent and liability lie?

The discussion around adequate informed consent evolved into a discussion
about institutional review boards (IRB) and how such a system might operate in the
age of the bio-citizen.

- What is the practicality of such a system?
- Are there different levels of approval that should be applied to the bio-citizen?
- Would such a system provide a level of legitimacy for the bio-citizen?
- Do rigid institutional governance frameworks prevent permissionless sandboxes?
- Do permissionless sandboxes hinder the establishment of a social license to operate for bio-citizens?

One idea was whether you could design a "peer-to-peer" IRB system or, more basically, provide access to the expertise and information that preserves the spirit of what a traditional IRB does. A similar type of project was developed around biosafety for the DIYbio community with its Ask a Biosafety Expert web portal (DIYbio 2013). Whether this type of system could work for issues that an IRB handles requires further thought and deliberation. For instance, could community IRBs lead to unconventional or non-traditional studies? Is approving unconventional and non-traditional experiments necessarily a sign of permissionless innovation?

One critical aspect is the liability associated with programs like this. Experience from the Ask a Biosafety Expert program suggests liability insurance is both needed and difficult to acquire without dedicated funding, which bio-citizens do not always have. How might bio-citizens who crowdfund the resources necessary to innovate acquire liability insurance? This type of program would also need some semblance of infrastructure and management in order for it to be useful for the community.

Other ideas that emerged from the discussion around intuitional review boards revolved around developing ethical and safety workshops/curriculums aimed at biocitizens, incubators, and community labs. These were also seen as potential capacity building opportunities for community biology labs and health incubators. The organization, Public Responsibility in Medicine and Research, was presented as a model that could be used. Their stated goals and activities focus on "creating a strong and vibrant community of ethics-minded research administration and oversight personnel, and providing educational and professional development opportunities that give that community the ongoing knowledge, support, and interaction it needs to raise the bar of research administration and oversight above regulatory compliance" (PRIMR 2019).

Ethical Innovation

There was a sense among the participants in the workshop, *Rise of the New Bio-Citizen* (Kuiken et al. 2018), that we need a better understanding of the underlying ethical issues associated with the bio-citizen and creating opportunities for inclusive innovation (de Beer and Jain 2018). Issues such as treatment vs enhancement or self-experimentation vs survival were discussed, and consensus was reached on the need for conceptual clarification. It was felt that we have little understanding on how to extrapolate health innovation "on the individual" to issues affecting society at large, particularly when discussed under the concept of social license to operate. A social license to operate "is an informal agreement that infers ongoing acceptance of…a project by a local community and the stakeholders affected by it" (Gallois et al. 2017). However, while many of the ethical issues focused on the individual, it was suggested that the issue be expanded beyond the individual to include public health, environmental health, and the impact on public science at large. This discussion led us to contemplate issues of power and control. Who gets to control another person's acts; who is the real villain or victim? The person who may engage in self-experimentation, the person who tries one of these innovations, or the person trying to stop any potential harm that might incur? The lines are fuzzy particularly when people, or the individual, think they are helping those who are seeking cures that do not currently exist or that they cannot afford.

One suggestion was for the community to address, or at least better understand, the underlying ethical issues associated with the bio-citizen. This would help to unpack how the regulatory structure affects the bio-citizen and evaluate how these ethical issues can guide what is happening, not stand in the way. It was felt that not meeting these ethical standards could cause others in society to reject what the bio-citizen might be doing and place societal roadblocks to the innovation platform or inclusive innovation.

It was suggested that innovators need to have some friction or speed bumps in the innovation process in order for them to see, or acknowledge, issues that are beyond the technical. Technologists and scientists typically focus on generating a specific kind of knowledge and are not well-suited, in the context of time, education, and influence, to assess and address potential ethical issues. By enabling ethicists, and other biosafety professionals, to work alongside scientists and technologists could provide this friction in order for the innovator to "take a step back" and think about the ethical and biosafety issues their projects raise. This type of reflection is evident in how the human practices and biosafety programs of iGEM operate. Interdependent issues encompassing ethics, social license to operate, and legitimacy were major underlying themes discussed throughout the workshop.

Though a social license to operate has typically been associated with industrial and energy industries (Ibid.), the concept elicits opinions about who/when/if you ask permission and whether acquiring a type of social license to operate establishes legitimacy. The "expression refers to mainly tacit [or, experiential] consent on the part of society toward the activities of business (or in our case the bio-citizen)…it constitutes grounds for the legitimacy of these activities" (Demuijnck and Fasterling 2016). A social license to operate does not necessitate or prevent permissionless innovation; rather, a social license to operate allows community bio-labs and bio-citizens to innovate in safe innovation spaces with ongoing risk and safety oversight. While establishing a social license to operate may help to break barriers to bio-citizen innovation, some questions remain in the social context. For instance, what is the entry point? Is it a social license, a market license, or an ethical or legal license? When do you ask for permission? Whom do you ask?

Finding the narrative story that shows the social good was suggested as a way to address this in part. You have to demonstrate the value of innovation for and by the bio-citizen. However, how do we establish communication between communities in order for them to understand what they are getting in return (particularly when sharing data)? How do we find the incremental value in bio-citizen innovation? How is that value or equity going back to the individual or community at large? Issues of equity and privilege are also important to recognize. For instance, some bio-citizens performing innovations with diseases, and innovations around those diseases, might not have the means to turn that into a business. Or gain access to the results of having participated if, for instance, those results were utilized by a company, resulting in therapies that the individual may not be able to afford.

The bio-citizen and their societies will need to define what a "social license to operate" means to them, particularly in a health context. There was a sense among the participants that we need to collectively shift the urgency toward these issues if we want to build an inclusive and trusted innovation platform for the bio-citizen, in part because our collective trust in institutions is declining. While clinicians are trusted, institutions are not, and there is even lower trust in government. At the same time, some participants in the workshop felt that people/publics may be scared of the bio-citizen and that increasing engagement channels (i.e., DIYbio days at local hospitals) could be an avenue to increase trust among these groups. Having bio-citizens coming in to answer the questions for themselves could help move toward a better understanding of the social good. Permissionless innovation can support experimenting in safe innovation spaces. However, how do we protect human rights in an ecosystem of permissionless innovation?

A Living Bio-citizen Tool Kit

Governance and ethical issues play a role in participatory health research and innovation—even if traditional regulatory approval does not. Traditionally, knowledge legitimacy has been tied to scientific knowledge; but citizen health innovators are beginning to change that paradigm and inject their experiential knowledge into biomedical research. Before bio-citizens can be seen as legitimate health innovators, they will need to gain the trust of other scientists and regulators.

Building off the ideas and discussions throughout the workshop, a living tool kit for future bio-citizens was developed that can evolve as the community of bio-citizens evolves. It provides engagement channels between patients-innovators, crowdfunders, ethicists, and regulators to design adaptive oversight mechanisms that will foster a culture of empowerment and responsibility.

Broadly, this tool kit seeks to address the following questions:

- How can we create a safe space for health innovators and community bio-labs to share and experiment with their data, value trade-offs, and ethical concerns in ongoing conversations with regulators?

- How can regulators and crowdfunding platforms help bio-citizens modernize practices that will give legitimacy to their research, devices, and treatments?
- Instead of trying to fit citizen-driven innovation into the existing regulatory framework, a more adaptive approach might help these citizens become literate in how to conduct research and help them identify the regulatory checkpoints. See the "Rise of the New Bio-citizen Tool Kit."

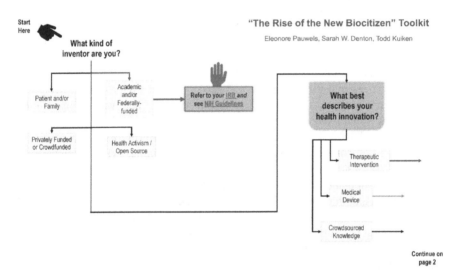

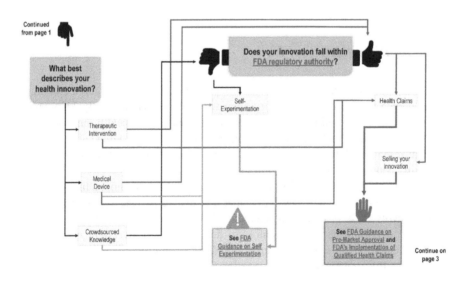

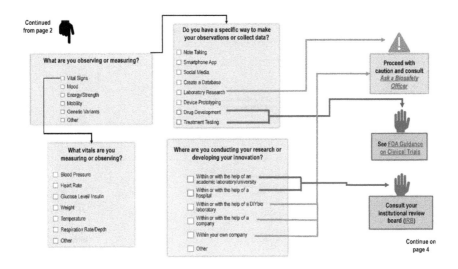

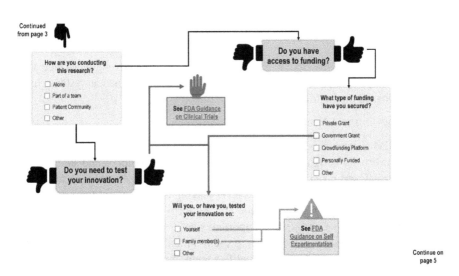

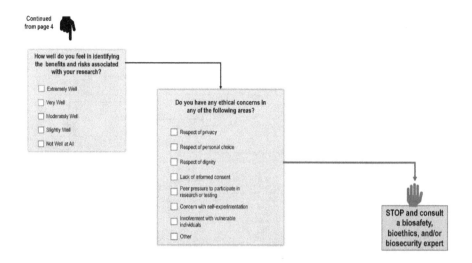

Thinking Forward: Is Collective Governance the Answer?

The examples discussed above represent a type of collective governance that involves multiple parties/stakeholders. These collective governance systems have functioned based on the following two principles:

1. *Direct buy-in from the community*

The iGEM participants, iGEM Foundation, and the larger iGEM community have all collectively bought into the need for the human practices and safety/security programs. These programs have captured most, if not all, country jurisdictional rules and universalized them to create a level playing field for all participants. Since the Safety and Security Committee has the authority to disqualify teams for not complying, they have a unique and important role in this type of collective governance, similar to legal consequences present in individual country regulations. While the goal is never to have to disqualify a team, this mechanism provides incentive for teams to comply with the program. In addition, providing awards for both human practices and safety provides additional incentives for teams to comply.

The DIYbio and citizen health communities have separate but similar reasons for buy-in. They have collectively recognized their responsibility to their own local communities and the larger global community, including the larger DIYbio and citizen health communities. Beyond the codes of conduct, which highlights the ethical and safety guidelines they follow, members of the DIYbio community understand that the actions, or missteps of one, will affect the entire community. In addition, the interactions and relationships built with both regulators and law enforcement from the beginning of the communities' development have provided the atmosphere and opportunities for this type of collective governance, without the typical top-down regulatory systems. The absence of a top-down regulatory system could be based on regulatory authorities' recognition of the responsible conduct the DIYbio and citizen health communities have developed and/or the inability for typical governance structures to realistically govern such a diverse and multinational community. In a sense, regulatory authorities and the DIYbio community need each other's buy-in to acquire a larger "social license" to operate.

2. *Flexibility in adapting to fast-changing technologies and applications*

The iGEM Safety and Security Program displays why flexibility in adapting to fast-changing technologies and applications is crucial for governing these types of communities. iGEM has the ability to adjust its rules and regulations annually and, if needed, during the working period leading up to the giant jamboree where the teams present their work. The Safety and Security Committee is similar to a governments' regulatory authority that reviews applications for permits. However, iGEM's Safety and Security Committee has more flexibility in terms of its authority and ability to adapt its rules based on the application/technology it encounters. Replicating this type of flexibility in a more traditional regulatory authority would be difficult, unless governments provided that flexibility in its overarching legal frameworks. The tool kit for citizen health innovators was developed in part with help from various US regulatory authorities. Similar to the buy-in discussed above, this partnership represents a "flexibility" in part by some US regulatory agencies and a recognition that the traditional regulatory structures are not capturing all that are participating. While they may not be adapting the regulations in real time, they are enabling these "outside" actors to maneuver through the system.

A Way Forward?

Synthetic biology and other evolving biotechnologies have given rise to a set of communities operating across countries and tied together by the technologies and purposes they ascribe. The iGEM community now represents over 40,000 students spread across 6 continents and over 40 countries. The DIYbio and citizen health innovator communities follow a similar trend in relation to geography with nearly 100 locations (see Fig. 3). Replicating the collective governance systems put in place at iGEM and within the DIYbio communities may require new legal and societal authorities to govern these and other emerging technologies. Yet, while the circumstances are unique, they provide guidance toward how a collective governance system could work.

Bibliography

Baumgaertner, E. (2018). As D.I.Y. gene editing gains popularity, 'someone is going to get hurt.' *New York Times.* https://www.nytimes.com/2018/05/14/science/biohackers-gene-editing-virus.html

Biosummit, Global Community. (2018). Global Community Biosummit. https://www.biosummit.org/

Bloomberg. (2018, September). This biohacker makes mutant frogs — And you can buy them on the internet. *Fortune.* http://fortune.com/2018/09/12/genetic-engineering-buy-frogs/

Carlson, R. (2005, May). Splice it yourself. *Wired.*

de Beer, J., & Jain, V. (2018). Inclusive innovation in biohacker spaces: The role of systems and networks. *Technology Innovation Management Review, 8*(2).

Demuijnck, G., & Fasterling, B. (2016). The social license to operate. *Journal of Business Ethics, 136*(4), 675–685. https://doi.org/10.1007/s10551-015-2976-7.

DIYbio. (2013). No title. Ask a biosafety professional. http://ask.diybio.org/

FDA. (2017). Information about self-administration of gene therapy. https://www.fda.gov/BiologicsBloodVaccines/CellularGeneTherapyProducts/ucm586343.htm

Gallois, C., Ashworth, P., Leach, J., & Moffat, K. (2017). The language of science and social licence to operate. *Journal of Language and Social Psychology, 36*(1), 45–60. https://doi.org/10.1177/0261927X16663254.

Grushkin, D., Kuiken, T., & Millet, P. (2013). Seven Myths & Realities about do-it-yourself biology. Washington, D.C.

IGEM. (2018). IGEM – previous competitions. http://igem.org/Previous_Competitions

iGEM. (2019). Human Practices Program. https://2019.igem.org/Human_Practices

iGEM Foundation. (2018a). 2018 exemplary projects in safety. http://2018.igem.org/Safety/Exemplary_Projects

iGEM Foundation. (2018b). IGEM animal use form. http://2018.igem.org/Safety/Animal_Use_Check_In

iGEM Foundation. (2018c). IGEM safety check in. http://2018.igem.org/Safety/Check_In

iGEM Foundation. (2018d). IGEM safety form. http://2018.igem.org/Safety/Final_Safety_Form

iGEM Foundation. (2018e). IGEM safety policies. http://2018.igem.org/Safety/Policies

iGEM Foundation. (2018f). IGEM safety rules. http://2018.igem.org/Safety/Rules

iGEM Foundation. (2018g). IGEM White List. http://2018.igem.org/Safety/White_List

IGEM, International Genetically Engineered Machines Competition. (2017). Registry of Standard Biological Parts.

International Genetically Engineered Machines Competition. (2018). IGEM human practices. http://igem.org/Human_Practices

Kuiken, T. (2016). Governance: Learn from DIY biologists. *Nature, 531*(7593), 167–168. https://doi.org/10.1038/531167a.

Kuiken, T., Pauwels, E., & Denton, S. W. (2018). The rise of the new bio-citizen workshop. *Washington, D.C.* https://www.wilsoncenter.org/article/the-rise-the-new-bio-citizen-workshop

Lunshof, J. E., & Birnbaum, A. (2017). Adaptive risk management of gene drive experiments. *Applied Biosafety, 22*(3), 97–103. https://doi.org/10.1177/1535676017721488.

Millett, P., Binz, T., Evans, S. W., Kuiken, T., Oye, K., Palmer, M. J., van der Vlugt, C., Yambao, K., & Yu, S. (2019). Developing a comprehensive, adaptive, and international biosafety and biosecurity program for advanced biotechnology: The IGEM experience. *Applied Biosafety*, 153567601983807, 64. https://doi.org/10.1177/1535676019838075.

Minnesota, Team. (2016). Shifting gene drives into reverse: Now mosquitoes are the yeast of our worries. *IGEM Foundation.* http://2016.igem.org/Team:Minnesota

Odin. (2018). The Odin. http://www.the-odin.com/about-us/

Open Philanthropy Project. (2019). Open philanthropy project. https://www.openphilanthropy.org/

Pauwels, E., & Denton, S. W. (2018). The rise of the new bio-citizen. *Washington, D.C.* https://www.wilsoncenter.org/article/the-rise-the-new-bio-citizen

PRIMR. (2019). Public responsibility in medicine and research. https://www.primr.org/about/

Sherman, R. E., Anderson, S. A., Pan, G. J. D., Gray, G. W., Gross, T., Hunter, N. L., LaVange, L., et al. (2016). Real-world evidence — What is it and what can it tell us? *New England Journal of Medicine, 375*(23), 2293–2297. https://doi.org/10.1056/NEJMsb1609216.

Smalley, E. (2018). FDA warns public of dangers of DIY gene therapy. *Nature Biotechnology, 36*(2), 119–120. https://doi.org/10.1038/nbt0218-119.

Swetlitz, I. (2016, December). College students try to hack a gene drive — And set a science fair abuzz. *STAT News*. https://www.statnews.com/2016/12/14/gene-drive-students-igem/

The Australia Group. (2019). List of human and animal pathogens and toxins for EXPORT control. https://australiagroup.net/en/human_animal_pathogens.html

U.S. FDA. (2017). Use of real world evidence to support decision-making for medical devices. Use of real world evidence to support decision-making for medical devices.

U.S. FDA. (2018). 21st century cures act. https://www.fda.gov/regulatoryinformation/lawsenforcedbyfda/significantamendmentstothefdcact/21stcenturycuresact/default.htm

Whalen, J. (2009). In attics and closets, 'biohackers' discover their inner Frankenstein. *The Wall Street Journal*. https://www.wsj.com/articles/SB124207326903607931

Zimmer, C. (2015, March 5). Amateurs are new fear in creating mutant virus. *New York Times*. https://www.nytimes.com/2012/03/06/health/amateur-biologists-are-new-fear-in-making-a-mutant-flu-virus.html

Synthetic Biology and Risk Regulation: The Case of Singapore

Benjamin D. Trump, George Siharulidze, and Christopher L. Cummings

Introduction[1]

The field of synthetic biology is rapidly expanding and is projected to become a \$38 billion-dollar industry by the year 2020. Today, this worldwide market is largely concentrated in North America and in Europe, but this international landscape is growing quickly, with some of the fastest-growing areas developing in East and Southeast Asia. Here, Singapore has quickly created a foothold, funneling increasing amounts of money into research and development as it aims to become a global leader in the field (Trump 2017).

Over the past 50 years, Singapore has grown from one of the poorest nations in the world in the 1960s to become a fully developed and scientifically advanced country (Davis and Gonzalez 2003; Wee 2007; Olds 2007). Due largely to its business-friendly stances and substantial funding for research and development, this progress has given the city state a reputation as a technological innovator, contributing to substantial funding for emerging technology research in nanotechnology, biotechnology, information system technology, and many other developing fields (Olds 2007; Altbach and Salmi 2011). Since its de jure independence in 1965, leadership of the "Lion City" has actively advocated for an aggressive research agenda in virtually all scientific fields, leading to collaborative relationships with most developed nations such as the United States and European Union alongside the development of

---

[1] This chapter includes material reproduced from Trump (2016).

B. D. Trump (✉) · G. Siharulidze
US Army Engineer Research and Development Center, Concord, MA, USA
e-mail: benjamin.d.trump@usace.army.mil

C. L. Cummings
Nanyang Technological University, Singapore, Singapore

© Springer Nature Switzerland AG 2020
B. D. Trump et al. (eds.), *Synthetic Biology 2020: Frontiers in Risk Analysis and Governance*, Risk, Systems and Decisions, https://doi.org/10.1007/978-3-030-27264-7_13

a number of industrial, governmental, and academic research ventures over the past several decades (Phan et al. 2005; Lee and Win 2004).

Among these areas of interest is synthetic biology, where two major universities (the National University of Singapore and Nanyang Technological University), a number of research groups, a biologics and therapeutics facility (Novartis), and a growing cohort of private companies and for-profit research ventures individually investigate various elements of synthetic biology research. Discussed more explicitly below, synthetic biology has emerged as a research venture in Singapore due to the potential for its scientists to engage with topical research questions and policy problems related to Southeast Asia such as with biofuels and combating tropical diseases such as dengue fever and malaria (Liang et al. 2011; Trump et al. 2019; Finkel et al. 2018) and its potential to galvanize the commercial industry.

Synthetic biology research has been formally explored and discussed in Singaporean universities since at least 2011 (Dhar Lab 2011). However, Singaporean researchers and laboratories have had connections with Western partners pertaining to synthetic biology research such as with the Massachusetts Institute of Technology since 2001. Singaporean students had begun to participate in the International Genetically Engineered Machine (iGEM) competition by 2008, with specific participation centered on the health track of the competition (NTU 2015).

By the end of 2011, Singaporean researchers at the Agency for Science, Technology and Research (A∗STAR) had begun to conduct research on DNA sequencing and metabolic engineering (Mitchell 2011), and by 2012, the National University of Singapore and Nanyang Technological University began to receive government grants to pursue metabolic and circuit engineering research (Oldham et al. 2012). Between 2012 and 2019, the Singaporean government has funded various projects at the two universities and has given grants worth up to half a million Singapore dollars to synthetic biology researchers collaborating with Chinese peers as part of a joint venture between Singapore's National Research Foundation and the National Natural Science Foundation of China. In 2016, Singapore established a consortium of research partners supported by the National Research Foundation and has expanded its research efforts. Since 2016, the government has invested nearly 60 million Singapore dollars in synthetic biology basic research. Most recently, in January of 2018, the Singaporean government launched a research and development program slated to receive 25 million Singapore dollars over 5 years. This program is focused on three primary areas of research: identifying genes for the sustainable production of synthetic cannabinoids, developing and producing rare fatty acids for industrial use, and establishing new, national strains of yeast and bacteria for commercial purposes.

Singapore is situated to further its leadership in synthetic biology research and is likely to continue to prolific growth in this sector. This chapter details how elements of risk culture can influence synthetic biology regulation in Singapore, thanks in part to the country's soft-authoritarianism yet cooperational and informal nature in facilitating synthetic biology regulation and governance reform. Specifically, we review (i) the history and practice of decision-making in Singapore with regard to its general regulatory culture and (ii) the current culture as it pertains to regulating synthetic biology risks.

Regulatory Culture and Regulatory Decision-Making in Singapore

Singapore's political and institutional identity centers on its status as a "soft-authoritarian state," which Turner (2015), Reilly (2016), and Olds (2007) describe as the situation where opposition parties are legally allowed to operate without significant fear of reprisal but are generally too weak or ineffective to seriously challenge power. For Singapore, the People's Action Party has served as the primary soft-authoritarian power, with effective control of the national government since 1959 (Hill and Lian 2013). A center-right party by nature, the People's Action Party has operated on the principles of pragmatism, meritocracy, multiracialism, and communitarianism, with the general motivation of the Party being to improve Singapore's economic position and the purchasing power of its citizenry through continual technological and economic investment within a racially and ethnically diverse population while also leveraging Singapore's historical advantages as a center for shipping and trade (Tremewan 1996).

The People's Action Party retains control over the three branches of government (executive, legislative, and judicial). Governmental structure is defined by the Westminster constitutional model (driven by Singapore's history prior to its independence from Britain), with a Prime Minister heading the national government and chosen from among the body of Parliamentary members (Sheehy 2004) and a President exercising largely ceremonial power as Head of State (Sheehy 2004; Hill and Lian 2013). Lastly, an independent judiciary checks executive and legislative actions that may be interpreted as violating the Singaporean Constitution, although judicial authority is limited and defers by law to the executive for instances where court authority is limited or uncertain.

Lawmaking in Singapore is carried out by Parliament, where ministers may propose bills (though most legislative proposals are initiated by a member of the Prime Minister's Cabinet) (Vasil 2004; Tan 2013). However, Parliamentary lawmaking is limited in cases of (i) those bills that seek to impose, increase, or abolish a tax, (ii) those bills that seek to borrow money on behalf of the government, and (iii) deposits or changes to the Singaporean Consolidated Fund (Constitution of Singapore; Tan 2013). Further, bills are screened for potential harms to minority rights, where those deemed to be explicitly harmful to a particular subsection of Singaporean society are removed from further consideration in Parliament (Constitution of Singapore) (Tan 2013; Vasil 2004). Bills allowed to circumvent screening for minority rights considerations include both those bills that the Prime Minister certifies as affecting the defense or security of Singapore or that relate to public safety, peace, or good order in Singapore and bills the Prime Minister certifies are so urgent that it is not in the public interest to delay enactment (Sheehy 2004; Hill and Lian 2013). In this way, both the President and Prime Minister exert control over the lawmaking process and retain the ability to (i) guide the legislative process by instructing a Cabinet member to propose and defend a bill or (ii) use their power to circumvent certain requirements of the deliberation of a bill in order to meet emerging public health and security concerns – factors that have been argued as enhancing the soft-authoritarian capabilities of Party leaders to influence lawmaking within Singapore's Parliament (Olds 2007; Tan 2013).

Specific to synthetic biology regulation and governance, Singapore captures the process of the field's development using existing hard and soft law previously crafted to govern genetically modified organisms (Malloy et al. 2016). Two such instruments include the Biological Agents and Toxins Act and the Singapore Biosafety Guidelines for Research on Genetically Modified Organisms. This is in lieu of using existing chemical regulation to capture elements of synthetic biology development such as within the United States, where such regulation has been used instead to cover biosecurity (the Strategic Goods (Control) Act) as well as product-driven regulation (the Medicines Act and The Health Products Act). Each of these will be discussed in turn below.

The Biological Agents and Toxins Act (2005) represents Singapore's key legislative instrument that shall most likely capture the process of synthetic biology development. Those with more knowledge of the law particularly referenced Chapter 24A, which was added as a revision to the original act in 2006. Chapter 24A is specifically geared to address regulatory policy to:

> prohibit or otherwise regulate the possession, use, import, transhipment, transfer and transportation of biological agents, inactivated biological agents and toxins, to provide for safe practices in the handling of such biological agents and toxins. (Biological Agents and Toxins Act 2006)

Administered by the Singaporean Ministry of Health, the Act states that those facilities which handle biological agents and toxins deemed "high risk" are required by law to acquire certification as "containment facilities," with inspection and recertification to occur on an annual basis.

This particular statute was directed at monitoring, reviewing, and assessing risk related to various elements of life sciences research, with coverage of synthetic biology research based upon the abilities of the Director of Medical Services and his or her appointed officers to monitor and review the possession, use, transportation, and production of biological agents. Biological agents are divided into a series of classes called "Schedules," with eight schedules referenced in Chapter 24A of the Biological Agents and Toxins Act (Biological Agents and Toxins Act 2006). Synthetic biology is not explicitly referenced within any of these schedules, although such products would most likely fall in the First, Second, or Third Schedule based upon the type of product that synthetic biology research would be conducted on. These schedules include some of the biological agents more tightly controlled and vigorously monitored by Singaporean officials, with explicit requirements of permitting and certification for most activities related to large-scale production, transport, possession, and use of such agents. Within the statute, these laws explicitly reference the Ministry of Health's ability to protect and preserve Singapore's biosafety with respect to biological agents and life sciences research generally speaking, with such language in line with other biosafety discussion related to synthetic biology and pharmaceuticals (Biological Agents and Toxins Act 2006).

A further legislative instrument includes is the Strategic Goods (Control) Act (Chapter 300) of 2002, which lists biological agents and toxins that are administered and reviewed by the Singaporean Customs Authority. This particular law addresses the preservation of both the nation's security relative to monitoring the brokering and the exchange of goods "capable of causing mass destruction," along with a review of the technologies moving in and out of the country that would otherwise be of interest to national security (Salerno and Gaudioso 2015). Geared more toward the biosecurity debates discussed since synthetic biology's modern inception in the early 2000s, these statutes seek to control the import and export of various materials that are potential threats to national security, with biological agents serving as one potential avenue of concern here.

Specific to the Strategic Goods (Control) Act, it is not explicitly clear how the Singaporean Customs Authority communicates with other bodies such as the Ministry of Health to identify and label certain technologies and products as being of concern for Customs agents at the nation's borders (Strategic Goods (Control) Act 2003). However, the statute does offer the Director-General of Singapore's Customs Authority the ability to, at their discretion:

> prescribe any military or dual-use technology as strategic goods technology for the purposes of [the] ACT.

This allows the Customs Authority to update their schedules and guidance regarding those technologies deemed strategic and of interest to the Singaporean government (Strategic Goods (Control) Act, Section 4a, 2003). Where greater flexibility was needed with respect to applying regulatory oversight to synthetic biology products, this Act offers a relatively adaptive and flexible approach to identifying biosecurity threats now and in the future and empowers Customs to regularly update their schedule of strategic goods and technologies based on notable threats and developments in areas ranging from energetics to life sciences. For purposes of synthetic biology and pharmaceuticals research, such flexibility would allow the Customs Authority's leadership to apply principles of soft law to include specific synthetic biology products on their list of materials that require permits and certification for travel and exchange.

Another important regulatory instrument related to synthetic biology development includes workplace safety considerations. Specifically, this includes the Ministry of Manpower (MOM) and the Workplace Safety and Health (WSH) Council. For the former, MOM includes 14 divisions (with 1 centered on Occupational Health and Safety) and is empowered by the Workplace Safety and Health Act to ensure workplace safety. Specifically, Part 4 of the Act asserts that:

> It shall be the duty of every occupier of any workplace to take, so far as is reasonably practicable, such measures to ensure that —

(a) the workplace;
(b) all means of access to or egress from the workplace; and
(c) any machinery, equipment, plant, article or substance kept on the workplace,

> are safe and without risks to health to every person within those premises, whether or not the person is at work or is an employee of the occupier.

Further, the WSH Council works in tandem with the MOM to review workplace safety concerns as laid out in the Workplace Safety and Health Act. This relationship is codified in Part 8 of the Act, where the WSH Part 8 Section 40a notes:

40A. The functions of the Council shall be:

(a) to develop or facilitate the development of acceptable practices relating to safety, health and welfare at work;
(b) to promote the adoption of acceptable practices relating to safety, health and welfare at work;
(c) to devise, organise and implement programmes and other activities for or related to providing support, assistance or advice to any person or organisation in preserving, improving and promoting safety, health and welfare at work;
(d) to facilitate and promote the development and upgrading of competencies, skills and expertise of the workforce relating to safety, health and welfare at work;
(e) to research into any matter relating to safety, health and welfare at work;
(f) to grant prizes and scholarships, and to establish and subsidise lectureships in universities and other educational institutions in subjects relating to safety, health and welfare at work;
(g) to provide practical guidance with respect to the requirements of this Act relating to safety, health and welfare at work; and
(h) to do all the things that it is authorised or required to do under this Act.

Such legal jurisdictions apply to genetically modified substances via the Fifth Schedule of the Act (Machinery, Equipment or Hazardous Substances). Discussed further below, the Singapore Biosafety Guidelines for Research on Genetically Modified Organisms also indicates how the MOM and WSH interact with the Genetic Modification Advisory Council and other agencies to explicitly cover workplace safety for genetically modified substances, where the legal authority of the two agencies derives from the Workplace Safety and Health Act.

With respect to soft law, various members of the academic, industry, and governmental axis referenced the Singapore Biosafety Guidelines for Research on Genetically Modified Organisms (GMOS) (2013), which serves as a more explicit connection to synthetic biology regulation than the hard law case of the Strategic Goods (Control) Act or the Biological Agents Control Act of 2005. While the term "synthetic biology" does not appear in the 2013 iteration of the Guidelines, the focus on genetically modified organisms and the various products that make use of such organisms drove most stakeholders knowledgeable of the document to argue for synthetic biology being thoroughly covered under the Guidelines. Within its contents, the Guidelines offer instruction on (i) the types of products and activities to be governed; (ii) the various governmental institutions and agencies with authority to review practices, adaptively improve regulatory mechanisms over time, and mete out consequences to those who defy best practices; and (iii) clear structure regarding governmental authority and workflows related to protecting Singaporean biosafety and biosecurity within the context of genetically modified organisms and their related research. Overall, the Guidelines do not carry the force of law as with the Biological Agents and Toxins Act, yet have been adopted by Singapore's research universities and are required for research organizations that receive funding from the Singaporean government (Tun et al. 2009; Asadulghani and Johnson 2015).

Looking first at the types of activities and products explicitly covered by the Singapore Biosafety Guidelines for Research on Genetically Modified Organisms, Section 2.1 (Extent of Guidelines) references the Guidelines' ability to offer guidance to:

> experiments that involve the construction and/or propagation of all biological entities (cells, prions, viroids, viruses or organisms) which have been made by genetic manipulation and are of a novel genotype and which are unlikely to occur naturally, or which could cause public health or environmental hazards.

While it is important to note that the Guidelines "do not cover work involving human subjects," the risk and hazard discussion centered on governing genetically modified organisms does consider both public health and environmental health outcomes as they arise from the research, manufacturing, use, and disposal of such materials. Further, the Guidelines note that they consider both the intentional and unintentional release of biological material deriving from genetically modified organisms yet also state that certain work or research may be subject to additional hard or soft law regulation depending on whether such work was able to get an exemption from external oversight or whether such work falls under a specific class of genetic modification research that the Singaporean government has denoted as not possessing significant biosafety risks to humans or the environment (Section 2.3).

Next, the Guidelines address at length the governmental institutions empowered to govern and regulate activities outlined in Section 2. Specifically, the Guidelines name eight agencies with some degree of authority to regulate research related to the genetic modification of biological material for predefined purposes, including:

- The Genetic Modification Advisory Committee of Singapore
- The Agri-Food and Veterinary Authority of Singapore
- The Ministry of Health, Singapore
- The National Environment Agency, Singapore
- The Ministry of Manpower, Singapore
- The Institutional Biosafety Committee
- The National Advisory Committee for Laboratory Animal Research
- The Bioethics Advisory Committee

Using this guidance, the Guidelines divide the regulation of genetically modified products or technologies into four general categories, including (i) the regulation of laboratories dealing with GMO research, involving animal pathogens and plant pests, (ii) the importation of organisms including GMOs, (iii) the certification or inspection of laboratories handling biological agents or toxins regulated under the Biological Agents and Toxins Act, and (iv) the regulation of workplace safety and health. Synthetic biology biosafety guidance is likely to currently fall under parts

(i) and (iii), where soft law provides guidance alongside notions of hard law certification and monitoring requirements of laboratories conducting genetic modification research. For these biosafety provisions (parts i and iii), the Guidelines state that both the Agri-Food and Veterinary Authority of Singapore and the Ministry of Health, respectively, serve as the two regulatory organizations empowered to govern such activities. With respect to part ii, the Guidelines note a collection of the Agri-Food and Veterinary Authority of Singapore, the MOH, and the National Environment Agency charged with the regulatory authority to oversee the importation of genetically modified organisms and products into Singapore and include a risk classification report regarding the proper shipping and assessment of such materials both as an import and with respect to internal transport within the country. Lastly, part (iv) is noted as being governed by the MOM, where occupational safety is an element of regulation that was on the horizon of synthetic biology and pharmaceutical research, yet greater consideration of imminent regulatory concerns remained both within the research and disposal stages of a generic pharmaceutical's life cycle.

Under this four-tiered framework of regulation and activity, the Genetic Modification Advisory Committee of Singapore is empowered to expand or add to such guidance where technologies emerge or research involving the genetic modification of biological material is uncertain or emerging (Section 5). As noted above, the nature of this guidance is nonbinding in a manner similar to the Biological Agents and Toxins Act yet is adopted within research universities and organizations receiving government funding in Singapore (Tun et al. 2009). After describing the types of experiments covered by the Guidelines as well as characteristics which make certain experiments exempt, Section 4 indicates that the Genetic Modification Advisory Committee is empowered with the ability to include further developments with research and experimentation to effectively expand the ability of the Guidelines to cover such projects as with synthetic biology – an important element in fostering an adaptive regulatory framework via iterative improvements to soft law regulation for the technology moving forward. Further, the Genetic Advisory Committee is empowered by the Biological Agents and Toxins Act to oversee, regulate, and approve of research related to genetic engineering.

To accomplish this goal, the Guidelines note that novel experimentation and genetic manipulation techniques may be reviewed by an Institutional Biosafety Council (IBC) relevant to the organization conducting synthetic biology research – the recommendations and observations of which may be submitted to the Genetic Modification Advisory Committee prior to the Committee's determination of how that particular product or experimental technique may be regulated.

Third, the Guidelines offer transparent workflows regarding which agency is responsible for monitoring a given activity alongside guidance for researchers and developers regarding how to identify the agency and regulations relevant to their vein of work. This is described in detail in Sections 3 and 5, respectively, where prospective researchers would be able to determine the degree of oversight their research requires as well as the various agencies involved in such oversight throughout research and development.

First, Section 3 indicates the "Summary of Procedures," which is a decision chart describing the assessment protocol and notification timelines for researchers engaging with work related to genetically modifying biological organisms. This includes considerations of self- regulation (IBCs) and external regulation (the Genetic Modification Advisory Committee and private government investigators). Next, Section 5 includes explicit notation of the roles and responsibilities held by the various government agencies throughout the regulatory process of genetic research. Collectively, the information found in Sections 3 and 5 serves as a measure of reducing uncertainty regarding the structure and actions taken by government for cases of research as with synthetic biology.

Generally speaking, Singapore's constitutional structure is an emulation of its colonial past under the British Empire, with modifications driven by paternalism and pragmatism that has pushed forward Singaporean regulation since the 1950s (Li-Ann 1993). The soft-authoritarian nature of regulation via the People's Action Party is enhanced by a centralization of power under the Prime Minister, who together with the President retains significant control over the legislative process (Tan 2013; Li-Ann 1993). This institutional, social, and political structure is significant to the formation of Singapore's regulatory risk culture – the characteristics of which are unpacked in the section below.

Synthetic Biology and Singapore's Risk Culture

With this general background on the functions of Singaporean government and the influence of the People's Action Party on the country's lawmaking process, it is important to next unpack considerations of Singapore's risk culture, or the institutional and political factors that influence their approach to regulation more generally, and with a specific focus on emerging science and technology. To cover this topic, this section begins by first discussing the historical path of synthetic biology regulation or noting how the regulation of genetically modified organisms in Singapore developed over time as well as noting the legal and regulatory mechanisms to cover related research, production, commercialization, and disposal of such material. The section then builds upon this by reviewing how elements of risk culture may influence Singaporean regulation of emerging technologies.

Historical Path of Synthetic Biology Regulation

In a manner similar to the European Union, Singapore's synthetic biology regulation is generally perceived to derive from the regulation of genetically modified organisms and includes a mixture of adherence to international regulation as well as domestic hard and soft law such as with the Biological Agents and Toxins Act

(Chapter 24) as well as the Biosafety Guidelines for GMOs (Genetic Modification Advisory Committee n.d.). Early regulation of such materials was driven by the need to govern food importation into Singapore, where genetically modified foods are viewed by the government as one avenue to improve food security and local nutrition for a nation with 90% of its food supply being imported from neighboring countries (Genetic Modification Advisory Committee n.d.; Tey et al. 2009). However, such regulation also covers other activities ranging from laboratory experimentation of genetically modified organisms to pharmaceuticals and other research ventures (Oriola 2002a, b). This section details both the adherence to international regulation and domestic hard and soft law that has been applied to cover synthetic biology research via the regulation of genetically modified organisms.

Prior to 2005, Singaporean regulatory authority over genetic modification was covered by formal legislation in specific research areas such as pharmaceutical development, agriculture, or food labeling (Ho 2011). Regulation specifically directed at genetic modification and emerging biotechnology research remained limited and informal until 2005, when Singapore's Parliament passed the Biological Agents and Toxins Act (Singapore Ministry of Health 2007: Ho 2011). It is important to note here that Chapter 24A of the Act was designed to govern the research, production, sale, distribution, transport, and disposal of genetically modified material (Singapore Ministry of Health 2007). The Act also included a system of approvals for those laboratories that sought to conduct such research – the process of which included biosafety risk considerations that researchers were required to discuss with regulators prior to receiving a permit for research (Singapore Ministry of Health 2007).

Later, Singapore's Genetic Modification Advisory Committee released the Singapore Biosafety Guidelines for Research on Genetically Modified Organisms in 2006, which served as a legally nonbinding approach to the regulation of research involving genetic modification that offered recommendations to counter biosafety and biosecurity risks for research involving genetic modification (Ho 2011; GMAC 2016). The Guidelines were further modified in 2008 and 2013 and included guidance on the biosafety and biosecurity concerns that researchers should work with their respective Internal Review Boards to address (GMAC 2016). As with the Biological Agents and Toxins Act, the Guidelines will be discussed in detail in below, yet it is important to emphasize their role in governing research on genetically modified organisms – including synthetic biology research.

Singapore's regulatory history for genetically modified organisms that currently covers synthetic biology research is a relatively limited one, with most guidance coming from soft law recommendations and applications from specific product development prior to 2005 (Trump 2017; Ho 2011). From 2008 to 2013, this was bolstered via hard law (Biological Agents and Toxins Act) and soft law (the Guidelines) geared explicitly to governing research related to genetic modification. Such guidance will likely continue to incorporate developments in domestic and international research pertaining to such modification and the potential biosafety and biosecurity concerns therein (Ho 2011).

Assessment of How Risk Culture Influences Regulation of Novel Compounds and Scientific Processes Including Synthetic Biology

After reviewing the historical path of regulation for genetically modified organisms and synthetic biology products in Singapore, this section reviews elements to consider including the institutional, social, and political values and behaviors that fashion Singaporean risk culture within the context of technology regulation. Specifically, risk culture considerations here include (i) the soft-authoritarianism and centralization of decision-making authority practiced by government leaders, and how this relates to technology regulation, (ii) the cooperational yet informal approach to overcoming regulatory disputes and driving technology regulation, and (iii) the more "proactionary" nature of Singaporean government leaders relative to innovation in order to strengthen the country's economic prospects (Olds 2007; Tan 2013; Ho 2011; Linkov et al. 2018). These three characteristics will explain the factors that local regulators must consider when reviewing options to govern specific emerging technologies as with the process of synthetic biology in general, as well as genetic modification and product development in particular (Trump et al. 2018; Cummings and Kuzma 2017).

For the first item, the soft-authoritarian nature of the Singaporean government's behavior serves as a pervasive characteristic that drives Singaporean lawmaking, regulatory behavior, and coordination of governmental and industry representatives (Olds 2007; Nasir and Turner 2013). Building off of the introductory discussion of soft-authoritarianism within Singapore in the previous sections, further characteristics that arise from this political and institutional arrangement include a centralization of decision-making authority within government alongside a lesser degree of transparency in the regulatory reform process than would be expected in a liberal democracy (Nasir and Turner 2013). More specifically, soft-authoritarian governments act in a manner where decision-making power is centralized among a powerful elite with limited checks on authority and little real competition in terms of election (Turner 2015). Such centralization includes the ability for government subject experts to introduce and implement regulatory reform in an efficient manner in comparison to a state where power is shared among more players (as is the case within the United States and the European Union) (Neo and Chen 2007; Merad and Trump 2020). An additional factor behind this includes Singapore's relatively small size in comparison to the United States and European Union, which limits democratizing factors and preserves the country's soft-authoritarian regime (Huat 2015; Lim and Lim 2016; Ufen 2015). Further, noting that the Prime Minister, as elected by the majority party in Parliament, also serves as the chief executive in a Westminster-style government further limits opportunities for regulatory reform to be hindered in passage (Tan 2013; Haque 2004).

However, even within an "imperfect democracy" and limited transparency in government, another characteristic of Singapore's soft-authoritarianism includes a general need to identify regulation that mitigates risk to the local population and the environment (Turner 2015). Such behavior can differ from a "full authoritarian" state that seeks to enrich elites and cadres often at the expense of the general public, where Roy (1994) and Olds (2007) argue instead that soft-authoritarian states like Singapore generally seek to represent the best interests of the general citizenry by promoting public safety, public health, and improved economic status. This is accomplished by the controlling political party's maneuvering within the Singaporean government and abiding by the Singaporean Constitution, although no serious challenge is raised by opposing political parties for those regulatory issues deemed higher priorities by elites in the People's Action Party (Roy 1994; Nasir and Turner 2013; Mauzy and Milne 2002). Within such a soft-authoritarian government, it is important to note that the Singaporean government is unlikely to use their controlling power to "force" hard or soft law through a resistant public but instead wait until international research developments or domestic necessity offers political and scientific reason to improve technological regulation in a specific manner.

For the second item, Singapore generally adopts a cooperational approach to resolve regulatory disputes and build regulation for emerging technologies like synthetic biology (Beng-Huat 1985; Srivastava and Teo 2009; Lim 2005). Such behavior is similar to that found within the European Union, where government stakeholders collaborate with stakeholders in industry, academia, and non-governmental institutions to construct regulation in a manner that is responsive to industry needs of promoting responsible innovation while balancing government requirements of upholding public health and safety (Kelemen 2011). However, the motivations for such behavior strongly differ in Singapore, where soft-authoritarianism limits the potential for significant resistance, dissent, or adversarial legalism in the process of technology regulation (with similar operational and political structures existing in examples such as Malaysia and Russia – Shevtsova 2014; Ufen 2015).

Instead, the "socially minded" approach to furthering the welfare of Singaporean citizens as described in Roy (1994), Nasir and Turner (2013), and Olds (2007) drives governmental elites to procure information about regulatory needs and innovation potential from members of industry and other stakeholders and use such information to make decisions about furthering the public good. With no real challenges to their political authority or significant threats of having their regulatory agendas seriously challenged in court, the People's Action Party can use informal regulation-building exercises to engage with concerned stakeholders in a manner that (i) allows them to acquire information on emerging technologies that allow them to balance risk and benefit in the regulatory process and (ii) identifies concerns and needs of industry researchers that would allow for a continued economic and technological pattern of growth within Singapore – a value central to the nation's identity (Beng-Huat 1985; Lim 2005; Mauzy and Milne 2002).

Third, Singapore's drive to innovate and grow economically allows it to take on a more proactionary nature (Li and Fang 2004). Specifically, Singaporean government agencies seek to empower universities and companies to conduct research related to emerging technologies in a less restrictive regulatory environment, with oversight driven both by internal mechanisms such as with internal review boards and informal contact with regulatory agencies such as with the Economic Development Board or the Genetic Modification Advisory Committee (Hobday 1995; Edquist and Hommen 2009; Peebles and Wilson 2002; Chieh 1999; Olds 2007). Such research is geared toward commercialization as soon as safely possible and toward benefitting the Singaporean economy and/or public health, with little government investment allocated without such intentions in mind (Williams and Narendran 1999). Overall, Singapore's adherence to technological proaction is driven by the wishes of the government to further boost its economic and technological capabilities in order to achieve greater development and promote the welfare of its citizens.

The factors described above represent the significant cultural, political, and institutional drivers that comprise Singaporean risk culture and influence regulation for emerging technologies as with synthetic biology. These factors must be considered by regulators when seeking to change or revise Singaporean hard or soft law for such technologies, where institutional and political norms help determine how regulatory change occurs, what legal requirements are needed to be met for such reform, and how political actors interact with one another as well as the lay citizenry in order to make such changes possible.

Synthetic Biology: Hard and Soft Law Regulation Within Singapore

Singapore's status as a growing economic power via capitalism and its subsequent ability to drive technological research is often viewed as a paradox based on the common scholarly discussion described by Rodan (2004), which note that such regulatory regimes rarely succeed in advancing successful capitalistic markets or a robust research base to drive innovation. Inglehart and Welzel (2005) note that such democratic and liberal economies tend to outperform more autocratic and closed-market regimes – leaving Singapore as something of an anomaly. Instead, the "soft-authoritarianism" in Singapore described by Olds (2007) allows the state to pursue global research and education opportunities in the spirit of forging new scientific opportunities beneficial to the country and its residents. Synthetic biology includes one of these opportunities, with particular academic attention and governmental resources paid to pharmaceutical and therapeutic product development. However, even with the growing degree of time, money, and manpower invested within Singapore on synthetic biology research, discussion regarding hard and soft governmental authority remains limited with respect to which regulations, statutes, and guidance mechanisms are functionally used to perceive, mitigate, and manage synthetic biology health risk along with whether or not new regulations or guidance would be effective at bolstering synthetic biology regulation in the near future (Greer and Trump 2019).

Singapore differs in political structure and smaller economic size from major synthetic biology developers such as the European Union and United States. This is driven by Singapore's status as a soft-authoritarian state yet continued economic success and capitalistic tendencies with respect to advancing technological innovation and development, making it difficult to ascertain the regulatory mechanisms in place to guide synthetic biology regulation or whether such mechanisms are de facto effective and valid (Lingle 1996; Rodan 2004).

However, it cannot be denied that Singapore has grown as a player in synthetic biology and other emerging technology research and scholarship and has growing connections to not only to Europe and America but also to Malaysia, the People's Republic of China, and other nations. As such, it will continue to grow in importance in the synthetic biology research area and will likely venture into new veins of biological research that organizations in the West may not yet have significant capabilities to develop. Such a divergence in research interest and capabilities will undoubtedly be of interest to governments, international companies, and public interest groups alike due to the increasing globalization of synthetic biology research and product development in a manner that differs from the risk assessment and governance principles of Western nations.

References

Altbach, P. G., & Salmi, J. (Eds.). (2011). *The road to academic excellence: The making of world-class research universities*. Washington, DC: World Bank Publications.

Asadulghani, M., & Johnson, B. (2015). Biosecurity in research laboratories, agriculture, and the food sector. *Foodborne Pathogens and Food Safety, 289*.

Beng-Huat, C. (1985). Pragmatism of the people's action party government in Singapore: A critical assessment. *Southeast Asian Journal of Social Science, 13*(2), 29.

Chieh, H. C. (1999). *What it takes to sustain research and development in a small, developed nation in the 21st century* (pp. 25–36). Singapore: Towards a Developed Status.

Cummings, C. L., & Kuzma, J. (2017). Societal risk evaluation scheme (SRES): Scenario-based multi-criteria evaluation of synthetic biology applications. *PLoS One, 12*(1), e0168564.

Davis, J. C., & Gonzalez, J. G. (2003). Scholarly journal articles about the Asian Tiger economies: Authors, journals and research fields, 1986-2001. *Asian-Pacific Economic Literature, 17*(2), 51–61.

Dhar Lab. (2011). *The untapped clinical potential – From combinatorial genomics to systems medicine*. National University of Singapore, Singapore.

Edquist, C., & Hommen, L. (2009). *Small country innovation systems: Globalization, change and policy in Asia and Europe*. Cheltenham: Edward Elgar Publishing.

Finkel, A. M., Trump, B. D., Bowman, D., & Maynard, A. (2018). A "solution-focused" comparative risk assessment of conventional and synthetic biology approaches to control mosquitoes carrying the dengue fever virus. *Environment Systems and Decisions, 38*(2), 177–197.

Genetic Modification Advisory Committee (GMAC). (2016). *Singapore biosafety guidelines for research on genetically modified organisms*. http://www.gmac.gov.sg/Index_Singapore_Biosafety_Guidelines_for_Research_on_GMOs.html

Genetic Modification Advisory Committee. (n.d.). *Government of Singapore*. http://www.gmac.gov.sg/Education/Index_FAQ_Genetically_Modified_Foods.html

Greer, S. L., & Trump, B. (2019). Regulation and regime: the comparative politics of adaptive regulation in synthetic biology. *Policy Sciences*, 1–20.

Haque, M. S. (2004). Governance and bureaucracy in Singapore: Contemporary reforms and implications. *International Political Science Review, 25*(2), 227–240.

Hill, M., & Lian, K. F. (2013). *The politics of nation building and citizenship in.* Singapore: Routledge.

Ho, W. C. (2011). Governance framework for biomedical research in Singapore: A risk-based Account. *Asia-Pacific Biotech News, 15*(5), 13–17.

Hobday, M. (1995). Innovation in East Asia: Diversity and development. *Technovation, 15*(2), 55–63.

Huat, C. B. (2015). Singapore: Growing wealth, poverty avoidance and management. In *Developmental pathways to poverty reduction* (pp. 201–229). Basingstoke: Palgrave Macmillan UK.

Inglehart, R., & Welzel, C. (2005). *Modernization, cultural change, and democracy: The human development sequence.* Cambridge: Cambridge University Press.

Kelemen, R. D. (2011). *Eurolegalism: The transformation of law and regulation in the European Union.* Cambridge, MA: Harvard University Press.

Lee, J., & Win, H. N. (2004). Technology transfer between university research centers and industry in Singapore. *Technovation, 24*(5), 433–442.

Li, S., & Fang, Y. (2004). Respondents in Asian cultures (e.g., Chinese) are more risk-seeking and more overconfident than respondents in other cultures (e.g., in United States) but the reciprocal predictions are in total opposition: How and why? *Journal of Cognition and Culture, 4*(2), 263–292.

Liang, J., Luo, Y., & Zhao, H. (2011). Synthetic biology: Putting synthesis into biology. *Wiley Interdisciplinary Reviews: Systems Biology and Medicine, 3*(1), 7–20.

Li-Ann, T. (1993). Post-colonial constitutional evolution of the Singapore legislature: A case study. *Singapore Journal sof Legal Studies, 80.*

Lim, M. K. (2005). Transforming Singapore health care: Public-private partnership. *Annals-Academy of Medicine Singapore, 34*(7), 461.

Lim, S. G., & Lim, J. H. (2016). *Part X: A study of Singapore, a study of honour-small city, small state.* World Scientific Publishing Co. Pte. Ltd.

Lingle, C. (1996). *Singapore's authoritarian capitalism.* Barcelona: Edicions Sirocco, SL.

Linkov, I., Trump, B. D., Anklam, E., Berube, D., Boisseasu, P., Cummings, C., et al. (2018). Comparative, collaborative, and integrative risk governance for emerging technologies. *Environment Systems and Decisions, 38*(2), 170–176.

Malloy, T., Trump, B. D., & Linkov, I. (2016). Risk-based and prevention-based governance for emerging materials. *Environmental Science & Technology, 50*(13), 6822–6824.

Mauzy, D. K., & Milne, R. S. (2002). *Singapore politics under the people's action party.* London: Psychology Press.

Merad, M., & Trump, B. D. (2020). *Expertise under scrutiny: 21st century decision making for environmental health and safety.* Cham: Springer. https://doi.org/10.1007/978-3-030-20532-4.

Mitchell, W. (2011). Natural products from synthetic biology. *Current Opinion in Chemical Biology, 15*(4), 505–515.

Nasir, M. K., & Turner, B. S. (2013). Governing as gardening: Reflections on soft authoritarianism in Singapore. *Citizenship Studies, 17*(3–4), 339–352.

Neo, B. S., & Chen, G. (2007). *Dynamic governance: Embedding culture, capabilities and change in Singapore.* River Edge, NJ: World Scientific.

NTU (Nanyang Technological University). (2015). *Synthetic biology.* http://www.scbe.ntu.edu.sg/Research/ResearchFocus/Pages/SB.aspx

Oldham, P., Hall, S., & Burton, G. (2012). Synthetic biology: Mapping the scientific landscape. *PLoS One, 7*(4), e34368.

Olds, K. (2007). Global assemblage: Singapore, foreign universities, and the construction of a "global education hub". *World Development, 35*(6), 959–975.

Oriola, T. A. (2002a). Ethical and legal issues in Singapore biomedical research. *The Pacific Rim Law & Policy Journal, 11*(3), 497–530.

Oriola, T. A. (2002b). Consumer dilemmas: The right to know, safety, ethics and policy of genetically modified food. *Singapore Journal of Legal Studies, 514.*

Peebles, G., & Wilson, P. (2002). *Economic growth and development in Singapore*. Cheltenham: Edward Elgar.

Phan, P. H., Siegel, D. S., & Wright, M. (2005). Science parks and incubators: Observations, synthesis and future research. *Journal of Business Venturing, 20*(2), 165–182.

Reilly, B. (2016). In the shadow of China: Geography, history and democracy in Southeast Asia. *Policy: A Journal of Public Policy and Ideas, 32*(1), 24.

Rodan, G. (2004). *Transparency and authoritarian rule in Southeast Asia: Singapore and Malaysia*. New York\London: Routledge.

Roy, D. (1994). Singapore, China, and the "soft authoritarian" challenge. *Asian Survey, 34*(3), 231–242.

Salerno, R. M., & Gaudioso, J. (Eds.). (2015). *Laboratory biorisk management: Biosafety and biosecurity*. Boca Raton: CRC Press.

Sheehy, B. (2004). Singapore, shared values and law: Non east versus west constitutional hermeneutic. *Hong Kong Law Journal, 34*, 67.

Shevtsova, L. (2014). The Russia factor. *Journal of Democracy, 25*(3), 74–82.

Singapore Ministry of Health. (2007). *Biological agents and toxins act 2005 (No. 36 of 2005)*. https://www.moh.gov.sg/content/moh_web/home/legislation/legislation_and_guidelines/biological_agentsandtoxinsact2005no36of2005.html

Srivastava, S. C., & Teo, T. S. (2009). Citizen trust development for e-government adoption and usage: Insights from young adults in Singapore. *Communications of the Association for Information Systems, 25*(1), 31.

Tan, K. P. (2013). The Singapore parliament. In *Parliaments in Asia: Institution Building and Political Development* (p. 27). Oxford: Routledge.

Tey, Y. S., Darham, S., Alias, E. F., & Ismail, I. (2009). Food consumption and expenditures in Singapore: Implications to Malaysia's agricultural exports. *International Food Research Journal, 16*(2), 119–126.

Tremewan, C. (1996). *The political economy of social control in Singapore (St. Anthony's Series)* (p. 105). London: Palgrave Macmillan.

Trump, B. (2016). *A comparative analysis of variations in synthetic biology regulation*. University of Michigan: Ann Arbor, Michigan.

Trump, B. D. (2017). Synthetic biology regulation and governance: Lessons from TAPIC for the United States, European Union, and Singapore. *Health Policy, 121*(11), 1139–1146.

Trump, B. D., Cegan, J., Wells, E., Poinsatte-Jones, K., Rycroft, T., Warner, C., et al. (2019). Co-evolution of physical and social sciences in synthetic biology. *Critical Reviews in Biotechnology, 39*(3), 351–365.

Trump, B., Cummings, C., Kuzma, J., & Linkov, I. (2018). A decision analytic model to guide early-stage government regulatory action: Applications for synthetic biology. *Regulation & Governance, 12*(1), 88–100.

Tun, T., Sai-Kit, A. L., & Sugrue, R. J. (2009). In-house BSL-3 user training: Development and implementation of programme at the Nanyang Technological University in Singapore. *Applied Biosafety, 14*(2), 89.

Turner, B. S. (2015). Soft authoritarianism, social diversity and legal pluralism: The case of Singapore. In *The sociology of Shari'a: Case studies from around the world* (pp. 69–81). Cham: Springer.

Ufen, A. (2015). Laissez-faire versus strict control of political finance: Hegemonic parties and developmental states in Malaysia and Singapore. *Critical Asian Studies, 47*(4), 564–586.

Vasil, R. K. (2004). *A citizen's guide to government and politics in Singapore*. Singapore: Talisman Pub.

Wee, C. L. (2007). *The Asian modern: Culture, capitalist development, Singapore*. Hong Kong: Hong Kong University Press.

Williams, S., & Narendran, S. (1999). Determinants of managerial risk: Exploring personality and cultural influences. *The Journal of Social Psychology, 139*(1), 102–125.

Effective and Comprehensive Governance of Biological Risks: A Network of Networks Approach for Sustainable Capacity Building

Tatyana Novossiolova, Lela Bakanidze, and Dana Perkins

Introduction

Natural outbreaks of disease could pose significant challenges to global security by undermining national economies, international trade and travel, public health and safety, and the trust of populace in its own government, potentially leading to ineffective governance or fragile state collapse (Bakanidze et al. 2010). The global biological threat environment is compounded by the possibility of rogue states and/or terrorists deliberately using biological agents as weapons of war. Laboratory-acquired infections (LAIs) have also attracted more attention in recent years, in particular with regard to high (biosafety level 3, or BSL-3) and maximum (BSL-4) containment laboratories. Poor personnel training increases the risk of a LAI or other biological accident in the laboratory and may also contribute to improper pathogen accounting, storage, and transportation, which in turn could contribute to the illicit acquisition of biological agents by terrorists or would-be bio-criminals. Any such use of a biological agent (whether overtly or covertly) could have potentially devastating consequences on public health or the environment.

The views expressed in this paper are those of the authors and may not reflect the official policy or position of the US Department of Health and Human Services or the US Government.

T. Novossiolova (✉)
Law Program, Center for the Study of Democracy, Sofia, Bulgaria
e-mail: tatyana.novossiolova@csd.bg

L. Bakanidze
On-Site Technical Assistance to the Chemical, Biological, Radiological and Nuclear Centres of Excellence Secretariats in Uzbekistan and Algeria, B&S Europe, Tashkent, Uzbekistan

D. Perkins
Office of the Assistant Secretary for Preparedness and Response, U.S. Department of Health and Human Services, Washington, DC, USA

© Springer Nature Switzerland AG 2020
B. D. Trump et al. (eds.), *Synthetic Biology 2020: Frontiers in Risk Analysis and Governance*, Risk, Systems and Decisions, https://doi.org/10.1007/978-3-030-27264-7_14

The expansion of biotechnology over the past several decades has been truly breathtaking, both in qualitative and quantitative terms (Novossiolova 2017). Forty years ago, scientists were fascinated by gene-splicing manipulations, while the tools and technologies available in the beginning of the twenty-first century have enabled them to *create* life forms from scratch (Cello et al. 2002; NAS 2005; NRC 2006; Wimmer 2006; Sample 2010; Noyce et al. 2018). Similarly, when initially conceived, the Human Genome Project (HGP) seemed a daunting undertaking, but within less than 10 years of its completion, the areas of genome-based diagnostics and therapeutics have been growing at a remarkable pace. While cutting-edge life science research was once confined to prestigious universities and state-of-the-art laboratories found in the highly industrialized countries in the global North, now studies involving highly dangerous microbes are conducted in research facilities scattered around the globe. Gene editing, synthetic genomics, NBIC (nano-bio-info-cognitive) technology convergence, and "do-it-yourself" (DIY) biology are just few examples of the scope, scale, and pace of the ongoing biotechnology (Daar 2002; Acharya et al. 2003; Roco et al. 2013; Crossley 2018; Seyfried et al. 2014).

Because infectious disease knows no borders, fostering biosafety and biosecurity capacities that aim to both ensure appropriate occupational health and safety procedures and practices and prevent unauthorized possession, loss, theft, misuse, diversion, or intentional release of biological agents and toxins is a shared responsibility at the international level. The effective national implementation of biological risk management regulations, policies, measures, and practices is a critical prerequisite for ensuring global health security and countering biological risks regardless of whether those are naturally occurring, accidental, or deliberate.

This chapter advances the argument that the effective and comprehensive governance of biological risks requires that all states take, in accordance with their constitutional processes, culture, and individual circumstances, relevant steps toward the effective and comprehensive national implementation of the existing internationally mandated or recommended measures in the area of biological security (Fig. 1). Sustainable capacity building underpinned by multi-stakeholder engagement and interagency cross-fertilization is an essential prerequisite for the achievement of this objective. The paper begins by an examination of the core aspects of the international governance of biological risks. It then looks into the ongoing efforts to develop consolidated national strategies and approaches for biological risk management. The paper concludes by highlighting the utility of capacity building through a "network of networks" approach for harmonizing the implementation of biological risk management policies and measures, in order to ensure sustainability and effectiveness.

Biological Risk Management: International Level

The spectrum of biological risks encompasses three categories: (1) naturally occurring disease outbreaks, including emerging threats such as antibiotic resistance; (2) accidental disease outbreaks resulting from, for example, laboratory lapses or neg-

Fig. 1 Biological risk management

ligence; and (3) deliberately caused disease outbreaks as a result of vandalism, sabotage, or the use of biological weapons (Stroot and Jenal 2011). Given the lack of a single focal point of threat, addressing this disparate array of risks falls within the remit of at least four regulatory frameworks: (1) health security, (2) prohibition of biological weapons, (3) biosafety and biosecurity, and (4) import and export controls, including the transport of dangerous goods and substances. The frameworks are mutually reinforcing and complementary, insofar as they seek to tackle different aspects of the biological risk spectrum.

The health security framework entails international regulations, guidelines, and initiatives that aim to promote and enhance the protection of global public, animal, and plant health. Human health and animal health are interdependent and bound to the health of the ecosystems in which they exist (OIE 2019a). In practice, this understanding of biological risks is commonly referred to as an integrated "One Health" approach. Health security requires effective capacities and mechanisms for disease prevention, epidemiological surveillance, early warning and diagnostics, and outbreak preparedness and response. It is closely linked to food security and economic development and cuts across virtually all sectors of social activity.

The framework for the prohibition of biological and toxin weapons comprises the international regulations and initiatives that aim to promote the peaceful use of life sciences and ensure that biological agents and toxins regardless of their origins are not misused for hostile purposes either by states or by non-state actors.

The biosafety and biosecurity framework encompasses international standards, guidelines, and initiatives that seek to foster safe, secure, and responsible practices within biological research facilities worldwide, in order to ensure the safe handling, including transfer, shipment, and transport and physical security of biological agents and toxins.

The import and export controls framework features the international arrangements and initiatives for the regulation of trade in sensitive dual-use materials, goods, and technology, in order to ensure that those are used only for peaceful, prophylactic, and protective purposes. This framework is traditionally considered in conjunction with the framework for the prohibition of biological and toxin weapons.

An indicative list of the international regulations, guidelines, and initiatives that pertain to each framework is presented in Table 1.

Biological Risk Management: National Level

The national implementation of biological risk management requires a synchronized effort in at least three key domains: (1) civil protection and preparedness, (2) law enforcement and counterterrorism, and (3) research oversight and responsible science culture. The domains are mutually reinforcing and complementary, insofar as they aim at the development of an integrated national approach to biological risks.

The domain of civil protection and preparedness entails policies, strategies, and measures for mobilizing resources and capacities for effective response to a disease outbreak regardless of its origins. Efficient interagency coordination, communication, and collaboration are essential for the adequate functioning of any national civil protection and preparedness system. Ongoing needs assessment and performance monitoring are key to maintaining the effectiveness and adaptive capacity of these systems, as well as to enhancing their sustainability and resilience.

The domain of law enforcement and counterterrorism encompasses policies, strategies, and measures for enhancing national capacities for preventing the hostile misuse of biological agents and toxins. This includes effective border and customs controls, adequate policing, engagement between the law enforcement community and life science and health communities, and public awareness of biological risks.

The domain of research oversight and responsible science culture (Perkins et al. 2018) comprises regulations, policies, and measures for promoting safe, secure, and responsible handling of biological agents and toxins. This includes the introduction of relevant legal rules, licensing and certification procedures, inspections and audit, training, education, and outreach programs and the development of institutional guidelines and codes of conduct.

An indicative list of national policies, strategies, and measures of relevance to each domain is presented in Table 2.

Table 1 International biological risk management

International regulation/ initiative	Implementing agency	Description	Supplementary tools
Health security			
International Health Regulations (IHR) – 2005	World Health Organization	Designed to help states build their capacities to detect, assess, and report public health events. The IHR also include specific measures at ports, airports, and ground crossings to limit the spread of health risks to neighboring countries and to prevent unwarranted travel and trade restrictions so that traffic and trade disruption is kept to a minimum (WHO 2018a)	*Health Security under the IHR (2005) Framework* – a training catalogue geared at both individual and institutional levels (WHO 2017) *Health Security Learning Platform* – educational resource (WHO 2018b)
OIE Terrestrial Animal Health Code	World Animal Health Organisation (OIE)	Sets out standards for the improvement of animal health and welfare and veterinary public health worldwide, including through standards for safe international trade in terrestrial animals (mammals, reptiles, birds, and bees) and their products (OIE 2018a)	*Performance of Veterinary Services (PVS) Pathway* – a global program for the sustainable improvement of a country's Veterinary Services' compliance with OIE standards (OIE 2018c). *OIE Tool for the Evaluation of Performance of Veterinary Services (OIE PVS Tool)* – a specific methodology which allows for evaluating the performance of Veterinary Services (OIE 2018c) *Biological Threat Reduction Strategy: Strengthening Global Biological Security* (OIE 2015). *OIE Guidelines for Investigation of Suspicious Biological Events* (OIE 2018d) *OIE Guidelines for Responsible Conduct in Veterinary Research: Identifying, Assessing and Managing Dual Use* (OIE 2019b)
OIE Aquatic Animal Health Code		Sets standards for the improvement of aquatic animal health and welfare of farmed fish worldwide, and for safe international trade in aquatic animals (amphibians, crustaceans, fish, and molluscs) and their products (OIE 2018b).	

(continued)

Table 1 (continued)

International regulation/initiative	Implementing agency	Description	Supplementary tools
International Plant Protection Convention (IPPC) – 1951	Food and Agricultural Organization of the United Nations (FAO)	Sets standards for the safe movement of plants and plant products to prevent the spread of plant pests and diseases internationally. Its governing body is the Commission on Phytosanitary Measures (CPM) comprising the IPPC Contracting Parties (183 states including the European Union) (IPCC 2018).	*International Standards for Phytosanitary Measures (ISPM)* – operate in conjunction with the 1995 World Trade Organization (WTO) Agreement on the Application of Sanitary and Phytosanitary Measures (SPS Agreement) and Codex Alimentarius (WTO 1998; FAO 2008, 2018)
Convention on Biological Diversity (CBD) – 1993	Conference of the Parties (COP)	Seeks to ensure (1) the conservation of biological diversity; (2) the sustainable use of the components of biological diversity; and (3) the fair and equitable sharing of the benefits arising out of the utilization of genetic resources (CBD 2018a). The Convention has 196 states parties including the European Union	*Strategic Plan for Biodiversity, 2011–2020* – provides an overarching framework for action by all countries and stakeholders to save biodiversity and enhance its benefits for people (CBD 2010); features 20 biodiversity targets, known as the *Aichi Targets* (CBD 2018b)
Transforming Our World: The 2030 Agenda for Sustainable Development	United Nations General Assembly	The Agenda is a plan of action for people, planet, and prosperity which seeks to strengthen universal peace in larger freedom. It features 17 Sustainable Development Goals and 169 targets (UN General Assembly 2015)	Goal 3: Good Health and Well-Being Goal 14: Life Below Water Goal 15: Life on Land
Global Health Security Agenda (GHSA) – 2014	Ad-hoc intergovernmental initiative	A partnership of nearly 50 nations, international organizations, and nongovernmental stakeholders which seeks to facilitate collaborative, capacity-building efforts to achieve specific and measurable targets around biological threats.	Structured around the thematic framework of Prevent–Detect–Respond with Action Packages covering specific relevant domains (GHSA 2018)

Global Health Security Initiative (GHSI) – 2001	Ad-hoc intergovernmental initiative	An informal, international partnership among like-minded countries to strengthen health preparedness and response globally to threats of biological, chemical, and radio-nuclear terrorism (CBRN) and pandemic influenza (GHSI 2018a). The GHSI operates through regular ministerial meetings at which progress is reviewed (GHSI 2018b)	
Prohibition of biological weapons			
Protocol for the Prohibition of the Use in War of Asphyxiating, Poisonous or Other Gases, and of Bacteriological Methods of Warfare (1925 Geneva Protocol)	140 states parties	Prohibits the use of chemical and biological weapons in war (UNODA 2018a)	
Biological and Toxin Weapons Convention (BTWC) – 1975	183 states parties 4 Signatories	Prohibits the development, production, stockpiling, or otherwise acquisition or retention of (1) "microbial or other biological agents, or toxins whatever their origin or method of production, of types and in quantities that have no justification for prophylactic, protective or other peaceful purposes; (2) weapons, equipment, or means of delivery designed to use such agents or toxins for hostile purposes or in armed conflict." The requirement is both to prohibit and to *prevent* the development of biological weapons (Rappert et al. 2006)	Article IV: "Each State Party to this Convention shall, in accordance with its constitutional processes, take any necessary measure to prohibit and prevent the development, production, stockpiling, acquisition or retention of agents, toxins, weapons, equipment and means of delivery specified in Article I of the Convention, within the territory of such state, under its jurisdiction or under its control anywhere."

(continued)

Table 1 (continued)

International regulation/initiative	Implementing agency	Description	Supplementary tools
Chemical Weapons Convention (CWC) – 1997	Organisation for the Prohibition of Chemical Weapons	Prohibits the development, production, acquisition, stockpiling, or retention of chemical weapons; their direct or indirect transfer; and the use thereof (OPCW 2018).	Article VII: "Each State Party shall, in accordance with its constitutional processes, adopt the necessary measures to implement its obligations under this Convention."
United Nations Security Council Resolution 1540 – 2004	1540 Committee	Calls upon all states to take and enforce effective measures to establish domestic controls to prevent the proliferation of nuclear, chemical, or biological weapons and their means of delivery, including by establishing appropriate controls over related materials (UNSC 2004)	
Global Counter-Terrorism Strategy	United Nations Office of Counter-Terrorism (UNOCT)	Adopted by the UN General Assembly in 2006 A unique international instrument to enhance national, regional, and international efforts to counter terrorism Comprises four pillars of action: (1) Addressing the conditions conducive to the spread of terrorism (2) Measures to prevent and combat terrorism (3) Measures to build states' capacity to prevent and combat terrorism and to strengthen the role of the United Nations system in that regard (4) Measures to ensure respect for human rights for all and the rule of law as the fundamental basis for the fight against terrorism (UNOCT 2006)	*Global Counter-Terrorism Plan of Action* (2) Measures to prevent and combat terrorism "To strengthen coordination and cooperation among States in combating crimes that might be connected with terrorism, including [...] smuggling of nuclear, chemical, biological, radiological and other potentially deadly materials. To step-up national efforts and bilateral, sub-regional, regional and international co-operation, as appropriate, to improve border and customs controls, in order to prevent and detect the movement of terrorists and to prevent and detect the illicit traffic in, inter alia, [...] nuclear, chemical, biological or radiological weapons and materials, while recognizing that States may require assistance to that effect" (UNOCT 2006)

| Securing Our Common Future: An Agenda for Disarmament | United Nations Office for Disarmament Affairs (UNODA) | "[…] concerns regarding the increasing risk of biological weapons have continued to grow as developments in science and technology lower barriers for their acquisition, access and use, including by non-State actors. There is therefore a need to strengthen the Biological Weapons Convention, which acts as a forum for consideration of preventative measures, such as strong national health systems, robust response capacities and effective counter-measures. The first step is to ensure more effective implementation of the Convention. This should be done by improving linkages with other relevant activities—for example in the domain of global health security—and oversight of dual-use research of concern, including in the context of Sustainable Development Goal 3 on health and well-being" (UNODA 2018b) |

(continued)

Table 1 (continued)

International regulation/ initiative	Implementing agency	Description	Supplementary tools
UN Secretary-General's Mechanism for Investigation of Alleged Use of Chemical and Biological Weapons (UNSGM)	United Nations Office for Disarmament Affairs (UNODA)	The purpose of the Secretary-General's Mechanism to carry out prompt investigations in response to allegations brought to his attention concerning the possible use of chemical and bacteriological (biological) and toxin weapons. The UNSGM is triggered by a request from any Member State which authorizes the Secretary-General to launch an investigation including dispatching a fact-finding team to the site(s) of the alleged incident(s) and to report to all UN member states. The UNSGM is intended to ascertain in an objective and scientific manner facts of alleged violations of the 1925 Geneva Protocol UNODA 2019)	Roster of experts and laboratories provided by UN Member States *UNSGM Guidelines and Procedures for the conduct of investigations*
INTERPOL Bioterrorism Projects	INTERPOL Bioterrorism Prevention Unit	Aims to facilitate targeted training and capacity building for law enforcement agencies on how to prevent, prepare, and respond to a bioterrorist attack as well as how to promote inter-agency coordination (INTERPOL 2018)	
Global Partnership against the Spread of Weapons and Materials of Mass Destruction – 2002	Ad-hoc intergovernmental initiative	An international initiative aimed at preventing the proliferation of chemical, biological, radiological, and nuclear weapons and related materials (Global Partnership 2018). With regard to biological risk management, the Global Partnership works in two thematic areas: mitigation of biological threats and support for the implementation of United Nations Security Council Resolution 1540	With regard to biological risk management, the Global Partnership works in two thematic areas: (1) mitigation of biological threats; and (2) support for the implementation of United Nations Security Council Resolution 1540

European Union Chemical Biological Radiological and Nuclear Risk Mitigation Centres of Excellence Initiative (EU CBRN CoE) – 2011	European Commission, EU CBRN CoE Partner Countries, UN Interregional Crime Research Institute (UNICRI)	Seeks to strengthen the institutional capacity of countries outside the European Union to mitigate CBRN risks by promoting regional security through the development of local ownership, local expertise, and long-term sustainability (EU CBRN CoE 2018).	Supports the development of national CBRN Teams; develops methodology and guidelines; enhances networking and cooperation both at a national and regional level (EU CBRN CoE 2018)
Recommendation on Science and Scientific Researchers	United Nations Educational, Scientific and Cultural Organization (UNESCO)	Recommended responsibilities and rights of scientific researchers: "to express themselves freely and openly on the ethical, human, scientific, social or ecological value of certain projects, and in those instances where the development of science and technology undermine human welfare, dignity and human rights or is 'dual use', they have the right to withdraw from those projects if their conscience so dictates and the right and responsibility to express themselves freely on and to report these concerns" (UNESCO 2017)	
Doing Global Science: A Guide to Responsible Conduct in the Global Research Enterprise, Chapter 3, "Preventing the Misuse of Research and Technology"	InterAcademy Partnership	Published in 2016 Contains educational material regarding the responsibilities of scientists including the duty to address issues of dual use and safeguard their work against hostile misuse (IAP 2016)	

(continued)

Table 1 (continued)

International regulation/ initiative	Implementing agency	Description	Supplementary tools
Biosafety and biosecurity			
ISO 15190:2003, Medical Laboratories – Requirements for Safety	International Organization for Standardization	Specifies requirements to establish and maintain a safe working environment in a medical laboratory (ISO 2003)	A new international standard titled ISO 35001, biorisk management for laboratories and other related organizations is currently under publication (ISO 2019)
ISO 15189:2012, Medical Laboratories – Requirements for Quality and Competence		Specifies requirements for quality and competence in medical laboratories. The Standard can be used by medical laboratories in developing their quality management systems and assessing their own competence (ISO 2012).	
Laboratory Biosafety Manual, 3rd Edition, 2003	World Health Organization	Provides practical guidance on biosafety techniques for use in laboratories at all levels, as well as guidelines for the commissioning and certification of laboratories (WHO 2003)	*Biorisk Management Advanced Trainer Programme* (WHO 2018c) *Laboratory Quality Management System Training Toolkit* (WHO 2018d) *Laboratory Issues for Epidemiologists Training Programme* (WHO 2018e)
Biorisk Management: Laboratory Biosecurity Guidance, 2006		Provides detailed guidance on biosecurity within a biological laboratory and addresses basic principles and best practices (WHO 2006)	
Responsible Life Sciences Research for Global Health Security, 2010		Aims at strengthening the culture of scientific integrity and excellence characterized by openness, honesty, accountability, and responsibility (WHO 2010).	
Laboratory Quality Management System: Handbook, 2011		Provides a comprehensive reference on laboratory quality management systems for all stakeholders in health laboratory processes, from management, to administration, to benchwork laboratorians (WHO 2011)	

Manual of Diagnostic Tests and Vaccines for Terrestrial Animals, 7th Edition 2012	World Animal Health Organisation	Aims to provide internationally agreed diagnostic laboratory methods and requirements for the production and control of vaccines and other biological products (OIE 2012)	
Manual of Diagnostic Tests for Aquatic Animals, 7th Edition, 2016		Aims to provide a standardized approach to the diagnosis of the diseases listed in the *Aquatic Code*, to facilitate health certification for trade in aquatic animals and aquatic animal products (OIE 2016)	
Cartagena Protocol on Biosafety – 2003	Intergovernmental Committee for the Cartagena Protocol on Biosafety (ICCP) – 171 states parties including the European Union (CBD 2018c)	Aims to ensure the safe handling, transport, and use of living modified organisms (LMOs) resulting from modern biotechnology that may have adverse effects on biological diversity, taking also into account risks to human health (CBD 2018c)	*Advance informed agreement (AIA)* procedure – ensures that countries are provided with the information necessary to make informed decisions before agreeing to the import of LMOs into their territory (CBD 2018c) Biosafety Clearing House (BCH) – facilitates the exchange of information on LMOs and assists States Parties in complying with their obligations under the *Protocol* (CBD 2018d)
Recommendations on the Transport of Dangerous Goods: Model Regulations (Rev.20)	United Nations Committee of Experts on the Transport of Dangerous Goods	Covers the classification of dangerous goods, their listing, the use, construction, testing, and approval of packaging and portable tanks, as well as consignment procedures such as marking, labelling, placarding, and documentation (UNECE 2017)	

(continued)

Table 1 (continued)

International regulation/ initiative	Implementing agency	Description	Supplementary tools
Guidance on Regulations for the Transport of Infectious Substances 2019-2020	World Health Organization	Provides information for identifying, classifying, marking, labelling, packaging, documenting, and refrigerating infectious substances for transportation by all modes of transport, both nationally and internationally, and ensuring their safe delivery (WHO 2019)	*Infectious Substances Shipping Training* – a course for shippers divided into modules addressing the classification, documentation, marking, labelling, packaging of infectious substances, and the preparation of shipments requiring the use of dry ice (WHO 2015b)
Guide for Shippers of Infectious Substances, 2015		Aims to assist shippers with classifying, documenting, marking, labelling, and packaging infectious substances (WHO 2015a)	
International Professional Certification Program	International Federation of Biosafety Associations (IFBA)	Based on a Body of Knowledge (BOK) developed by international subject matter experts and validated through a Job Task Analysis (JTA) process in accordance with international standards. IFBA certifications are valid for a period of 5 years after which the certificants must undergo a recertification process (IFBA 2018)	Technical disciplines: (1) Biorisk management (2) Biological waste management (3) Biocontainment facility design, operations, and maintenance (4) Biosafety cabinet selection, installation, and safe use (5) Biosecurity

Biological and Toxin Weapons Convention (BTWC) – 1975	183 states parties 4 signatories	Article III: "Each State Party to this Convention undertakes not to transfer to any recipient whatsoever, directly or indirectly, and not in any way to assist, encourage, or induce any State, group of States or international organisations to manufacture or otherwise acquire any of the agents, toxins, weapons, equipment or means of delivery specified in Article I of the Convention."	Article X: (1) The States Parties to this Convention undertake to facilitate, and have the right to participate in, the fullest possible exchange of equipment, materials and scientific and technological information for the use of bacteriological (biological) agents and toxins for peaceful purposes. Parties to the Convention in a position to do so shall also co-operate in contributing individually or together with other States or international organisations to the further development and application of scientific discoveries in the field of bacteriology (biology) for the prevention of disease, or for other peaceful purposes. (2) This Convention shall be implemented in a manner designed to avoid hampering the economic or technological development of States Parties to the Convention or international co-operation in the field of peaceful bacteriological (biological) activities, including the international exchange of bacteriological (biological) agents and toxins and equipment for the processing, use or production of bacteriological (biological) agents and toxins for peaceful purposes in accordance with the provisions of the Convention.

(continued)

Table 1 (continued)

International regulation/ initiative	Implementing agency	Description	Supplementary tools
United Nations Security Council Resolution 1540 – 2004	1540 Committee	"[…] All States shall take and enforce effective measures to establish domestic controls to prevent the proliferation of […] biological weapons and their means of delivery" (UNSC 2004)	The domestic controls required by UNSCR 1540 include "effective border controls and law enforcement efforts to detect, deter, prevent and combat […] the illicit trafficking and brokering in such items"; "appropriate effective national export and trans-shipment controls over such items, including appropriate laws and regulations to control export, transit, trans-shipment and re-export and controls on providing funds and services related to such export and trans-shipment such as financing, and transporting that would contribute to proliferation, as well as establishing end-user controls."
Wassenaar Arrangement on Export Controls for Conventional Arms and Dual-Use Goods and Technologies – 1995	Ad-hoc intergovernmental initiative	Seeks to contribute to regional and international security and stability, by promoting transparency and greater responsibility in transfers of conventional arms and dual-use goods and technologies (Wassenaar Agreement 2018a)	*List of Dual-Use Goods and Technologies and the Munitions List* – refers to biological agents and related protective equipment and detection systems (Wassenaar Agreement 2018b)
Regulatory System of Dual-Use Trade Controls	European Union	Aimed at complementing international efforts to prevent the proliferation of weapons of mass destruction (WMD) and related technology	Common export control rules. A common EU list of dual-use items. A "catch-all clause" for nonlisted items Controls on brokering dual-use items and transit thereof. Record-keeping and registers. Information exchange (European Commission 2018)

Australia Group	Ad-hoc intergovernmental initiative	Aims to promote the use of licensing measures to ensure that exports of certain chemicals, biological agents, and dual-use chemical and biological manufacturing facilities and equipment do not contribute to the spread of CBW (Australia Group 2007a)	*Australia Group Common Control List Handbook* – a commodity-oriented training material to enhance the capabilities of enforcement officers to identify dual-use materials and equipment in cargo shipment (Australia Group 2007b)
Green Customs Initiative	Partnership of international organizations	Seeks to enhance the capacity of customs and other relevant border control officers to monitor and facilitate the legal trade and to detect and prevent illegal trade in environmentally sensitive commodities covered by relevant trade-related Multilateral Environmental Agreements (MEAs) and international conventions (Green Customs 2018)	

Table 2 Biological risk management at a national level

Country	Initiative	Type of initiative	Stakeholder	Description
Civil protection and preparedness				
USA	*National Biodefense Strategy*	National strategy	US Government	Seeks to ensure that the United States actively and effectively prevents, prepares for, responds to, recovers from, and mitigates risk from natural, accidental, or deliberate biological threats
				Acknowledges that biological threats are global in nature, persistent, and originate from multiple sources, underscoring the need for a multi-sectoral cooperation and interdisciplinary approach for effective disease prevention and response
				Core goals:
				1. Enable risk awareness to inform decision-making across the biodefense enterprise.
				2. Ensure biodefense enterprise capabilities to prevent bioincidents.
				3. Ensure biodefense enterprise preparedness to reduce the impacts of bioincidents.
				4. Rapidly respond to limit the impacts of bioincidents.
				5. Facilitate recovery to restore the community, the economy, and the environment after a bioincident (The White House 2018a)
USA	*National Health Security Strategy*	National strategy	US Government, DHHS	Published every 4 years to establish a strategic approach to enhance health security in times of crisis
				Accompanied by a *NHSS Implementation Plan 2019–2022* which presents a road map for promoting health security through enhanced readiness and response capabilities (DHHS 2019b)
				Core objectives:
				1. Prepare, mobilize, and coordinate the Whole-of-Government to bring the full spectrum of federal medical and public health capabilities to support state, local, tribal, and territorial authorities in the event of a public health emergency, disaster, or attack.
				2. Protect the nation from the health effects of emerging and pandemic infectious diseases and chemical, biological, radiological, and nuclear (CBRN) threats.
				3. Leverage the capabilities of the private sector (DHHS 2019a)

UK	*UK Biological Security Strategy*	National strategy	UK Government	Intended as a consolidated document that draws together the work carried out by different arms of the government for the purpose of protecting the country from biological risks, no matter how these risks occur and no matter who or what they affect Acknowledges that an all-hazards approach addressing natural, accidental, and deliberate risks to public, animal, and plant health is required to meet emerging biosecurity challenges in an era of rapid globalization Four pillars: 1. Understanding biological risks 2. Preventing biological risks 3. Detecting, characterizing, and reporting biological risks 4. Responding to biological risks Two additional aspects: 1. That the effective response to biological risks requires certain scientific capabilities and capacities 2. That the biological sectors offer opportunities that need to be fully leveraged (UK Government 2018)
Law enforcement and counterterrorism				
Australia	*Biosecurity Act 2015*	Regulation	Government, Department of Agriculture and Water Resources; Department of Health	Addresses managing diseases and pests that may cause harm to human, animal, or plant health or the environment (Australian Government 2015)
Canada	*Human Pathogens and Toxins Act*	Regulation	Government, Public Health Agency of Canada	Addresses risks posed by domestically acquired and/or domestically produced human pathogens and toxins, as well as by imported human pathogens and toxins (Almquist et al. 2015)

(continued)

Table 2 (continued)

Country	Initiative	Type of initiative	Stakeholder	Description
Canada	*Human Pathogens and Toxin Regulations*	Regulation	Government, Public Health Agency of Canada	Support the implementation of a national licensing program for Canadian laboratories working with human pathogens and toxins and a security clearance program for researchers and individuals with access to a list of high-consequence human pathogens and toxins (known as "security-sensitive biological agents") in Canada) Outline the functions to be performed by a designated biological safety officer at a licensed organization (Almquist et al. 2015)
Denmark	*Biosecurity Act 2008*	Regulation	Government, Centre for Biosecurity and Biopreparedness	Addresses the securing of specific biological substances, delivery systems, and related materials (CBB 2017)
USA	*Federal Select Agent Program*	Regulation	Government, Centers for Disease Control and Prevention, Animal and Plant Health Inspection Service, Federal Bureau of Investigation	Develops, implements, and enforces the Select Agent Regulations Oversees the possession, use, and transfer of biological select agents and toxins, which have the potential to pose a severe threat to public, animal, or plant health or to animal or plant products (FSAP 2019)
USA	*FBI Biosecurity Program*	Program	Government, Federal Bureau of Investigation	1. Enforcement of the regulations related to the Federal Select Agent Program (detailed in the following section) – research institutions that carry out activities involving any of the 65 agents and toxins featured on the Biological Select Agents and Toxins List are required to register if they possess, use, or transfer these agents and to report instances of misplacement, theft, and loss thereof.
				2. Personnel reliability programs (PPRs) – work with certain select agents and toxins require that staff are subject to background (preemployment suitability) and continuous reliability checks in order to help prevent the misuse of highly dangerous agents and toxins.

Country	Type	Organization	Document	Description
				3. Monitoring and analysis of relevant advances in science and technology.
				4. Scientist engagement – fostering partnerships with institutional biosafety committees, practicing researchers, and the "do-it-yourself" biology community in order to promote a trust-based approach to the governance of biological risks (So 2015)
USA	Guidance	Federal Bureau of Investigation, Centers for Disease Control and Prevention	Joint Criminal and Epidemiological Investigations Handbook (2016)	Facilitates the use of resources and to maximize communication and interaction among law enforcement and public health officials in an effort to minimize potential barriers to communication and information sharing during a response to biological threat. Provides an overview of criminal and epidemiological investigational procedures and methodologies for a response to a biological threat, identifies potential challenges to data sharing, proposes possible solutions, and demonstrates effective law enforcement and public health collaboration (FBI 2016)
USA	Guidance	Department of Agriculture, Food and Drug Administration, Federal Bureau of Investigation	Criminal Investigation Handbook for Agroterrorism (2008)	Aims to facilitate communication and interaction among officials and representatives from law enforcement, animal health, plant health, and public health who become involved in a joint investigation of a potential or actual agroterrorism event. Aims to foster a greater understanding about the food and agriculture sector among law enforcement officials to minimize potential communication barriers during an agroterrorism event (FBI 2008)
USA	National strategy	Government	National Strategy for Counterterrorism	Aims to ensure that "terrorists are unable to acquire or use WMDs, including chemical, biological, radiological, and nuclear weapons, and other advanced weaponry." (The White House 2018b)
USA	National strategy	Government	National Strategy for Countering Weapons of Mass Destruction (WMD) Terrorism	Aims to reduce the probability that extremist groups and individuals will conduct attacks using WMD through the achievement of the following strategic objectives: 1. That the agents, precursors, and materials needed to acquire WMD are placed beyond the reach of terrorists and other malicious non-state actors, and the global quantity of WMD and related materials is reduced.

(continued)

Table 2 (continued)

Country	Initiative	Type of initiative	Stakeholder	Description
				2. That states and individuals are deterred from providing support to would-be WMD terrorists.
				3. That an effective architecture is in place to detect and defeat terrorist WMD networks.
				4. That national defenses against WMD terrorism are strengthened, and state, local, tribal, and territorial preparedness to contend with WMD threats is enhanced.
				5. That the United States is able to identify and respond to technological trends that may enable terrorist development, acquisition, or use of WMD (The White House 2018c)
Research oversight and responsible science culture				
Canada	_Biosafety and Biosecurity Programme_	Program	Government, Public Health Agency of Canada	Key elements: 1. Facility licensing – All persons that conduct controlled activities in Canada with human pathogens or toxins (i.e., possessing, handling, using, producing, storing, permitting access to, transferring, importing, exporting, releasing, abandoning, or disposing) are required to obtain a license from the Public Health Agency of Canada. The organization where activities are taking place needs to apply for a license to ensure that all people and laboratories under its purview are authorized to handle and store these agents.
				2. Biological safety officers – The biological safety officer is the main point of contact between the licensed organization and the Public Health Agency of Canada and a key compliance and monitoring resource for the license holder.
				3. Security clearances – Individuals that work with, or have access to, security-sensitive biological agents are required to obtain a security clearance from the Public Health Agency of Canada. This includes individuals conducting research, those who require access to storage space, workers or researchers that open shipments containing security-sensitive biological agents, and personnel that handle or care for animals experimentally infected.

		4. Incident reporting – A license holder has a legal obligation to inform the Public Health Agency of Canada if any of the following incidents with human pathogens and/or toxins occur: Inadvertent release from a laboratory, inadvertent production of a human pathogen or toxin, an incident that has or may have caused a disease, and stolen or otherwise missing human pathogens or toxins.
		5. Institutional biosafety program – A licensed organization is required to have a biosafety program in place that is supported by a Biosafety Manual that documents the program and describes how the organization will achieve the goals and objectives of the program. Common safety measures in a biosafety program include good microbiological laboratory practices, appropriate primary containment equipment, and proper physical design of the containment zone.
		6. Institutional biosecurity plan – All licensed organizations are required to develop, implement, monitor, and review a biosecurity plan that describes mitigation strategies for the risk associated with biological assets (i.e., human pathogens and toxins) in a facility's possession. Biosecurity plans must cover five areas: physical security, personnel suitability and reliability, pathogen and toxin accountability and inventory control, incident and emergency response, and information security.
		7. National compliance verification program – Canada's approach to biosafety and biosecurity includes a national compliance and enforcement program that uses different tools and strategies to promote, monitor, and verify compliance, including the delivery of training and educational resources and conducting audits and inspections (Almquist et al. 2015)
Denmark	Policy	Centre for Biosecurity and Biopreparedness	Key elements: 1. Licensing – research institutions, pharmaceutical companies, hospital laboratories, etc. wishing to work with biological dual-use components should hold a license from the CBB as part of the country's implementation of obligations under United Nations Security Council Resolution 1540 to prevent the diversion and acquisition of biological materials for nefarious purposes.
	Biosecurity Licensing Policy		

(continued)

Table 2 (continued)

Country	Initiative	Type of initiative	Stakeholder	Description
				2. Biosecurity officer – all entities with a license need to have a biosecurity officer who is responsible for registering stocks and training personnel who have access to controlled materials. Biosecurity officers undergo a formal training procedure at the CBB which provides a theoretical introduction to the background of the legislation on biosecurity and addresses all aspects of daily administration of biosecurity rules and regulations.
				3. Training and awareness-raising – CBB offers courses for different stakeholders with the goal of strengthening biosecurity. *International Biosecurity Course* provides detailed guidance on how to establish and implement a national biosecurity system (CBB 2015). *Responsible science, ethics, and biosecurity in the life sciences* is targeted at graduate students and aims to facilitate discussion on the scientific responsibility of preventing misuse of research results and methods in life science (Petersen 2015; CBB 2018a, b)
Germany	*Biosecurity – Freedom and Responsibility of Research (2014)*	High-level review	German Ethics Council	Examines the issue of dual-use research and its legal, ethical, and policy context and the need for basic, advanced, and continuing education in the area of biosecurity. Among the recommendations of the report are: 1. Raising the level of awareness for questions of biosecurity in the scientific community; and
				2. Elaboration of a national biosecurity code of conduct for responsible research that defines what constitutes a responsible manner of dealing with biosecurity questions (German Ethics Council 2014)

Germany	Scientific Freedom and Scientific Responsibility: Recommendations for Handling Security-Relevant Research (2014)	Guidelines	German Research Foundation, National Academy of Sciences Leopoldina	Directed at individual scientists and research institutions, respectively, and seek to "raise awareness of risks, provide ethical guidelines to assist with answering ethical questions, and minimise risks through self-regulation." The "General Recommendations on Ethically Responsible Research" cover the following areas: (1) risk analysis, (2) minimizing risk, (3) evaluating publications, (4) forgoing research as a last resort, (5) documentation and communication of risks, (6) training and information, and (7) responsible persons. The "Supplementary Organisational Recommendations for Research Institutions" cover the following areas: (1) legal provisions and compliance units, (2) ethics rules and research ethics committees; and (3) education and training (German Academy of Sciences 2014)
The Netherlands	Biosecurity Policy	Policy	Ministry of Health, Welfare, and Sport National Institute for Public Health and the Environment Biosecurity Office	Seeks to secure high-risk biological material within organizations by focusing on eight biosecurity priority areas: (1) biosecurity awareness, (2) personnel reliability, (3) transport security and export control, (4) information security, (5) accountability for materials, (6) emergency response, (7) management, and (8) physical security. Self-Scan Toolkit and Biosecurity Vulnerability Scan have been developed to assist practitioners in conducting self-checks on the state of biosecurity within organizations (Biosecurity Office 2018a). Both tools are structured around the eight priority areas outlined above. The Self-Scan Toolkit is a relatively fast scan with a limited number of closed questions that can easily form an indication of strong and weak biosecurity aspects within an organization (Biosecurity Office 2018b). The Biosecurity Vulnerability Scan is a more extended scan featuring questions, scenarios, and best practices
The Netherlands	Code of Conduct for Biosecurity (2008)	Guidance	Royal Netherlands Academy of Arts and Sciences (KNAW)	Aims "to prevent life sciences research or its application from directly or indirectly contributing to the development, production or stockpiling of biological weapons, as described in the Biological and Toxin Weapons Convention (BTWC), or to any other misuse of biological agents and toxins." Intended for those in the life sciences whether in government, industry, or academia. The code defines the following rules of conduct: (1) raising awareness, (2) research and publication policy, (3) accountability and oversight, (4) internal and external communication, (5) accessibility, and (6) shipment and transport (KNAW 2008)

(continued)

Table 2 (continued)

Country	Initiative	Type of initiative	Stakeholder	Description
The Netherlands	*Improving Biosecurity – Assessment of Dual-Use Research (2013)*	High-level review	Royal Netherlands Academy of Arts and Sciences (KNAW)	With regard to the institutional oversight of dual-use research, the report recommended that: "The Code of Conduct for Biosecurity should be an ongoing topic of interest in education and researcher training and for research team heads and funding bodies. Drawing attention to the Code will raise awareness of possible dilemmas in dual-use research and may encourage stakeholders to be more active and vigilant" (KNAW 2013)
UK	*Position Statement on Dual Use Research of Concern and Research Misuse (2015)*	Position statement	Biotechnology and Biological Sciences Research Council (BBSRC), Medical Research Council, and the Wellcome Trust	11. "As a condition of funding institutions in receipt of BBSRC, MRC and WT funds are responsible for ensuring that they comply fully with the requirements of all regulatory authorities for the storage, use and transfer of all potentially harmful materials, including pathogenic organisms, and any additional provisions to safeguard security that may be specified by such authorities. Institutions also accept full responsibility for the management, monitoring and control of all research work funded by grants, and for ensuring that permanent and temporary staff and students undertaking such work receive training appropriate to their duties. The funding bodies continue to work with their sponsored institutes to ensure high levels of compliance" (BBSRC et al. 2015)
USA	*United States Government Policy for Oversight of Life Sciences Dual-Use Research of Concern (2012)*	Policy	Government Department of Health and Human Services	Applies to all federally funded research involving 15 select agents and toxins and 7 types of experiments (US Government 2012)

USA	*United States Government Policy for Institutional Oversight of Life Sciences Dual Use Research of Concern (2014)*	Policy	Government Department of Health and Human Services	Designed to assist funding agencies and research institutions that are subject to the 2012 Policy. Institutional oversight of dual-use research of concern is considered a critical component of a comprehensive oversight system because institutions are most familiar with the life sciences research conducted in their facilities and are in the best position to promote and strengthen the responsible conduct and communication (US Government 2014)
USA	*Screening Framework Guidance for Providers of Synthetic Double-Stranded DNA (2010)*	Guidance	Government Department of Health and Human Services	Sets forth recommended baseline standards for the gene and genome synthesis industry and other providers of synthetic double-stranded DNA products regarding the screening of orders, so that they are filled in compliance with current US regulations and to encourage best practices in addressing biosecurity concerns associated with the potential misuse of their products to bypass existing regulatory controls. Comprises customer screening (customer legitimacy check), sequence screening ("sequences of concern" check), and follow-up screening (end-use legitimacy check). Provides recommendations regarding proper records retention protocols and screening software (DHHS 2010)
USA	*Neuroethics Guiding Principles (2018)*	Guidance	National Institutes of Health – BRAIN Initiative	4. "Attend to possible malign uses of neuroscience tools and neurotechnologies Novel tools and technologies, including neurotechnologies, can be used both for good ends and bad. [...] Researchers have a responsibility to try to predict plausible misuses and ensure that foreseeable risks are understood, as appropriate, by research participants, IRBs, ethicists, and government officials." (Greely et al. 2018)
USA	*Recommendations for the Evaluation and Oversight of Proposed Gain-of-Function Research (2016)*	Guidance	National Science Advisory Board for Biosecurity	Recommendation 1: "Research proposals involving GOF research of concern entail significant potential risks and should receive an additional, multidisciplinary review, prior to determining whether they are acceptable for funding. If funded, such projects should be subject to ongoing oversight at the federal and institutional levels" (NSABB 2016)

(continued)

Table 2 (continued)

Country	Initiative	Type of initiative	Stakeholder	Description
USA	*Framework for Conducting Risk and Benefit Assessments of Gain-of-Function Research (2015)*	Guidance	National Science Advisory Board for Biosecurity	Contains guiding principles for designing and conducting risk and benefit assessment of gain-of-function experiments (NSABB 2015)
USA	*Guidance for Enhancing Personnel Reliability and Strengthening the Culture of Responsibility (2011)*	Guidance	National Science Advisory Board for Biosecurity	Contains recommendations on (1) hiring and employment practices, (2) encouraging biosecurity awareness and promoting responsible conduct, and (3) assessing the effectiveness of practices aimed at enhancing personnel reliability and the culture of responsibility (NSABB 2011)
USA	*Governance of Dual Use Research in the Life Sciences: Advancing Global Consensus on Research Oversight: Proceedings of a Workshop (2018)*	High-level review	National Academy of Sciences	Contains a comprehensive overview of activities across the governance landscape and regional and international forums, organizations, and bodies (NAS 2018)

Biological Risk Management: A "Network of Networks" Approach to Capacity Building

In basic terms, capacity building refers to the development of the human and infrastructural aspects that are deemed essential for the effective management of biological risks. That is, the availability of adequately trained personnel capable of performing required duties has to be matched with the availability of technical equipment that corresponds to local needs. Infrastructural requirements have to be met in a cost-effective manner that allows for the continued maintenance of the installed equipment. Relevant skills and competence are necessary for the development of both top-down (e.g., regulations, policies, guidelines, inspections, audit, etc.) and bottom-up approaches (education, training, outreach, awareness-raising, codes of conduct, standard operating procedures – SOPs, professional certification, etc.) for biological risk management. Thus, capacity building for biological risk management is both a precondition for and a consequence of the proper national implementation and effective functioning of biological risk management policies, strategies, measures, and approaches. As such, capacity building requires the active engagement of multiple stakeholders including government agencies, law enforcement services, relevant professional associations, industry, learned societies, funding bodies, publishers, academic and research institutions, and civil society organizations. In short, it requires a network of networks.

The concept of a network of networks has its origins in the interdisciplinary field of network science and has been used to describe the interdependency between different critical infrastructure systems in modern societies:

> In reality, diverse critical infrastructures are coupled together and depend on each other, including systems such as water and food supply, communications, fuel, financial transactions and power stations. [...] In interdependent networks, the failure of nodes in one network leads to the failure of dependent nodes in other networks, which in turn may cause further damage to the first network, leading to cascading failures and possible catastrophic consequences. (Gao et al. 2014)

Within the context of biological risk management, the concept of a network of networks signifies the interrelationship of different stakeholders highlighting the mutually reinforcing nature of their roles and activities for generating a continuous positive feedback on the level of the entire system (Pearson 1992, 2015; Sture et al. 2013; Millett 2010; Trump et al. 2019). The underlying assumption here is that due to its inherent complexity and versatile nature, biological risk management requires a multilayered coordinated action from a multitude of agents operating in disparate spheres (Fig. 2). Each agent operates within their established network. Each network has a duty as far as the management of biological risks is concerned, but none is sufficient on its own to meet the challenge. Therefore, a concerted effort by all stakeholders involved is needed to ensure the effective functioning of the network of networks.

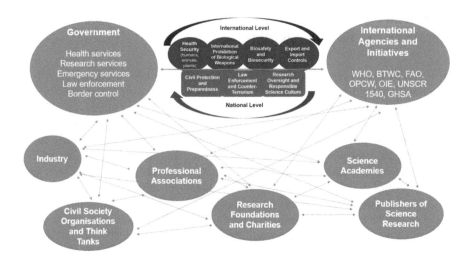

Fig. 2 A network of networks model for biological risk management

Nuclear Security Education Through a Network of Networks Approach

The practical manifestations of the network of networks approach to capacity building are notable in the area of nuclear security.

The International Atomic Energy Agency (IAEA) provides a framework for promoting nuclear security in a harmonized and coordinated manner. Since 2002, the IAEA's Board of Governors has been adopting *Nuclear Security Plans* which outline the main activities to be carried out, in order to address Member States' nuclear security priorities. The *Nuclear Security Plan 2018–2021* lays an emphasis on the importance of capacity building, including education and training (IAEA 2017). The *Plan* further underscores the need for international cooperation through the maintenance and strengthening of existing partnership networks, such as the International Nuclear Security Education Network (INSEN) and the International Network for Nuclear Security Training and Support Centres (NSSCs). Set up in 2010, INSEN seeks to enhance global nuclear security by developing, sharing, and promoting excellence in nuclear security education (IAEA 2018a). The NSSCs Network contributes to sustaining the national nuclear security regime by supporting competent authorities, authorized persons, and other organizations with nuclear security responsibilities in terms of human resource development and the provision of technical and scientific support services (IAEA 2018b).

The Master's Program in Nuclear Security offered by the University of National and World Economy (UNWE), Bulgaria, is an example of capacity-building model that is based on multi-stakeholder engagement featuring the IAEA, government authorities, industry, and academia (UNWE 2018). The program is a result of an Agreement of Cooperation in Education and Research in the Field of Nuclear Security between the IAEA and UNWE signed in 2014 and is supported by the Nuclear Regulatory Agency – the principal national authority for nuclear safety and security oversight. The program is structured in accordance with the IAEA's "Technical Guidance" titled *Educational Programme in Nuclear Security* (*Nuclear Security Series No.12*) and comprises a set of 12 compulsory and 10 elective modules (IAEA 2010). The compulsory courses cover such aspects as legal frameworks, nuclear technologies and applications, radiation protection, as well as prevention, detection, and response. Theoretical courses are combined with practical sessions including demonstrations, laboratory exercises, technical visits, and simulations. The program aims to prepare qualified managerial personnel for the needs of the nuclear industry and as such is open to candidates from all over the world.

Fostering a Network of Networks for Biological Risk Management

Each state has a responsibility, within its jurisdiction, to ensure that any biological risks that may occur regardless of their origin are effectively addressed. At the same time, no state exists in a vacuum, and hence the options for international cooperation through, for example, assistance programs and mechanisms for data sharing, peer-learning, and exchange of experience and lessons learned need to be further harmonized in order to maximize effectiveness and achieve greater efficiency. A comprehensive national system for biological risk management comprises at least three main elements:

- Civil protection and preparedness.
- Law enforcement and counterterrorism.
- Research oversight and responsible science culture.

Each element requires a set of relevant capacities in the form of national laws and regulations; policies, strategies, and plans; technical equipment and infrastructure that correspond to local needs; licensing, certification, and accreditation procedures; audit and inspection procedures; institutional guidelines and codes of conduct; education and training; and outreach activities. Fostering such capacities and ensuring their sustainability is a long-term, dynamic process that rests upon the engagement of multiple stakeholders in a synchronized network of networks. Connecting networks means building relationships. And relationships are based on trust. Trust, for its own part, requires mutual understanding. Communities of stakeholders that have little experience of direct collaboration among one another need to develop the habit of working together in a systematic manner. Overcoming this barrier in a way that promotes mutual understanding among stakeholders is by far the most significant challenge to the implementation of biological risk management, and it is on this task that the bulk of effort needs to be invested.

References

Acharya, T., et al. (2003). Biotechnology and the UN's millennium development goals. *Nature Biotechnology, 21*(12), 1434–1436.

Almquist, K., et al. (2015). National implementation of biosecurity in Canada. In S. Whitby et al. (Eds.), *Preventing biological threats: What you can do*. University of Bradford.

Australia Group. (2007a). Available at https://australiagroup.net/en/introduction.html. Accessed 19 Dec 2018.

Australia Group. (2007b). *Volume II: Biological weapon-related common control lists*. Available at https://australiagroup.net/en/controllisthandbooks.html. Accessed 19 Dec 2018.

Australian Government. (2015). *Biosecurity Act 2015. No. 61*. Available at https://www.legislation.gov.au/Details/C2018C00363. Accessed 6 Feb 2019.

Bakanidze, L., et al. (2010). Biosafety and biosecurity are essential pillars of international health security and cross-cutting elements of biological non-proliferation. *BMC Public Health, 10*(Suppl.1), S12.

Biosecurity Office. (2018a). *User manual for biosecurity vulnerability scan*. https://www.bureaubiosecurity.nl/en/toolkit/user-manual-for-biosecurity-vulnerability-scan. Accessed 20 Dec 2018.

Biosecurity Office. (2018b). *Toolkit*. Available at https://www.bureaubiosecurity.nl/en/toolkit. Accessed 20 Dec 2018.

Biotechnology and Biological Sciences Research Council; et al. (2015). *Position statement on dual use research of concern and research misuse*. Available at https://wellcome.ac.uk/sites/default/files/wtp059491.pdf. Accessed 18 Feb 2019.

Cello, J., et al. (2002). Chemical synthesis of poliovirus cDNA: Generation of infectious virus in the absence of natural template. *Science, 297*(5583), 1016–1018.

Centre for Biosecurity and Biopreparedness. (2015). *An efficient and practical approach to biosecurity*. https://www.biosecurity.dk/biosecuritybook/. Accessed 20 Dec 2018.

Centre for Biosecurity and Biopreparedness. (2017). The *Danish biosecurity legislation*. Available at https://www.biosecurity.dk/613/. Accessed 6 Feb 2019.

Centre for Biosecurity and Biopreparedness. (2018a). Available at https://www.biosecurity.dk/home/. Accessed 20 Dec 2018.

Centre for Biosecurity and Biopreparedness. (2018b). *Educational activities*. Available at https://www.biosecurity.dk/educationalactivities/. Accessed 20 Dec 2018.

Convention on Biological Diversity. (2010). *Decision adopted by the Conference of the Parties to the Convention on Biological Diversity at its tenth meeting*. UNEP/CBD/COP/DEC/X/2, 29 October. Available at https://www.cbd.int/doc/decisions/cop-10/cop-10-dec-02-en.pdf. Accessed 19 Dec 2018.

Convention on Biological Diversity. (2018a). Available at https://www.cbd.int/intro/default.shtml. Accessed 19 Dec 2018.

Convention on Biological Diversity. (2018b). *Quick guides for the aichi biodiversity targets*. Available at https://www.cbd.int/nbsap/training/quick-guides/. Accessed 19 Dec 2018.

Convention on Biological Diversity. (2018c). *The cartagena protocol on biosafety*. Available at http://bch.cbd.int/protocol/. Accessed 19 Dec 2018.

Convention on Biological Diversity. (2018d). *Biosafety clearing house*. Available at http://bch.cbd.int/. Accessed 19 Dec 2018.

Crossley, M. (2018). *What is CRISPR gene editing and how does it work?*. The Conversation, 31 January. http://theconversation.com/what-is-crispr-gene-editing-and-how-does-it-work-84591. Accessed 19 Dec 2018.

Daar, A. (2002). Top ten biotechnologies for improving health in developing countries. *Nature Genetics, 32*, 229–232.

Department of Health and Human Services. (2010). *Screening framework guidance for providers of synthetic double-stranded DNA*. Available at https://www.phe.gov/preparedness/legal/guidance/syndna/documents/syndna-guidance.pdf. Accessed 18 Jan 2019.

Department of Health and Human Services. (2019a). *National health security strategy*. Available at https://www.phe.gov/Preparedness/planning/authority/nhss/Pages/default.aspx. Accessed 20 Jan 2019.

Department of Health and Human Services. (2019b). *National health security strategy implementation plan, 2019–2022*. Available at https://www.phe.gov/Preparedness/planning/authority/nhss/Pages/overview.aspx. Accessed 20 Jan 2019.

EU CBRN Centres of Excellence. (2018). *Addressing regional CBRN risk mitigation needs*. Available at http://www.cbrn-coe.eu/Projects.aspx. Accessed 19 Dec 2018.

European Commission. (2018). *Dual-use trade controls*. Available at http://ec.europa.eu/trade/import-and-export-rules/export-from-eu/dual-use-controls/. Accessed 19 Dec 2018.

Federal Bureau of Investigation. (2008). *Criminal investigation handbook for agroterrorism*. Available at https://www.fsis.usda.gov/shared/PDF/Investigation_Handbook_Agroterrorism.pdf. Accessed 19 Dec 2018.

Federal Bureau of Investigation. (2016). *Joint criminal and epidemiological investigations handbook*. Available at https://www.fbi.gov/file-repository/joint-criminal-and-epidemiological-investigations-handbook-2016-international-edition/view. Accessed 19 Dec 2018.

Federal Select Agent Programme. (2019). Available at https://www.selectagents.gov/index.html. Accessed 18 Jan 2019.

Gao, J., et al. (2014). From a single network to a network of networks. *National Science Review, 1*, 346–356.

German Academy of Sciences (Leopoldina). (2014). *Scientific freedom and scientific responsibility: Recommendations for handling security-relevant research*. Available at https://www.leopoldina.org/uploads/tx_leopublication/2014_06_DFG-Leopoldina_Scientific_Freedom_Responsibility_EN.pdf. Accessed 20 Dec 2018.

German Ethics Council. (2014). *Biosecurity – Freedom and responsibility of research*. Berlin: German Ethics Council. Available at https://www.ethikrat.org/fileadmin/Publikationen/Stellungnahmen/englisch/opinion-biosecurity.pdf. Accessed 20 Dec 2018.

Global Health Security Agenda. (2018). *GHSA action packages: Renewal under GHSA 2024*. Available at https://www.ghsagenda.org/docs/default-source/default-document-library/ghsa-2024-files/action-package-renewal%2D%2D-2019-action-packages.pdf. Accessed 19 Dec 2018.

Global Health Security Initiative. (2018a). Available at http://www.ghsi.ca/english/index.asp. Accessed 19 Dec 2018.

Global Health Security Initiative. (2018b). *Ministerial statements*. Available at http://www.ghsi.ca/english/statements.asp. Accessed 19 Dec 2018.

Global Partnership against the Spread of Weapons of Mass Destruction. (2018). Available at https://www.gpwmd.com/about. Accessed 19 Dec 2018.

Greely, H., et al. (2018). Neuroethics guiding principles for the NIH BRAIN initiative. *The Journal of Neuroscience, 38*(50), 10586–10588.

Green Customs Initiative. (2018). Available at https://www.greencustoms.org/. Accessed 19 Dec 2018.

International Atomic Energy Agency. (2010). *Educational programme in nuclear security*. Technical Guidance. IAEA Nuclear Security Series. 12. Available at https://www-pub.iaea.org/MTCD/Publications/PDF/Pub1439_web.pdf. Accessed 20 Dec 2018.

International Atomic Energy Agency. (2018a). *International nuclear security education network*. Available at http://www-ns.iaea.org/security/workshops/insen-wshop.asp. Accessed 20 Dec 2018.

International Atomic Energy Agency. (2018b). *International network for nuclear security training and support centres*. https://www.iaea.org/services/networks/nssc. Accessed 20 Dec 2018.

International Atomic Energy Agency; Board of Governors. (2017). *Nuclear security plan, 2018–2021*. GC(61)/24. Vienna, Austria. Available at https://www-legacy.iaea.org/About/Policy/GC/GC61/GC61Documents/English/gc61-24_en.pdf. Accessed 20 Dec 2018.

International Federation of Biosafety Associations. (2018). *Certification: About the program*. Available at https://www.internationalbiosafety.org/index.php/professional-certification/ifba-professional-certifications/about-the-program. Accessed 20 Dec 2018.

International Organisation for Standardisation (ISO). (2019). Biorisk Management for Laboratories and Other Related Organisations. ISO 35001. Available at https://www.iso.org/standard/71293. html. [accessed 4/10/2019].

International Organisation for Standardisation (ISO). (2003). *Medical laboratories – Requirements for safety*. ISO 15190:2003. Available at https://www.iso.org/obp/ui/#iso:std:iso:15190:ed-1:v1:en. Accessed 19 Dec 2018.

International Organisation for Standardisation (ISO). (2012). *Medical laboratories – Requirements for quality and competence*. ISO 15189:2012. Available at https://www.iso.org/standard/56115. html. Accessed 19 Dec 2018.

International Plat Protection Convention. (2018). Available at https://www.ippc.int/en/. Accessed 19 Dec 2018.

INTERPOL. (2018). *Bioterrorism*. Available at https://www.interpol.int/Crime-areas/CBRNE/ Bioterrorism/Introduction. Accessed 19 Dec 2018.

Millett, P. (2010). The biological weapons convention: Securing biology in the twenty-first century. *Journal of Conflict and Security Law, 15*(1), 25–43.

National Academies of Sciences. (2018). *Governance of dual use research in the life sciences: Advancing global consensus on research oversight: Proceedings of a workshop*. Washington DC: National Academies Press. Available at https://www.nap.edu/catalog/25154/. Accessed 6 Dec 2019.

National Academy of Sciences. (2005). *An international perspective on advancing technologies and strategies for managing dual-use risks: Report of a workshop*. Washington, DC: National Academies Press.

National Research Council. (2006). *Biosecurity, globalisation and the future of the life sciences*. Washington, DC: National Academy Press.

National Science Advisory Board for Biosecurity. (2011). *Guidance for enhancing personnel reliability and strengthening the culture of responsibility*. Available at https://osp.od.nih.gov/biotechnology/nsabb-reports-and-recommendations/. Accessed 6 Feb 2019.

National Science Advisory Board for Biosecurity. (2015). *Framework for conducting risk and benefit assessments of gain-of-function research*. Available at https://osp.od.nih.gov/biotechnology/nsabb-reports-and-recommendations/. Accessed 6 Feb 2019.

National Science Advisory Board for Biosecurity. (2016). *Recommendations for the evaluation and oversight of proposed gain-of-function research*. Available at https://osp.od.nih.gov/wp-content/uploads/2016/06/NSABB_Final_Report_Recommendations_Evaluation_Oversight_Proposed_Gain_of_Function_Research.pdf. Accessed 6 Feb 2019.

Novossiolova, T. (2017). *Governance of biotechnology in post-soviet Russia*. Basingstoke: Palgrave.

Noyce, R. S., et al. (2018). Construction of an infectious horsepox virus vaccine from chemically synthesized DNA fragments. *PLoS One, 13*(1), e0188453.

Organisation for the Prohibition of Chemical Weapons. (2018). *Chemical weapons convention*. Available at https://www.opcw.org/chemical-weapons-convention. Accessed 19 Dec 2018.

Pearson, G. (1992). *Talking point: Preventing biological warfare*. New Scientist. 21 March. Available at https://www.newscientist.com/article/mg13318130-100-talking-point-preventing-biological-warfare/. Accessed 6 Feb 2019.

Pearson, G. (2015). The idea of a web of prevention. In S. Whitby et al. (Eds.), *Preventing biological threats: What you can do*. University of Bradford.

Perkins, D., et al. (2018). The culture of biosafety, biosecurity, and responsible conduct in the life sciences: A comprehensive literature review. *Applied Biosafety, 24*, 34. https://doi.org/10.1177/1535676018778538.

Petersen, R. (2015). The Danish biosecurity system. In S. Whitby et al. (Eds.), *Preventing biological threats: What you can do*. University of Bradford.

Rappert, B. et al. (2006). *In-depth implementation of the BTWC: Education and outreach*. Review Conference Paper No.18, University of Bradford. Available at http://www.opbw.org/sbtwc/ RCP_18.pdf. Accessed 19 Dec 2018.

Roco, M. et al. (2013). *Convergence of knowledge, technology, and society: Beyond convergence of nano-bio-info-cognitive technologies*. WTEC, Lancaster PA. Available at http://www.wtec. org/NBIC2/Docs/FinalReport/Pdf-secured/NBIC2-FinalReport-WEB.pdf. Accessed 19 Dec 2018.

Royal Netherlands Academy of Arts and Sciences (KNAW). (2008). *A code of conduct for biosecurity*. Amsterdam: Royal Netherlands Academy of Arts and Sciences. Available at https:// www.knaw.nl/en/news/publications/a-code-of-conduct-for-biosecurity. Accessed 20 Dec 2018.

Royal Netherlands Academy of Arts and Sciences (KNAW). (2013). *Improving biosecurity: Assessment of dual-use research*. Amsterdam: Royal Netherlands Academy of Arts and Sciences. Available at https://www.knaw.nl/en/news/publications/improving-biosecurity. Accessed 20 Dec 2018.

Sample, I. (2010). *Craig Venter creates synthetic life form*. The Guardian, 20 May, available at http://www.theguardian.com/science/2010/may/20/craig-venter-synthetic-life-form. Accessed 19 Dec 2018.

Seyfried, G., et al. (2014). European do-it-yourself (DIY) biology: Beyond the hope, hype and horror. *BioEssays, 36*(6), 548–551.

So, W. (2015). The Federal Bureau of Investigation Biosecurity Program: A case study of law enforcement and outreach. In S. Whitby et al. (Eds.), *Preventing biological threats: What you can do*. University of Bradford.

Stroot, P., & Jenal, U. (2011). A new approach: Contributing to BWC compliance via biosafety, biosecurity and biorisk management. *Nonproliferation Review, 18*(3), 545–555.

Sture, J., et al. (2013). Biosafety, biosecurity and internationally mandated regimes: Compliance mechanisms for education and global health security. *Medicine, Conflict and Survival, 29*(4), 289–321.

The InterAcademy Partnership. (2016). *Doing global science: A guide to responsible conduct in the global research enterprise*. Princeton NJ: Princeton University Press. Available at http:// www.interacademies.org/33345/Doing-Global-Science-A-Guide-to-Responsible-Conduct-in-the-Global-Research-Enterprise. Accessed 20 Dec 2018.

The White House. (2018a). *National biodefense strategy*. Available at https://www.whitehouse. gov/wp-content/uploads/2018/09/National-Biodefense-Strategy.pdf. Accessed 19 Dec 2018.

The White House. (2018b). *National strategy for counterterrorism of the United States of America*. Available at https://www.whitehouse.gov/wp-content/uploads/2018/10/NSCT.pdf. Accessed 19 Dec 2018.

The White House. (2018c). *National strategy for countering weapons of mass destruction Terrorism*. Available at https://www.whitehouse.gov/wp-content/uploads/2018/12/20181210_ National-Strategy-for-Countering-WMD-Terrorism.pdf. Accessed 19 Dec 2018.

Trump, B. D., et al. (2019). Co-evolution of physical and social sciences in synthetic biology. *Critical Reviews in Biotechnology, 39*(3), 351–365.

UK Government. (2018). *UK biological security strategy*. Available at https://assets.publishing. service.gov.uk/government/uploads/system/uploads/attachment_data/file/730213/2018_UK_ Biological_Security_Strategy.pdf. Accessed 19 Dec 2018.

United Nations Economic Commission for Europe (UNECE). (2017). *Recommendations on the transport of dangerous goods – Model regulations (Rev.20)*. Available at http://www.unece.org/ index.php?id=46066&L=0. Accessed 19 Dec 2018.

United Nations Educational, Scientific, and Cultural Organisation. (2017). *Recommendation on science and scientific researchers*. http://portal.unesco.org/en/ev.php-URL_ID=49455&URL_ DO=DO_TOPIC&URL_SECTION=201.html. Accessed 6 Feb 2019.

United Nations Food and Agriculture Organisation (FAO). (2008). *The international plant protection convention*. Available at http://www.fao.org/docs/up/easypol/785/international_plant_pro-tection_convention_slides_078en.pdf. Accessed 19 Dec 2018.

United Nations Food and Agriculture Organisation (FAO). (2018). *Codex alimentarius: International food standards*. Available at http://www.fao.org/fao-who-codexalimentarius/ home/en/. Accessed 19 Dec 2018.

United Nations Office for Disarmament Affairs. (2018a). *1925 Geneva protocol: Protocol for the prohibition of the use in war of asphyxiating, poisonous or other gases, and of bacteriological methods of warfare.* Available at https://www.un.org/disarmament/wmd/bio/1925-geneva-protocol/. Accessed 19 Dec 2018.

United Nations Office for Disarmament Affairs. (2018b). *Securing our common future: An agenda for disarmament.* Available at https://www.un.org/disarmament/sg-agenda/en/. Accessed 25 Feb 2019.

United Nations General Assembly. (2015). Transforming Our World: The 2030 Agenda for Sustainable Development. A/RES/70/1. 21 October 2015. Available at https://www.un.org/ga/search/view_doc.asp?symbol=A/RES/70/1&Lang=E. [accessed on 29/09/2019].

United Nations Office for Disarmament Affairs. (2019). Secretary-General's Mechanism for Investigation of Alleged Use of Chemical and Biological Weapons. Available at https://www.un.org/disarmament/wmd/secretary-general-mechanism/. [accessed 29/09/2019].

United Nations Office of Counter Terrorism. (2006). UN *Global counter terrorism strategy.* Available at https://www.un.org/counterterrorism/ctitf/en/un-global-counter-terrorism-strategy. Accessed 19 Dec 2018.

United Nations Security Council. (2004). *United Nations security council resolution 1540.* Available at http://www.un.org/en/ga/search/view_doc.asp?symbol=S/RES/1540(2004). Accessed 19 Dec 2018.

University of National and World Economy. (2018). *Master's programme in nuclear security.* Available at https://www.unwe.bg/nuclear-security/en. Accessed 20 Dec 2018.

US Government. (2012). *United States government policy for oversight of life sciences DURC.* Available at https://www.phe.gov/s3/dualuse/Pages/USGOversightPolicy.aspx. Accessed 19 Dec 2018.

US Government. (2014). *United States government policy for institutional oversight of life sciences dual use research of concern.* Available at https://www.phe.gov/s3/dualuse/Pages/InstitutionalOversight.aspx. Accessed 19 Dec 2018.

Wassenaar Agreement. (2018a). *The Wassenaar arrangement on export controls for conventional arms and dual-use goods and technologies.* Available at https://www.wassenaar.org/. Accessed 19 Dec 2018.

Wassenaar Agreement. (2018b). *Control lists.* Available at https://www.wassenaar.org/control-lists/. Accessed 19 Dec 2018.

Wimmer, E. (2006). The test-tube synthesis of a chemical called poliovirus: The simple synthesis of a virus has far-reaching societal implications. *EMBO Reports, 7,* S3–S9.

World Animal Health Organisation (OIE). (2012). *Manual of diagnostic tests and vaccines for terrestrial animals* (7th ed.). Available at http://www.oie.int/standard-setting/terrestrial-manual/. Accessed 19 Dec 2018.

World Animal Health Organisation (OIE). (2015). *Biological threat reduction strategy: Strengthening global biological security.* Available at http://www.oie.int/fileadmin/Home/eng/Our_scientific_expertise/docs/pdf/A_Biological_Threat_Reduction_Strategy_jan2012.pdf. Accessed 19 Dec 2018.

World Animal Health Organisation (OIE). (2016). *Manual of diagnostic tests for aquatic animals* (7th ed.). Available at http://www.oie.int/standard-setting/aquatic-manual/. Accessed 19 Dec 2018.

World Animal Health Organisation (OIE). (2018a). *Terrestrial animal health code.* Available at http://www.oie.int/en/standard-setting/terrestrial-code/. Accessed 19 Dec 2018.

World Animal Health Organisation (OIE). (2018b). *Aquatic animal health code.* Available at http://www.oie.int/en/standard-setting/aquatic-code/. Accessed 19 Dec 2018.

World Animal Health Organisation (OIE). (2018c). *The OIE PVS pathway.* Available at http://www.oie.int/solidarity/pvs-pathway/. Accessed 19 Dec 2018.

World Animal Health Organisation (OIE). (2018d). Guidelines for Investigation of Suspicious Biological Events. Available at https://www.oie.int/scientific-expertise/biological-threat-reduction/. [accessed 29/09/2019].

World Animal Health Organisation (OIE). (2019a). *One health "at a Glance"*. Available at http://www.oie.int/en/for-the-media/onehealth/. Accessed 6 Feb 2019.

World Animal Health Organisation (OIE). (2019b). Guidelines for Responsible Conduct in Veterinary Research: Identifying, Assessing and Managing Dual Use. Available at https://www.oie.int/scientific-expertise/biological-threat-reduction/. [accessed 29/09/2019].

World Health Organisation. (2003). *Laboratory biosafety manual* (3th ed.). Available at https://www.who.int/csr/resources/publications/biosafety/WHO_CDS_CSR_LYO_2004_11/en/. Accessed 19 Dec 2018.

World Health Organisation. (2006). *Biorisk management: Laboratory biosecurity guidance*. Available at http://www.who.int/ihr/publications/WHO_CDS_EPR_2006_6/en/. Accessed 19 Dec 2018.

World Health Organisation. (2010). *Responsible life sciences research for global health security*. Available at http://apps.who.int/iris/bitstream/handle/10665/70507/WHO_HSE_GAR_BDP_2010.2_eng.pdf:jsessionid=286DF456CB6A91593902951FBCD0ED05?sequence=1. Accessed 19 Dec 2018.

World Health Organisation. (2011). *Laboratory quality management system: Handbook*. Available at https://www.who.int/ihr/publications/lqms/en/. Accessed 19 Dec 2018.

World Health Organisation. (2015a). *Guide for shippers of infectious substances*. Available at http://www.who.int/ihr/infectious_substances/en/. Accessed 19 Dec 2018.

World Health Organisation. (2015b). *Infectious substances shipping training*. Available at http://www.who.int/ihr/i_s_shipping_training/en/. Accessed 19 Dec 2018.

World Health Organisation. (2017). *Health security and the International Health Regulations (2005)*. Available at https://www.who.int/ihr/publications/WHO-WHE-CPI-2016.11/en/. Accessed 19 Dec 2018.

World Health Organisation. (2018a). *Strengthening health security by implementing the International Health Regulations (2005): About IHR*. Available at https://www.who.int/ihr/about/en/. Accessed 19 Dec 2018.

World Health Organisation. (2018b). *Health security learning platform*. Available at https://extranet.who.int/hslp/training/. Accessed 19 Dec 2018.

World Health Organisation. (2018c). *Strengthening health security by implementing the International Health Regulations (2005): Biorisk management advanced training programme (BRM ATP)*. Available at https://www.who.int/ihr/training/biorisk_management/en/. Accessed 19 Dec 2018.

World Health Organisation. (2018d). *Strengthening health security by implementing the International Health Regulations (2005): Laboratory quality management system training toolkit*. Available at https://www.who.int/ihr/training/laboratory_quality/en/. Accessed 19 Dec 2018.

World Health Organisation. (2018e). *Strengthening health security by implementing the International Health Regulations (2005): Laboratory issues for epidemiologists*. Available at https://www.who.int/ihr/lyon/surveillance/laboratory/en/. Accessed 19 Dec 2018.

World Trade Organisation. (1998). *Understanding the WTO agreement on sanitary and phytosanitary measures*. Available at https://www.wto.org/english/tratop_e/sps_e/spsund_e.htm. Accessed 19 Dec 2018.

World Health Organisation. (2019). Guidance on Regulations for the Transport of Infectious Substances 2019-2020. Available at https://www.who.int/ihr/publications/WHO-WHE-CPI-2019.20/en/. [accessed 29/09/2019].

The Role of Expert Disciplinary Cultures in Assessing Risks and Benefits of Synthetic Biology

Christina Ndoh, Christopher L. Cummings, and Jennifer Kuzma

Like other technological fields before it, synthetic biology (SB) has been ascribed different definitions by different scholars (Pauwels 2013; Smith 2013; Wang et al. 2013). One commonly used definition of SB is the extraction of living parts for organisms that are then inserted into other organisms to create a "new" organism with parts from the donor and recipient (Benner and Sismour 2005). Synthetic biology has also been described as "the use of computer assisted, biological engineering to design and construct new synthetic biological part" (Hoffman and Newman 2012). Others like the National Science Foundation and the Engineering and Physical Sciences Research Council have noted that synthetic biology is the identification and application of biology in the design of biological parts and systems for use in the creation or redesign of natural biological systems for useful purposes (Engineering and Physical Sciences Research Council 2009).

At first glance, the term "SB" appears somewhat of an oxymoronic label. The word biology is usually defined as the study of life and living organisms, whereas synthetic is often defined as something not of natural origin or alternately as something that is fake or not genuine. A lay understanding of the term could lead one to believe that SB is a combination of living and artificial or unnatural components. However, if instead a definition of synthetic that looks at "synthesis of parts" is used, a more common scientific understanding of SB can be achieved. Such definition differences may be due to distinct "expert cultures" who view the field, its products, and subsequent risks in distinct ways. This work explores these potential

C. Ndoh
Kumon, Morrisville, NC, USA

C. L. Cummings (✉)
Nanyang Technological University, Singapore, Singapore
e-mail: CCUMMINGS@ntu.edu.sg

J. Kuzma
North Carolina State University, Raleigh, NC, USA

© Springer Nature Switzerland AG 2020
B. D. Trump et al. (eds.), *Synthetic Biology 2020: Frontiers in Risk Analysis and Governance*, Risk, Systems and Decisions, https://doi.org/10.1007/978-3-030-27264-7_15

cultural perceptions and focuses on potential differences in expert groups' beliefs and attitudes regarding risk analysis and governance needs for SB.

This chapter takes a case study approach similar to related work (see Cummings and Kuzma 2017; Trump et al. 2018a; Valdez et al. 2019) to examine a case study in SB, the planned enhancement, and use of the bacterium species *Mesorhizobium loti* (*M. loti;* formerly known as *Rhizobium loti*). The non-SB form of *M. loti* was studied extensively in the 1990s and had its complete genome sequence identified in 2000 (Kaneko et al. 2000). *M. loti* is a Gram-negative bacterium commonly found in the root nodules of many plant species and serves a symbiotic relationship with the plant in nitrogen fixation.

The SB-enhanced version is planned to improve the natural nitrogen-fixing qualities of the bacterium among nonlegumes, thus potentially increasing plant health and crop yield. While the bacterium is being speculated for widespread use, it is still under development. Christopher Voigt, at the Massachusetts Institute of Technology, researches this technology as a method for minimizing fertilizer application by allowing crops that previously relied on nitrogenous fertilizer applications to now rely on nitrogen production from engineered bacteria (Charpentier and Oldroyd 2010). The Voigt lab is also researching two additional pathways for achieving nitrogen fixation in nonleguminous plants, both of which involve engineering the plant instead of the bacteria. Since the two different genetic manipulations (plant or bacteria) would likely have different risk governance issues, we focus on the genetic manipulation of the soil bacteria.

Using this case study, we examine the boundaries and differences between groups of experts who may approach the case from distinct disciplines. We define disciplinarity as a form of cultural similitude between experts who may share similar ontological and epistemological structures, methods, organizations, and assumptions in their professional perspectives (Becher 1994). From this premise, we propose two axes that help us to chart boundaries between "expert cultures."

The first axis is the broad disciplinary grouping of either "natural scientists" or "social scientist," and the second axis is expert positioning of either "upstream" or "downstream" in relation to the technological development (Trump et al. 2018b; Trump et al. 2019). Stemming from the early work of Mary Douglas, Kahan and Braman (2006) identify risk perceptions from their cultural theory of risk that evaluates individuals' worldviews on the basis of "group" and "grid" preferences. In their framework, a "group" represents an individual's beliefs on how individualistic or communal a society should be, while the "grid" represents the individual's beliefs on the organization and durability of authority within society. A "high-group" perspective exhibits desires for a high degree of collective control, while those among the "low group" maintain much lower desires for authority and demonstrate preference for individual self-sufficiency. A "high-grid" perspective is characterized by desires for durable and conspicuous roles and authority structures within society, while the "low grid" is notably more egalitarian in its role orientation. In this chapter, we adapt cultural theory of risk to include academic disciplines as cultures, and we position expert cultures to their relationship with the development of the SB technology to assess potential patterns of cultural perspectives regarding the potential risks associated with the enhanced *M. loti* bacterium.

In order to study this case among expert cultures, we conducted a Policy Delphi study among 48 experts. To guide our inquiry, we posed the following research question: Does the expert group culture affect the views on riskiness of the genetically modified *M. loti*?

This chapter next reviews relevant literature from the fields of anticipatory governance and risk perception; then outlines our methodology and methods used for case study selection, expert elicitation, and data analysis; then reports results from this mixed-method inquiry; and identifies patterns and implications of the differences in expert culture perspectives of *M. loti*. We conclude the chapter discussing the implications for this work in helping to inform risk and governance discussions for emerging technologies in the area of SB.

Literature Review

Risk Analysis and Governance

Risk analysis (RA) includes the "traditional model" of scientific risk assessment employed by many federal agencies. It generally involves human dose-response metrics that are used to determine acceptable levels of risk based on exposure to a particular concern (DeSesso and Watson 2006; National Research Council Staff 1993). In risk assessment methods for genetically engineered plant microbes, human exposure and environmental sensitivities are given a static analysis, and subsequently, a determination on risk is made. Traditionally, RA has come after the technology development process when products are nearing the market for widespread use (Wareham and Nardini 2015).

In the case of emerging technologies, many scholars have called for more proactive governance approaches (Gutmann 2011; Mandel and Marchant 2014; Tait 2012). Anticipatory governance seeks to evaluate and potentially make recommendations on best practices for managing and governing emerging technologies prior to the technology being widely introduced into the public sphere (Guston 2014; Guston and Sarewitz 2002; Kuzma et al. 2008b; Quay 2010). In addition, anticipatory governance strategies have been used as a tool to promote early public engagement around a technology (Macnaghten 2008). One goal of anticipatory governance is, through upstream discussion and analysis, to prepare for emerging technologies and thus minimize potential negative externalities that could occur based on unknown risks associated with the technology's deployment. Given that there may be considerable uncertainty in a pre-dissemination technology, anticipatory governance often involves evaluating multiple factors that could affect society and similarly will likely involve evaluating multiple endpoint scenarios for the technology (Kuzma and Tanji 2010).

Anticipatory governance can be seen as an umbrella concept, under which many practical tools are held. One such tool, real-time technology assessment (RTTA), is an argument made for moving beyond the traditional static risk assessment model and, instead, adopting a more adaptive system that allows for "real-time" evaluation of societal and ethical implications of a technology under development (Guston and Sarewitz 2002). This reorientation provides new risk evaluation structures to be placed "upstream" where experts can incorporate this feedback into the design of a technology which may allow developers and researchers to craft a product that better maximizes benefits and minimizes potential risks. RTTA also provides a mechanism for making incremental changes, thus providing iterative feedback on the effectiveness of each step.

Similarly, the use of upstream oversight assessment (UOA) encourages experts to think beyond the traditional RA framework when considering potential data, information, and regulatory oversight needs of an emerging technology (Kuzma et al. 2008a). Both UOA and RTTA can be considered anticipatory governance approaches. This work's practical framing uses UOA, which has been defined as the advanced consideration of technology case studies to explore risk, regulatory, and societal issues (Kuzma et al. 2008a). Defining the emerging technology is a critical step in conducting a UOA. Once the emerging technology is defined, selecting a representative case study within the technology helps to ground conversations. Whereas a rigid definition is not required, boundaries for the technology that help determine potential oversight needs must be developed in order to have a fruitful discussion of options. Upstream oversight assessments are conducted by analyzing a case study from an emerging technology to "highlight oversight issues" by thinking through the potential deployment of the technology in society and how that technology fits into the current regulatory landscape (Kuzma et al. 2008b). Case studies are a way for anticipatory governance strategies such as UOA to proceed with discussions of specifics and to make progress in highlighting issues associated with SB applications (Kuzma and Tanji 2010).

Anticipatory governance seeks to capture a wider range of voices earlier in technological development. For example, Stirling's (2008) *Science, Precaution, and the Politics of Technical Risk* argues that including precautionary and participatory approaches complements the traditional "science-based" risk assessment. Furthermore, the incorporation of precautionary and participatory principles, such as expanded RA methods and increased public engagement, can improve democratic legitimacy and provide a more comprehensive decision-making process. Stirling notes "deliberate attention to potential blind spots" as one of several key features of a precautionary approach. While our assessment of *M. loti* does not explicitly seek to promote a precautionary approach, we do deliberately investigate potential blind spots in RA and governance needs assessments by evaluating commonalities in perceptions that are shared among expert cultures.

We employed expert elicitation methods to collect data on expert group risk perceptions and governance preferences. These methods are particularly germane in emerging areas, such as SB, for which data and information are scarce and uncertainty is high (Otway and von Winterfeldt 1992). Whereas these methods have been used and accepted in policy decision-making (Morgan 2005; Swor and Canter 2011), we must be cautious to recognize potential cognitive biases and overconfidence of the expert group (Morgan 2014).

Risk Perception

For many cases similar to *M. loti*, little data or information exists regarding risk governance needs. Thus, expert judgments become a valuable source of governance strategies and potential outcomes related to the technology (Cummings and Kuzma 2017). However, experts themselves are influenced by their disciplinary cultures, as well as their individual life experiences, interests, motivations, and specific knowledge. Mile's law (Miles Jr. 1978), which notes that "where you stand depends on where you sit," captures the potential unassuming bias likely to influence decision-making processes of all types, including those regarding risks. In looking at perceptions of risk, scholars have explored several theories to explain perceptions that are based on factors other than the physical riskiness (harm from a dose or exposure) of an application.

One prominent theory in this area is the cultural theory of risk. This theory posits that risk perceptions are influenced by our cultural worldviews. Cultural cognition looks at the characteristics of a group to which an individual belongs to explain part of that individual's worldview. In turn, this worldview influences the perception of risk that the individual holds (Kahan and Braman 2006). The cultural theory of risk thus proposes that an individual's affiliation with cultural groups will determine which types of hazards resonate with that individual. Kahan and Braman's model of cultural cognition uses continuous attitudinal scales to measure where an individual falls on two measures: (1) the hierarchy-egalitarianism scale and (2) the individualism-communitarianism (solidarism) scale. Those who have a low-group (individualistic) worldview expect individuals to be self-resourceful and have little expectation of group support. Those who have a high-grid worldview (hierarchical) expect that resources be divided based on characteristics specific to a given social order; conversely, low-grid (egalitarian) individuals expect resource allocation to be equitable and not consider any social ordering.

Case Study: Genetically Modified *M. loti*

Kuzma and Tanji (2010) argue that discussions regarding SB in general are too broad for evaluation of anticipatory governance options. As such, this work uses a specific application within SB, genetically modified *M. loti* for extending nitrogen fixation to nonlegumes, as a method for grounding conversation in UOA in SB. The use of genetic engineering applications on plant microbes to extend nitrogen fixation to nonleguminous plants has been studied for many decades (Charpentier and

Oldroyd 2010; Wang et al. 2013). The process involves multiple genetic manipulations before a successful new symbiotic relationship between microbes that already has the ability to fix nitrogen can interact with a plant that does not already pose the ability to host the microbe (Wang et al. 2013).

Currently, *M. loti* is the only bacterium known to have a naturally occurring symbiotic relationship with legumes (Charpentier and Oldroyd 2010). It is this natural symbiotic relationship between *M. loti* and the plant that has interested researchers to attempt to modify *M. loti* to similarly be able to fix nitrogen among nonlegumes including rice cereals. The benefits of the relationship to both organisms are readily available food supplies and host sites for the bacteria. For the symbiotic relationship to be established, there is a multistep process that must take place between the two organisms (Oldroyd and Dixon 2014; Santi et al. 2013). First the legume will secrete flavonoids into the soil, which are detected by *M. loti*. The bacteria will secret Nod factor in response to the recognition of the flavonoids. The Nod factor, once recognized by the legume, leads to the creation of nodules in the plant root hairs. These nodules become the host site for the bacteria, and the *M. loti* colonize within the nodules. Once the bacteria have infected the plant host, and colonized in the root nodules, the symbiotic exchange of essential nutrients begins (Oldroyd and Dixon 2014). The plants provide the bacteria with needed organic matter, and the bacteria provide the plant host with ammonium. In contrast to the atmospheric nitrogen, ammonium provides a readily available source of nitrogen for the plant. An overview of this symbiotic relationship is shown in Fig. 1.

The genetic engineering goal for symbiosis is to extend the abilities to cereal crops, including rice, wheat, and maize. Some benefits that have been discussed around this technology include decreasing global nitrogenous fertilizer demands

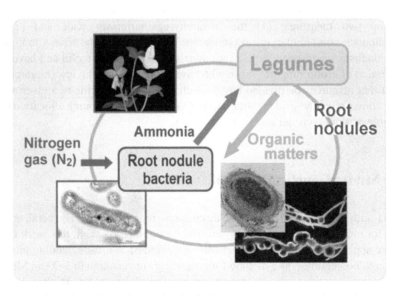

Fig. 1 Overview of symbiotic nitrogen fixation in legumes

and increasing crop yields for cereals, which could potentially lead to less environmental degradation due to fertilizer application and a partial solution to address global hunger needs. Having a readily available source of nitrogen has been shown to be a limiting factor in climates that would otherwise support cereal crop growth (Oldroyd and Dixon 2014).

There are at least eight genes involved in root nodule symbiosis (RNS) once the Nod factor has been recognized by the host plant (Oldroyd and Dixon 2014). These eight common symbiotic components (SYM) are thought to have common ancestry with arbuscular mycorrhizal (AM) symbiosis. Cereal crops possess AM genetic materials and are thought to have had an ancient symbiotic relationship with other soil bacteria that also provided essential nutrients (Charpentier and Oldroyd 2010). Given that cereals already possess genetic material that could be used for SYM pathways and that the bacteria needed for the symbiotic relationship already colonize the soil where cereals exist, there has been considerable hope that a series of genetic modifications could lead to the successful symbiosis of cereals and nitrogen-fixing bacteria. Also, given the global significance of rice crops in diets, the potential for extending the symbiosis to rice has been studied by many scholars (Oldroyd and Dixon 2014). This case study represents an emerging technology in SB that is nearer-term and where similar genetic engineering technologies have been developed (U.S. Environmental Protection Agency 2013), thus making it a good candidate for our overall project's aims.

Methodology

We use a mixed-method design to evaluate the case study among distinct expert cultures as part of a larger Alfred P. Sloan grant (PIs Kuzma and Cummings, *Looking Forward to Synthetic Biology Governance: Convergent Research Cases to Promote Policy-Making and Dialogue* [#556583]).

The project first reviewed multiple SB applications in early development and identified four case studies: biomining using engineered microbes, cyberplasm, de-extinction, and, the currently explored case, nitrogen fixation using engineered plant microbes. After the case studies were selected, more detailed descriptions were written by research staff using interview data from the technology's developers and available literature. These case study descriptions were then shared with the recruited expert panel to elicit feedback on governance needs for each case study specifically, as well as for SB holistically using Policy Delphi approach.

The Delphi method has been used for many decades to obtain group consensus from experts using iterative controlled intensive questionnaires (Landeta 2006). Originally named after the Oracle of Delphi, this method attempts to forecast potential risks and evaluate myriad policy options to maximize benefits and identify risks and areas of uncertainty that warrant greater information and attention. In this way, Delphi studies can serve an agenda-building function to promote areas of concern and create new goals and objectives for study in areas of need. The method is best used "when accurate information is unavailable or expensive to obtain, or evaluation models require subjective inputs to the point where they become the dominating parameters" (Linstone and Turoff 1975).

Delphi methods have been used extensively in social science research to reduce hindrances to group processes like group think, dominant personalities, and inhibition. The Delphi method has been used often in natural and social sciences, and scholars have upheld its validity for forecasting and supporting decision-making (Landeta 2006). A Policy Delphi differs from the traditional Delphi methods by seeking to uncover a range of policy options and pros and cons of those options (Turoff 1970). Our Policy Delphi study aimed to elicit expert-stakeholder perceptions regarding potential risks and benefits of a technology, as well as the potential ethical, legal, and societal (ELSI) issues associated with the case studies under evaluation to uncover a range of SB risk analysis and governance needs from participants through iterative individual and group reflections.

The overall project's Delphi consisted of four rounds. The first round consisted of a standardized open-ended interview, which is a form of qualitative data collection that is more structured than most other interview methodologies and thus [a]increases comparability of responses (Patton 2002). The second round included an online quantitative survey that was drafted from results drawn from Round One. Within the survey, panel members were asked to respond to a variety of scale items regarding risk and governance issues pertaining to each cases. The third round consisted of a face-to-face workshop where the goal was to envision ideal governance for SB in coming generations. The final round consisted of a second short online survey used to assess general trends among the sample. Data presented in this chapter come primarily from the first and second rounds of the project, and the interview protocol and survey are detailed in the following section.

Interview and Survey Methods

Prior to the beginning of Round One, experts were introduced to the *M. loti* case in the form of a two-page dossier that summarized the scientific goals of the developing technology and outlined the current state of the research. This summary was vetted for accuracy by a set of SB experts who were not participants in this study. Participants were asked to refer to the summary and any other information they had gathered as they completed the Round One qualitative interview.

In the interview, participants were primarily asked the following three questions during the interview with regard to RA of *M. loti* and the other case studies: (1) What are the types of data and information needed to assess the risks and benefits? (2) What are the associated uncertainties with this application, and how might they affect oversight? (3) How can risk analysis methods be used in the face of such uncertainties? The interviews were conducted via Skype and telephone, and audio files from the interviews were sent to a service for transcription. Transcribed data were then imported into NVivo software for coding and analysis.

A few weeks following the completion of the Round One interviews, we conducted the second round of the Policy Delphi study which consisted of an online survey with quantitative and open-ended questions that were based on initial themes that emerged from the Round One interviews. The Round Two survey posed questions to the expert panel on potential risks and uncertainty associated with each of the four case studies and asked experts to provide scaled responses to questions such as "How beneficial are engineered plant microbes to the environment?" Experts were also able to give feedback on the governance structures that they deemed most appropriate and which entities are, or should be, primarily responsible for oversight of the technology. These questions helped to further elicit expert opinions of governance, risk analysis, and data needs for regulating engineered plant microbes, such as *M. loti*.

The following background information was provided to the experts prior to the Round Two survey:

> Background: Many plant microbes are being researched for their ability to assist in crop production. One such example, the bacterium Mesorhizobium loti [M. loti] is being engineered to improve nodulation signaling for rice crops, thus allowing the two to enter into a symbiotic relationship where the M. loti colonize the newly formed nodules of the rice crop and provide a readily usable form of nitrogen. For the following survey questions, please assume that Engineered Plant Microbes are to be used in situ with open-release for agricultural purposes.

Respondents were then given the option to answer on a Likert scale of 1–10, with 1 representing the lowest response possible and 10 representing the highest.

1. How certain or uncertain are the risks of engineered plant microbes?
2. How likely is engineered plant microbes to be commercially developed and used in the next 15 years?
3. How potentially hazardous is engineered plant microbes to human health?
4. How potentially hazardous is engineered plant microbes to the environment?
5. How manageable are the potential hazards of engineered plant microbes?
6. How beneficial is engineered plant microbes to human health?
7. How beneficial is engineered plant microbes to the environment?
8. What might be the level of public concern regarding the risk of engineered plant microbes?
9. To what degree are the potential hazards of engineered plant microbes irreversible?

In addition, experts were asked to give an ordinal ranking to a list of potential issues for risk research concerning engineered plant microbes that could fix nitrogen for cereal crops. The list of options included the following:

1. Biopersistence
2. Competitiveness with other organisms
3. Disposal
4. Ecological system effects
5. Economic trade-offs with using other technologies

6. Environmental trade-offs with using other technologies
7. Genetic stability
8. Horizontal gene transfer
9. Life cycle
10. Organism tracking in situ
11. Pathogenicity
12. Regulation of tools
13. Regulatory approval process
14. Route of exposure to humans
15. Route of exposure to other organisms
16. Toxicity and biogeochemical cycling
17. Other

Experts could also select "other" and provide additional items to be considered in the ranking. Items were ranked between 1 and 17, with 1 being a risk consideration with highest priority and 17 being a risk consideration with lowest priority. Following the data collection in Round Two, we shifted efforts into classifying our expert panel into distinct cultural groups that would support our inquiry in answering our research question.

Expert Disciplinary Culture Classification

For the current analysis, the expert sample was classified into "expert cultures" from two data points. First, self-reports were used to classify individuals by their academic area of expertise into either "natural" or "social" sciences. Natural sciences have been found to promote more linear methods than social sciences and also have been found to promote hierarchical methods more readily than social sciences (Neumann and Becher 2002). Second, expert position of either upstream or downstream was determined by conducting searches of CVs or other published information in order to determine the expert role in *M. loti* development. Downstream experts are those involved with evaluation of the technology, policy, or societal concerns involved with the technological application. Examples of downstream scientists would include lawyers and risk scientists shown in Fig. 2. In contrast, upstream experts are involved with the technology or social science innovation or creation, like the ethicists and natural scientists shown in Fig. 2.

Survey data from 38 participants of the Policy Delphi were available for analysis of our defined expert cultures. Given the nature of such small expert studies, we compared expert cultures along the two axes, of upstream vs. downstream and natural science vs. social science, but did not group the experts into dual-axes categories, such as "upstream-natural scientists." In our final counts, we identified relatively equal expert culture groups between downstream ($N = 20$) and upstream ($N = 18$) groups and natural science ($N = 19$) and social science ($N = 19$) groups.

Fig. 2 Expert grouping by position and academic discipline

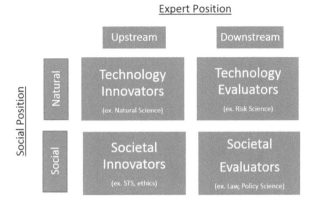

Results

Round One: Interview Data

The Round One interview asked individual experts to consider the data and information needs for genetically modified *M. loti*. The risk analysis need stated most by expert regardless of their disciplinary culture concerned gene flow and gene transfer from genetically modified *M. loti* after introduction into the natural environment or microbial population. Other risk assessment needs that were repeatedly mentioned by the experts included the need for human health metrics of toxicity, allergenicity, and pathogenicity. Another risk criteria that were highly mentioned among experts were competitiveness of the microbe in the environment. There were also three experts who voiced opinions that the existing risk assessment framework for traditional organisms is sufficient for governing all GMOs. A cross-tabs analysis of sub-themed risk analysis needs that emerged from interview data categorized by disciplinary culture is shown in Fig. 3.

A diversity of opinions around risk analysis needs emerged and were distinct between disciplinary cultures. In total count of references for top RA needs, "downstream experts" (technology and policy evaluators) listed 32 concerns compared with 21 needs identified by "upstream experts" (technology and policy innovators). When looking at natural scientists, we see 36 RA needs were identified compared to 17 from the social scientists. The RA criterion most mentioned was gene transfer, with a total of 28 references. This was mentioned most by natural scientist with 10 total references and least by social scientist with only 4 mentioned. For upstream and downstream experts, both groups mentioned gene transfer seven times. When comparing downstream to upstream expert preferences for human health metrics, downstream experts made 15 references to allergenicity, toxicity, and pathogenicity compared to only 3 from the upstream experts. Similarly, natural science experts mentioned human health concerns more than social science experts with 14 references compared to 4.

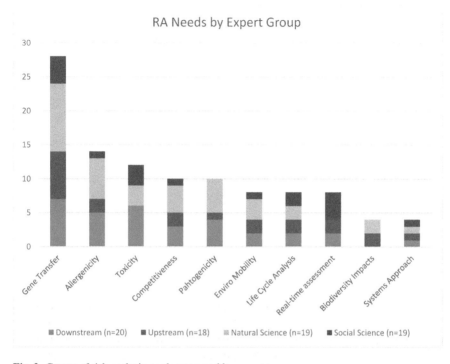

Fig. 3 Counts of risk analysis needs expressed by expert types

When asked about at RA needs, three alternative methods of assessment were mentioned by the experts: life cycle analysis, real-time assessments, and systems approaches. Life cycle approaches and system approaches were mentioned the same number of times by experts in all four groups, with two and one references per group, respectively. Real-time assessment was referenced most by social scientists with a total of four references, followed by two references from both upstream and downstream experts, and no references from natural scientists.

Round Two: Risk Perception Data

The Round Two survey consisted of multiple segments. The first asked our expert panel to answer nine questions pertaining to their risk perceptions for engineered *M. loti*. Responses were reported on a 10-point semantic differential scale where the poles reflected the core content of each question, and higher scores indicate increased risk perceptions. Figure 4 visualizes the results for each expert culture, and Table 1 reviews the descriptive statistics among each group and item.

When looking at uncertainty of risk for engineered plant microbes, we find that social scientists believe the risks to be slightly less certain with a score of 5.9 and

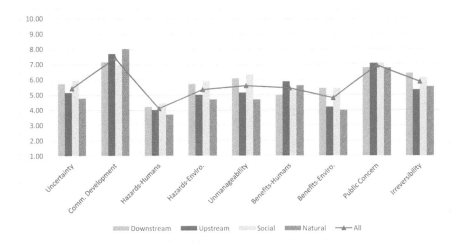

Fig. 4 Risks and benefits ratings by expert group

that natural scientists have slightly lower perceptions of uncertainty with a score of 4.8. We also find downstream scientists to be slightly less certain of risk than upstream scientists.

All four expert groups see this technology as likely to be commercially developed within the next 15 years, with natural and upstream scientists expressing highest scoring of likely development with score of 8.0 and 7.7, respectively. Upstream experts showed higher expectation of near-term development than downstream scientists, and natural showed higher expectation than social. This higher expectation from upstream and natural scientists could be due to greater experience with other genetically engineered microbes in the environment.

When looking at hazards and benefits of this technology to humans and the environment, we see the entire expert group perceives the hazards to environment as greater than the hazards to human health, but the benefits to human health greater than the benefits to the environment. Downstream scientists perceive hazards to both the environment and humans to be slightly greater than hazard perceptions from the upstream scientists, and social scientists perceive hazards in both areas as greater than their natural science counterparts. In contrast, environmental benefits were perceived as higher by downstream and social scientists than by the upstream and natural counterparts. However, upstream scientists assigned slightly higher human health benefits than did downstream scientists, and natural scientists gave slightly higher human health benefits than social scientist.

Downstream scientist expressed greater concern over manageability of this technology than did upstream scientists, and social scientist expressed greater concern than the natural scientist counterparts. Similarly, downstream and social scientists expressed greater concern than upstream and natural science groups regarding concerns over irreversibility of effects from this technology.

Table 1 Risk data by expert groups

	All experts (n = 29)			Downstream (n = 14)			Upstream (n = 15)			Social (n = 16)			Natural (n = 13)		
	Mean	Std. Dev	Range	Mean	Std. Dev	Range	Mean	Std. Dev	Range	Mean	Std. Dev	Range	Mean	Std. Dev	Range
Uncertainty	5.4	2.0	8	5.7	2.2	8	5.1	1.7	5	5.9	2.1	8	4.8	1.6	5
Commercial development	7.4	1.7	7	7.1	1.9	7	7.7	1.4	5	6.9	1.9	7	8.0	1.2	4
Hazardous to humans	4.1	1.8	7	4.2	1.9	7	4.0	1.6	6	4.4	1.9	7	3.7	1.5	6
Hazardous to the environment	5.3	1.4	6	5.7	1.6	6	5.0	1.0	4	5.9	1.3	6	4.7	1.2	4
Unmanageability	5.6	2.1	9	6.1	2.1	9	5.1	2.0	6	6.3	2.0	9	4.7	1.9	6
Benefit to humans	5.4	2.0	8	5.0	1.4	5	5.9	2.3	8	5.3	1.5	6	5.6	2.4	8
Benefit to the environment	4.8	2.0	8	5.4	1.4	5	4.2	2.3	8	5.4	1.7	6	4.0	2.1	7
Public concern	6.9	1.7	6	6.8	1.8	6	7.1	1.7	6	7.1	1.4	5	6.8	2.0	6
Irreversibility	5.9	1.6	6	6.4	1.5	6	5.3	1.4	5	6.1	1.6	6	5.5	1.5	5

When asked about public perceptions of risk for this technology, all groups expected moderate public concern. Downstream experts expected slightly greater perceptions of public concern than did upstream experts, and social scientist expected greater concern than natural scientists.

Round Two: Ordinal Ranking Data

In Round Two, we also explored the ordinal ranking of RA needs data that were provided by experts. To give an overall group ranking to risk assessment items ranked individually by experts, values within each expert group were averaged, and overall rank for each criteria was shown for each expert group. The results are shown in Table 2.

When looking at the top three RA needs identified by each of the expert groups, we see consistency among groups for the top three priorities. Biopersistence was ranked as the most important criteria for all expert groups, with the social scientists ranking both biopersistence and ecological system effects as equally most important. Other RA considerations ranked in the top three for any expert group include horizontal gene transfer and competitiveness with other organisms. The same four RA needs were ranked in the top three for each expert group; additionally, these four needs are the highest ranked overall when looking at the combined dataset for all experts.

Overall, the disciplinary groups showed fairly consistent ranking of RA needs. Ranked values were highlighted in instances where an expert group's ranking of RA needs differed from the overall group ranking by more than two. For downstream experts, we find that 12 of the 17 RA needs are similar to the overall group ranking, with differences of 2 or less. For upstream and social science experts, we see similarity in 15 of 17 of the RA needs ranking compared to the overall group and, for natural science experts, similarity across 16 of the 17 RA needs.

We also see some agreement between the ranking of RA needs from this ordinal data and the most frequently mentioned RA needs from the interview data. "Gene transfer" was the most mentioned RA theme during interviews and also was ranked in the top three RA needs. Competitiveness, another RA need ranked in the top three, was also mentioned frequently during the expert interviews.

Based on theories of disciplinary cultures (Becher 1994; Valimaa 1998; Trump 2017), it might be hypothesized that factors that involve expanding the traditional RA framework, such as considering environmental and economic trade-offs, would be favored by social sciences that are more accepting of multiple frameworks. However, those considerations were ranked the same for both natural and social scientists. However, the downstream scientists of both groups ranked the concerns considerably lower than the upstream scientists. Not surprisingly, the "route of exposure to other organisms" and "route of exposure to humans" were ranked higher for downstream scientist than for upstream scientists. In fact, the ranking for "route of exposure to other organism" from downstream scientists differed from the

Table 2 Ordinal ranking of RA needs by expert groups

Potential issues for risk research	Priority ranking based on averaged ranking scores				
	All experts[a] (n = 29)	Downstream (n = 14)	Upstream (n = 15)	Social (n = 16)	Natural (n = 13)
1. Biopersistence	1	1	1	1	1
4. Ecological system effects	2	2	3	1	4
2. Competitiveness	3	3	4	3	3
8. Horizontal gene transfer	4	5	2	5	2
7. Genetic stability	5	4	6	4	5
6. Environmental trade-offs	6	9	5	6	6
5. Economic trade-offs	7	11	7	7	7
11. Pathogenicity	8	7	8	9	9
10. Organism tracking in situ	9	12	9	14	8
3. Disposal	9	7	12	8	12
9. Life cycle	11	13	11	10	11
13. Regulatory approval process	12	14	9	13	10
15. Exposure to other organisms	13	6	15	11	15
14. Exposure to humans	14	10	14	12	14
12. Regulation of tools	15	16	13	15	13
16. Toxicity and biogeochemical cycling	16	15	16	16	16
17. Other	17	17	17	17	17

[a]The numbers shown by each "RA need" correspond with the numbering shown in the list of RA needs provided in the methodology section for this article, with full descriptions of each need. Ordinal rankings where the expert disciplinary group deviated from the overall group ranking are highlighted

overall ranking more than any other RA need ranking. This differential in upstream and downstream scientists' prioritization of exposure supports previous findings that downstream scientists are more concerned about potential environmental and human health effects from emerging technologies than their upstream counterparts (Powell 2007). Disposal is another RA need where we see some variation in expert rankings. Upstream and natural scientists have the lowest ranking of this consideration (12 of 17), while downstream and social scientists have higher rankings (7 of 17 and 8 of 17, respectively).

Discussion

There has been limited work on the study of "disciplinary culture" as a factor in risk perception. Through interviews of a small sample of scientists ($n = 20$), Powell (2007) was able to show preliminary findings that disciplinary cultures of "upstream" and "downstream" expert position can influence risk perceptions. Specifically, downstream scientists are generally more concerned with human health and environmental risks from nanotechnology, another field of emerging technologies. Powell also suggests that experts who are "upstream" in the developmental process perceive less uncertainty with the technology than their "downstream" counterparts. When evaluating the expert responses to the risk questions, we see results that support Powell's findings. In this study, downstream experts also had greater uncertainty in risk perceptions as well as greater perceptions of potential human and environmental hazards than did upstream scientists.

In addition, this work also found differences in risk perceptions between natural science disciplines and social science disciplines. Our natural science experts stated lower expectations of human and environmental hazards than did our social science experts but also stated lower concerns regarding irreversibility of environmental effects and unmanageability of this technology. This dataset, though limited in size, supports "disciplinary cultures," as a component affecting risk perceptions, similar to our "cultural cognition" groups.

This also suggests that the two "axes" for "disciplinary culture" studied in this work of "discipline" and "position" both influence perceptions of risk and governance needs for this technology. Future research could evaluate the relative influence of each of these axes in overall perceptions of risks. Additionally, more specific measures of cultural cognition could be examined by administering a scaled test specific to cultural cognition among area experts.

One limitation of this work is the lack of targeted testing of the standard cultural cognition paradigm of group and grid preferences. Future studies that include specific measure of cultural cognition factors, in addition to disciplinary group factors, could be tested to explore the relative influence of each component in risk perception. Corley et al. (2009) tested nanotechnology policy opinions of expert nonscientists using explicit measures of some cultural cognition factors. They found that academic disciplinary grouping may influence experts' opinions regarding regulatory needs for nanotechnology, thus providing some support for cultural cognition influences in risk perception of emerging technologies.

From a theoretical perspective, this work seeks to begin a discussion on whether expert opinions of RA needs for SB are more aligned with the standard cultural cognition paradigms of group and grid preferences or with disciplinary culture. Practically, this work can help provide a framework for understanding how inclusion or exclusion of expert groups may bias or limit strategies for anticipatory governance. It shows how having a diverse group of upstream and downstream experts and natural and social scientists is likely to expand the conversation during deliberative assessments of technology and its oversight so that a full range of options is considered. According to postnormal science (Funtowicz and Ravetz 1992) and responsible research and innovation (Stilgoe et al. 2013), as well as recent National Academies of Science (2017) and International Risk Governance Council reports (2015), a broad inclusion of these perspectives is important for appropriate governance of synthetic biology which is accompanied by high complexity, novelty, and uncertainty.

References

Becher, T. (1994). The significance of disciplinary differences. *Studies in Higher Education, 19*(2), 151–161. https://doi.org/10.1080/03075079412331382007.

Benner, S. A., & Sismour, A. (2005). Synthetic biology. *Nature Reviews: Genetics, 6*(7), 533–543. https://doi.org/10.1038/nrg1637.

Charpentier, M., & Oldroyd, G. (2010). How close are we to nitrogen-fixing cereals? *Current Opinion in Plant Biology, 13*(5), 556–564. https://doi.org/10.1016/j.pbi.2010.08.003.

Corley, E. A., Scheufele, D. A., & Hu, Q. (2009). Of risks and regulations: How leading U.S. nanoscientists form policy stances about nanotechnology. *Journal of Nanoparticle Research, 11*(7), 1573–1585. https://doi.org/10.1007/s11051-009-9671-5.

Cummings, C., & Kuzma, J. (2017). Societal Risk Evaluation Scheme (SRES): Scenario-based multi-criteria evaluation of synthetic biology applications. *PLoS One.* https://doi.org/10.1371/journal.pone.0168564.

DeSesso, J. M., & Watson, R. E. (2006). The case for integrating low dose, beneficial responses into US EPA risk assessments. *Human and Experimental Toxicology, 25*(1), 7–10. https://doi.org/10.1191/0960327106ht578oa.

Engineering and Physical Sciences Research Council and National Science Foundation. (2009). *New directions in synthetic biology: A call for participants to take part in a five-day sandpit to look for innovative ways to explore future developments in synthetic biology.* Warrenton, VA: Airlie Conference Center.

Guston, D. H. (2014). Understanding 'anticipatory governance'. *Social Studies of Science, 44*(2), 218–242. https://doi.org/10.1177/0306312713508669.

Guston, D. H., & Sarewitz, D. (2002). Real-time technology assessment. *Technology in Society, 24*(1–2), 93–109. https://doi.org/10.1016/S0160-791X(01)00047-1.

Gutmann, A. (2011). The ethics of synthetic biology: Guiding principles for emerging technologies. *The Hastings Center Report, 41*(4), 17–22. https://doi.org/10.1002/j.1552-146X.2011.tb00118.x.

Funtowicz, S. O., & Ravetz, J. R. (1992). Risk management as a postnormal science2. *Risk Analysis, 12*(1), 95–97. https://doi.org/10.1111/j.1539-6924.1992.tb01311.x.

Hoffman, E., & Newman, S. (2012). Big promises backed by bad theory. *Genetic Engineering & Biotechnology News, 32*(10), 6–7. https://doi.org/10.1089/gen.32.10.01.

International Risk Governance Council. (2015). *Guidelines for emerging risk governance.* Switzerland: Lausanne.

Kahan, D. M., & Braman, D. (2006). Cultural cognition and public policy. *Yale Law & Policy Review, 24*(1), 149–172.

Kaneko, T., Nakamura, Y., Sato, S., Asamizu, E., Kato, T., Sasamoto, S., et al. (2000). Complete genome structure of the nitrogen-fixing symbiotic bacterium mesorhizobium loti. *DNA Research, 7*(6), 331–338. https://doi.org/10.1093/dnares/7.6.331.

Kuzma, J., Paradise, J., Ramachandran, G., Kim, J., Kokotovich, A., & Wolf, S. M. (2008b). An integrated approach to oversight assessment for emerging technologies. *Risk Analysis, 28*(5), 1197–1220. https://doi.org/10.1111/j.1539-6924.2008.01086.x.

Kuzma, J., Romanchek, J., & Kokotovich, A. (2008a). Upstream oversight assessment for agri-food nanotechnology: A case studies approach. *Risk Analysis, 28*(4), 1081–1098. https://doi.org/10.1111/j.1539-6924.2008.01071.x.

Kuzma, J., & Tanji, T. (2010). Unpackaging synthetic biology: Identification of oversight policy problems and options. *Regulation & Governance, 4*(1), 92–112. https://doi.org/10.1111/j.1748-5991.2010.01071.x.

Landeta, J. (2006). Current validity of the delphi method in social sciences. *Technological Forecasting and Social Change, 73*(5), 467–482. https://doi.org/10.1016/j.techfore.2005.09.002.

Linstone, H. A., & Turoff, M. (Eds.). (1975). *The Delphi method: Techniques and applications.* Boston, MA: Addison-Wesley.

Macnaghten, P. (2008). Nanotechnology, risk and upstream public engagement. *Geography, 93*(2), 108–113.

Mandel, G. N., & Marchant, G. E. (2014). The living regulatory challenges of synthetic biology. *Iowa Law Review, 100*(1), 155–200.

Miles, R. E., Jr. (1978). The origin and meaning of Miles' Law. *Public Administration Review, 38*(5), 399–403. https://doi.org/10.2307/975497.

Morgan, K. (2005). Development of a preliminary framework for informing the risk analysis and risk management of nanoparticles. *Risk Analysis: An Official Publication of the Society for Risk Analysis, 25*(6), 1621–1635. https://doi.org/10.1111/j.1539-6924.2005.00681.x.

Morgan, M. G. (2014). Use (and abuse) of expert elicitation in support of decision making for public policy. *Proceedings of the National Academy of Sciences of the United States of America, 111*(20), 7176–7184. https://doi.org/10.1073/pnas.1319946111.

National Academies of Science, Engineering, and Medicine. (2017). *Preparing for future products of biotechnology*. Washington, DC: National Academies Press.

National Research Council Staff. (1993). *Issues in risk assessment*. Washington, DC: National Academies Press.

Neumann, R., & Becher, T. (2002). Teaching and learning in their disciplinary contexts: A conceptual analysis. *Studies in Higher Education, 27*(4), 405–417. https://doi.org/10.1080/0307507022000011525.

Oldroyd, G. E., & Dixon, R. (2014). Biotechnological solutions to the nitrogen problem. *Current Opinion in Biotechnology, 26*, 19–24. https://doi.org/10.1016/j.copbio.2013.08.006.

Otway, H., & von Winterfeldt, D. (1992). Expert judgment in risk analysis and management: Process, context, and pitfalls. *Risk Analysis, 12*(1), 83–93. https://doi.org/10.1111/j.1539-6924.1992.tb01310.x.

Patton, M. Q. (Ed.). (2002). *Qualitative research & evaluation methods* (3rd ed.). Thousand Oaks, CA: SAGE Publications.

Pauwels, E. (2013). Public understanding of synthetic biology. *Bioscience, 63*(2), 79–89. https://doi.org/10.1525/bio.2013.63.2.4.

Powell, M. C. (2007). New risk or old risk, high risk or no risk? How scientists' standpoints shape their nanotechnology risk frames. *Health, Risk & Society, 9*(2), 173–190. https://doi.org/10.1080/13698570701306872.

Quay, R. (2010). Anticipatory governance. *Journal of the American Planning Association, 76*(4), 496–511. https://doi.org/10.1080/01944363.2010.508428.

Santi, C., Bogusz, D., & Franche, C. (2013). Biological nitrogen fixation in non-legume plants. *Annals of Botany, 111*(5), 743–767. https://doi.org/10.1093/aob/mct048.

Smith, K. (2013). Synthetic biology: A utilitarian perspective. *Bioethics, 27*(8), 453–463. https://doi.org/10.1111/bioe.12050.

Stilgoe, J., Owen, R., & Macnaghten, P. (2013). Developing a framework for responsible innovation. *Research Policy, 42*(9), 1568–1580. https://doi.org/10.1016/j.respol.2013.05.008.

Stirling, A. (2008). Science, precaution, and the politics of technological risk. *Annals of the New York Academy of Sciences, 1128*(1), 95–110. https://doi.org/10.1196/annals.1399.011.

Swor, T., & Canter, L. (2011). Promoting environmental sustainability via an expert elicitation process. *Environmental Impact Assessment Review, 31*(5), 506–514. https://doi.org/10.1016/j.eiar.2011.01.014.

Tait, J. (2012). Adaptive governance of synthetic biology. *EMBO Reports, 13*(7), 579–579. https://doi.org/10.1038/embor.2012.76.

Trump, B. D., Cegan, J., Wells, E., Poinsatte-Jones, K., Rycroft, T., Warner, C., et al. (2019). Co-evolution of physical and social sciences in synthetic biology. *Critical Reviews in Biotechnology, 39*(3), 351–365.

Trump, B. D., Cegan, J. C., Wells, E., Keisler, J., & Linkov, I. (2018b). A critical juncture for synthetic biology: Lessons from nanotechnology could inform public discourse and further development of synthetic biology. *EMBO Reports, 19*(7), e46153. https://doi.org/10.15252/embr.201846153.

Trump, B., Cummings, C., Kuzma, J., & Linkov, I. (2018a). A decision analytic model to guide early-stage government regulatory action: Applications for synthetic biology. *Regulation & Governance, 12*(1), 88–100. https://doi.org/10.1111/rego.12142.

Trump, B. D. (2017). Synthetic biology regulation and governance: Lessons from TAPIC for the United States, European Union, and Singapore. *Health Policy, 121*(11), 1139–1146.

Turoff, M. (1970). The design of a policy delphi. *Technological Forecasting and Social Change, 2*(2), 149–171. https://doi.org/10.1016/0040-1625(70)90161-7.

U.S. Environmental Protection Agency. (2013). *TSCA biotechnology notifications, FY 1998 to present.* Retrieved 03/02, 2014, from http://www.epa.gov/biotech_rule/pubs/submiss.htm

Valdez, R., Kuzma, J., Cummings, C., & Peterson, N. (2019). Anticipating risks, governance needs, and public perceptions of de-extinction. *Journal of Responsible Innovation, 6*(2), 1–21.

Valimaa, J. (1998). Culture and identity in higher education research. *Higher Education, 36*(2), 119–138. https://doi.org/10.1023/A:1003248918874.

Wang, X., Yang, J., Chen, L., Wang, J., Cheng, Q., Dixon, R., & Wang, Y. (2013). Using synthetic biology to distinguish and overcome regulatory and functional barriers related to nitrogen fixation. *PLoS One, 8*(7), e68677. https://doi.org/10.1371/journal.pone.0068677.

Wareham, C., & Nardini, C. (2015). Policy on synthetic biology: Deliberation, probability, and the precautionary paradox. *Bioethics, 29*(2), 118–125. https://doi.org/10.1111/bioe.12068.

Scientists' and the Publics' Views of Synthetic Biology

Emily L. Howell, Dietram A. Scheufele, Dominique Brossard,
Michael A. Xenos, Seokbeom Kwon, Jan Youtie, and Philip Shapira

Introduction: Synthetic Biology at the Intersection of Science and Society

In the past decade, advances in synthetic biology research and applications have raised important questions about the role of humans in shaping the natural world. A broad field combining multiple disciplines, synthetic biology involves the engineering of biological components and systems to create novel organisms or to change the makeup of existing organisms in novel ways. It has applications for medicine, energy, sustainability, security, and agriculture, among others. Although some argue that humans have always been changing nature (Kaebnick 2013; Kaebnick and Murray 2013), or that categorizing humans as separate from nature is misconceived,

E. L. Howell (✉) · M. A. Xenos
Department of Life Sciences Communication, University of Wisconsin-Madison,
Madison, WI, USA
e-mail: elhowell@wisc.edu

D. A. Scheufele · D. Brossard
Department of Life Sciences Communication, University of Wisconsin-Madison,
Madison, WI, USA

Morgridge Institute for Research, Madison, WI, USA

S. Kwon
School of Public Policy, Georgia Institute of Technology, Atlanta, GA, USA

J. Youtie
School of Public Policy, Georgia Institute of Technology, Atlanta, GA, USA

Enterprise Innovation Institute, Georgia Institute of Technology, Atlanta, GA, USA

P. Shapira
School of Public Policy, Georgia Institute of Technology, Atlanta, GA, USA

Manchester Institute for Innovation Research, The University of Manchester, Manchester, UK

© Springer Nature Switzerland AG 2020   371
B. D. Trump et al. (eds.), *Synthetic Biology 2020: Frontiers in Risk Analysis and Governance*, Risk, Systems and Decisions, https://doi.org/10.1007/978-3-030-27264-7_16

discussion surrounding synthetic biology reveals that many see synthetic biology as increasingly blurring the lines between natural and man-made (Boldt 2013; European Commission Directorate General for Health and Consumers 2010; Jennings 2013). Specific concerns differ depending on the particular field and applications, with a range of ecological, environmental, bio- and cybersecurity, health, regulatory, infrastructural, societal, and equity uncertainties raised (Cummings and Kuzma 2017; Goodman and Hessel 2013; Hoffman et al. 2017; Wintle et al. 2017).

An underpinning set of moral and ethical concerns about synthetic biology focus on power and humans' place in nature. This includes concerns over how much power we should have over other life forms, how much power certain people or groups should have over decisions affecting other people, and what risks exist if we do not recognize or respect the limits of our own foresight (Boldt 2013; International Risk Governance Council 2010; Jennings 2013; Kaebnick and Murray 2013; Lustig 2013). At the same time, synthetic biology research exists because of beliefs in using nature and science to better ourselves and the world. From this viewpoint, not conducting research that could further these goals has ethical implications as well (European Commission Directorate General for Health and Consumers 2010; Fauci 2010; International Risk Governance Council 2010; Kaebnick and Murray 2013).

As with many complex issues, there is no clear "right" path for future research and development in synthetic biology. Uncertainty clouds the potential societal implications, and value-based decisions and considerations will necessarily play a role in shaping what research is done, what applications are developed, and what is socially acceptable. Decisions about these potential implications and about what is socially acceptable, therefore, are not purely based on questions that scientific and technical expertise can answer.

Nonetheless, with some exceptions (see, e.g., Bhattachary et al. 2010), much of the discussions about synthetic biology's implications have taken place within the scientific community (European Commission Directorate General for Health and Consumers 2010; International Risk Governance Council 2010; Presidential Commission for the Study of Bioethical Issues 2010; Sarewitz 2015; Vincent 2013) without explicit involvement of publics. Additionally, in the few available public polls, only a small portion of respondents report being even somewhat familiar with synthetic biology (Akin et al. 2017; Hart Research Associates 2013). Indeed, as Marris (2015) observes, institutions of science tend to discount public views about synthetic biology as uninformed and fear that "misunderstandings" among publics about synthetic biology will hinder its development.

Such views of the public and lack of public awareness and involvement are likely to limit what relevant considerations shape decision-making on development and applications of synthetic biology. As a large body of science and technology studies research emphasizes, there are serious limitations to making decisions based on only the views of those who work within the particular scientific field of focus (Brossard and Lewenstein 2010; Evans 2013; Jasanoff 1990; Pielke Jr. 2007; Sarewitz 2015; Vincent 2013). Science does not remove the appearance or existence of values shaping policy decisions, nor does it produce a "perfect, objectively verifiable truth," nor carry enough authority to end controversial societal debates (Jasanoff 1990). Scientists can also fall into the role of "stealth advocate" – seen as implicitly representing or advocating for a particular stance, either in appearance or reality. This can occur if participating scientists do not make explicit the values and assumptions shaping their own viewpoints on the development of applications of synthetic biology and if there is no space for discussion and decision-making to incorporate public values (Jasanoff 1990; Pielke Jr. 2007).

To be fair, many scientists are aware of the limitations of including only those with scientific and technical expertise in the governance of the synthetic biology field and its future development. Numerous advisory committees and workshops have determined that public dialogue will be necessary to avoid the polarization seen around similar biotechnology issues such as genetically modified crops in Europe and stem cell research in the US (European Commission Directorate General for Health and Consumers 2010; International Risk Governance Council 2010; Presidential Commission for the Study of Bioethical Issues 2010). Less common, however, is recognition of the need for involvement of publics beyond just for increasing public acceptability of any resulting decisions (Jasanoff et al. 2015). Experts in synthetic biology are experts because they have particular technical skills and familiarity with synthetic biology. Even experts trained in bioethics or similar ethical, legal, and social studies are inclined to look at the world in a particular way with a particular set of language and ideas. At the same time, academics and scientists do not, and cannot, reflect wide-ranging demographics, so excluding public views or including them only within bounds preset by experts could produce a deceptively narrow set of considerations and options. In a domain such as synthetic biology, in which a large and varied array of future benefits and consequences is possible that could affect a diversity of groups and systems, an exclusionary approach would be not only inequitable and undemocratic but also risky, subjective, and inadequate.

This chapter examines similarities and differences between scientists' and non-scientists' views of synthetic biology and the factors that shape them, as well as limitations of available research and the need for more focus on the views of both groups. We combine data from a survey of researchers in synthetic biology and a nationally representative survey of US adults on synthetic biology to compare the characteristics of respondents in each group and how those general characteristics could shape each group's views. We end with a call for more, and more detailed, social science research to facilitate effective public engagement that creates space for the variety of views and concerns that will shape synthetic biology and its governance.

For the remainder of this chapter, our use of "expert" will indicate a person who has professional knowledge and in-depth familiarity with the scientific field or technology of concern. Rowe and Wright (2001) argued that the title of "expert" should indicate "known accuracy of [the individual's] risk judgements" for the issue of interest rather than the individual's particular role. As we will discuss, however, there are a wide variety of risk considerations that shape perceptions of, and decisions concerning, science and technology, many of which are impossible to cleanly categorize as "accurate" or "inaccurate." Synthetic biology is a new and developing field where uncertainty is still so high and assessing much of the risk involved requires judgments and value positions about what is acceptable. Even when assessing technical risk, finding an "accurate" measure is difficult. In this study, therefore, we are more concerned with how the people who will have a say in determining progress in and acceptability of synthetic biology view the risks and benefits. The goal is to understand how to develop public engagements that address risk and benefit perceptions, whatever they might be, as well as what values and experiences shape those perceptions as research and applications in synthetic biology continue to develop.

Where Experts and the Public Differ: Science, Experience, and Risk

Limited research exists on what shapes public views of synthetic biology (Akin et al. 2017), and no research that we are aware of has systematically captured expert views of synthetic biology. Research on the extent to which public and expert views of science and technology differ on other issues indicates that experts, in general, have lower risk perceptions and more neutral or positive associations with the technologies they work with than do publics (Flynn et al. 1993; Ho et al. 2011; Savadori et al. 2004; Siegrist et al. 2007). Furthermore, experts believe the risks to be more precisely known and relatively controllable than do people who work outside the respective field (Wright et al. 2000).

There is a high degree of variability, however, in terms of the methodological rigor surrounding many of these studies (see Ho et al. 2011 for overviews; Rowe and Wright 2001; Scheufele et al. 2009). The most common weaknesses are not controlling for relevant demographic factors such as gender or age (the case for Flynn et al. 1993; Wright et al. 2000) or using small and/or not statistically representative samples of either the public or the relevant groups of experts (the case for Flynn et al. 1993; Savadori et al. 2004; Wright et al. 2000). This is not to discount findings from this earlier work. These findings from previous research do suggest that different experiences (or different people drawn to different fields and sectors) play important roles in understanding risk perceptions among experts.

Perhaps explaining the differences in expert and public views, several studies suggest that people are less likely to rely on heuristic processing or rely on different pathways for processing for issues that they have expertise in. A study of anesthetists concluded that experts interpret risk differently than the general public for issues in which they are more familiar (or are, in fact, experts) but not for issues in which they are less familiar due to "a reduced reliance on a low-effort heuristic for experts given an expertise-relevant context" (Fleming et al. 2012). Supporting this, a study comparing experts in nanotechnology with the public found significant differences in experts' and publics' value-related considerations (which the authors referred to as (pre)dispositions, such as trust and religiosity) related to each group's assessment of the risks and benefits of nanotechnology (Ho et al. 2011). Although experts' views on the risks and benefits of nanotechnology related to predispositions such as deference to science and trust in scientists, they did so to a lesser extent than did the publics. Additionally, public views of risk were significantly related to their level of religious guidance, which was not true of the nanotechnology researchers.

Similar findings highlight differences in the role of value-laden personal characteristics, or predispositions, for members of the public and experts, with political ideology and religiosity playing a critical role in predicting public support of regulation of nanotechnology research (Su et al. 2016). Scientists working in nanotechnology, in comparison, were less likely to rely on such predispositions and more likely to have attitudes that related to their perceptions of how regulation could impact scientific progress in the field. The studies above attribute these differences in scientists' and publics' views to heuristic processing. It is worth noting, however, that the differences could also be due to systematic processing, with values and experience structuring more deliberate reasoning (Dragojlovic and Einsiedel 2013).

These studies, as well as this chapter, do not categorize one view of risk, expert or public, as correct. As a large body of risk research highlights, technical and measured risk is only one aspect of a wide range of recognized risk considerations (for an overview, see Renn 1992). Other factors shape the perceived severity and acceptability of potential risks, such as how controllable exposure to a particular risk seems, how familiar a person is with the risk, how horrible (or dreadful) the consequences would be if the risk does become reality, and how the risk information spreads among people through interpersonal communication and media systems (Fischhoff et al. 1979; Kasperson et al. 1988; Slovic et al. 1979). These qualitative aspects of risk play an important role in public acceptance of technology (e.g., in public opinion on nuclear power: Slovic et al. 1991) and in assigning damages in legal cases (Dowie 1994; Strobel 1994; Tesh 1988). Stakeholders involved in decision-making can better characterize the variety of potential aspects of risk, and the assumptions and weights people use to assess risks, when diverse groups of public and expert stakeholders coordinate in decision-making.

Values as Perceptual Filters: Religiosity and Political Ideology

As mentioned above, values shape how we weigh different considerations as we assess information (Festinger 1962; Kunda 1990; Yeo et al. 2015). The result can be biased information processing, such as motivated reasoning, in which reasoners are more likely to reach conclusions that are consonant with their held beliefs (Festinger 1962; Kunda 1990; Lord et al. 1979). Because of this mental processing, values and value-based considerations often predict attitudes across a range of emergent science and risk by providing mental structures through which we process incoming information on the issue (Dragojlovic and Einsiedel 2013; Ho et al. 2010, 2011; Leeper and Slothuus 2014; Malka et al. 2009; Su et al. 2016; Yeo et al. 2014). Research regarding public views of synthetic biology is just starting to untangle how different values relate to particular views, and research has not yet examined how values relate to expert views. We can expect aspects of synthetic biology to overlap with ethical and religious views concerning the role of humans in changing life, which have shaped views of other biotechnologies (Ho et al. 2008; Scheufele and Beier 2017; Shih et al. 2012).

In the synthetic biology domain, there has been conversation around ethics that includes concerns of synthetic biology offering "too much" power to humans. These concerns are related to religious views or voiced in religious terms capturing the risk of human limitations such as hubris and lack of foresight (Dragojlovic and Einsiedel 2013; Jennings 2013; Kaebnick and Murray 2013; Lustig 2013; Schmidt et al. 2009). Previous studies on similar biotechnology-related issues such as applications of stem cell research found that religion was a significant predictor of attitudes toward the technology (Dragojlovic and Einsiedel 2013; Shih et al. 2012; Ho et al. 2008). Additionally, polling on public views of synthetic biology finds that individuals who belong to religious denominations or regularly attend church are more likely to perceive risk from synthetic biology and support banning synthetic biology research (Akin et al. 2017; Hart Research Associates 2010, 2013).

Political ideology is another characteristic that encapsulates a wide range of values shaping individuals' opinions on controversial or emergent technology, especially on views toward regulation. For synthetic biology in particular, political ideology appears to be relevant for shaping views of regulating the field. US polling data finds that Democrats are more likely to support government regulation of research in synthetic biology while Republicans are more likely to support voluntary guidelines (Hart Research Associates 2013). Additional research found that political ideology did not significantly relate to general support for synthetic biology research (Akin et al. 2017), so it is possible that the findings concerning regulation of the science are reflecting respondents' views on regulation in general, perhaps more so than their support for or views on synthetic biology itself.

Religiosity and political ideology are only two characteristics that relate to how people form opinions on issues, and they are very broad characteristics at that. We use them here as an example, however, of two characteristics on which experts in synthetic biology and members of the US public greatly differ. In the next section, we will examine how these differences could overlap with different views and concerns associated with synthetic biology.

Capturing Expert and Public Views

To examine public and expert views of synthetic biology, we combined two separate surveys. The first is of US scientists who have published research in an area of synthetic biology. The sample selection used a systematic publication keyword search of synthetic biology-related terms (Shapira et al. 2017) to compile a list of US-based scientists who have published synthetic biology research in the Web of Science database between January 2000 and October 2015. The study contacted 1748 researchers, and 46.1 percent (or 806 researchers) completed the survey, primarily online between November 2015 and January 2016. Respondents who reported that they did not work in the field were removed, resulting in a final sample of 790 respondents, 8 percent of whom completed a mail survey instead of the online survey.

The second survey captures a nationally representative sample of US adults and was conducted online by the GfK Group between July and August 2014. The final stage completion rate was 48.0 percent, or 3145 respondents. Within the survey, respondents were assigned to one of four situations, answering questions on synthetic biology, nanotechnology, hydraulic fracturing, or climate change. The respondents who answered questions regarding synthetic biology are the sample of the public that this study focuses on ($N = 808$). We applied a weight GfK recommended and supplied to address demographic differences between sample respondents and the US public.

In terms of basic demographics, political ideology, and levels of religious guidance, experts greatly differ from members of the US public. As seen in Table 1, respondents in the expert sample are more likely to be male and white. *Religiosity* is a single-item measure on an 11-point scale asking respondents, "How much guidance does religion provide in your everyday life" ("0" = "no guidance at all"; "10" = "a great deal of guidance")? Half responded "0" on the item asking how much they rely on religious guidance in their daily lives while only 5 percent responded with the highest level indicating "a great deal of guidance" from religion. The nationally representative survey, in contrast to experts, is less than half male, a smaller majority white, and only 16 percent reported relying on "no guidance at all" from religion compared to 20 percent reporting they rely on a "great deal of guidance."

Conservate and liberal political ideologies (the terms typically used to characterize the major US political alignments) can manifest in different ways depending on whether the topic is viewed as mostly an economic or a social issue. We thus treat economic ideology and social ideology separately and asked respondents to indicate "In terms of [economic or social] issues, would you say you are..." on a scale from "very liberal" to "very conservative" with "moderate" in the middle.

Table 1 Descriptives of US expert and public samples

	Scientists (%)	Public (%)
Gender (male)	80.7	48.1
Race (white)	71.1	66.2
Level of religious guidance		
"No guidance at all"	50.0	16.0
"A great deal of guidance"	5.4	20.5
Economic political ideology		
Conservative	13.6	41.3
Moderate	34.1	35.0
Liberal	52.3	23.8
Social political ideology		
Conservative	5.9	20.8
Moderate	17.8	24.5
Liberal	76.3	54.7

Public respondents are also more likely to be conservative than are the respondents in the expert sample, with 41 percent conservative for economic issues and 20.8 percent for social issues.

What Do These Differences Mean for Views of Synthetic Biology?

The point of these differences is not to argue that scientists should look different than they do or hold different religious and political views than they do. These demographic statistics also do not give insight into the variety of experiences or views that could be present among either experts or the public. They do illustrate, however, that scientists and the public differ in some basic characteristics – ones that often overlap with particular experiences and perspectives relevant to views of biotechnology and emerging scientific research. Research consistently finds significant differences in risk perceptions between men and women, for example, with men perceiving less risk from emerging technologies, especially those with potential health and environmental risks (Barke et al. 1997; Bord and O'Connor 1987; Dohmen et al. 2011; Eisler et al. 2003; Flynn et al. 1994; Ho et al. 2010). Gender differences appear in studies of expert risk perceptions as well (Barke et al. 1997; Ho et al. 2011; Slovic et al. 1995), although the studies do not control for differences potentially arising from the specific fields and industries men and women tend to work in.

For attitudes toward synthetic biology, the differences between men and women are less studied or reported even when controlled for. Available results indicate that women are less likely to accept synthetic biology (Akin et al. 2017; Dragojlovic and Einsiedel 2013; Finucane et al. 2000; Hart Research Associates 2010, 2013). Similarly, race can have significant relationships with risk perceptions and support for different technologies, with nonwhite respondents in the US often likely to see more risk and be less supportive of new technology (Ansolabehere and Konisky 2009; Finucane et al. 2000; Flynn et al. 1994; Whitfield et al. 2009). In polls on public opinion of synthetic biology, non-white respondents were more likely to see greater risk (Hart Research Associates 2010, 2013).

That scientists not only have a particular experience with and knowledge of synthetic biology but also differ in some key demographic characteristics could, therefore, have implications for how their views and decisions concerning synthetic biology compare to those of publics. In particular, the main demographic characteristics of the scientists in this sample overlap with the characteristics of those who, in more general US samples, tend to be more supportive of and less concerned with risks associated with new technologies. Of course, within samples of men and women and people with different racial identities, there is a wide range of experiences and perspectives on technologies and risk. Because there are systematic patterns in how people across gender and race perceive risk and benefit from technology, however, it is likely that a sample, such as the experts here, that contains fewer women and nonwhite minorities could have systematically views of synthetic biology. Decisions among only expert groups, then, risk being narrow not only because of the scientific experience of the experts but also because of more general experiences with risk, benefit, and technology that overlap with gender and racial identity.

Beyond the demographic differences, the differences between experts and the US public in religiosity and political ideology are especially striking. To examine the extent to which these differences in religiosity and political ideology relate to expert and public views, we ran a series of regression models predicting perceptions of risks and benefits of synthetic biology. We included several of the basic demographic variables and the measures of religiosity and political ideology. *Gender* and *race* are both coded as dummy variables (1 = "female" and 1 = "nonwhite," respectively; 63.5 percent male and 67.2 percent white). We acknowledge that the diverse groups within the "nonwhite" category are likely to have varied perspectives. Because relatively few respondents marked each "nonwhite" group (black, Hispanic, Asian, and other), however, we do not have a large enough sample to analyze each separately. So race is simplified as "white" and "nonwhite" to capture effects from majority and minority race. *Age* is how old the respondent is in years and ranges from 18 to 100 years old ($M = 47$; SD = 15.17).

Perceived risk and *perceived benefit* are different items in the expert and public surveys, which is a difficultly of having limited data on expert views, not to mention data that directly mirrors measures in public surveys. For the public survey, *perceived risk and benefit* are both single items with 11-point scales, asking "How risky (beneficial) do you think [synthetic biology] is for society as a whole" ("0" = "not risky (beneficial) at all" to "10" = "very risky (beneficial)")? For the expert survey, *perceived risk and benefit* are single items with a 5-point scale, asking respondents to indicate the extent to which they agree or disagree with the statement "Synthetic biology is risky/beneficial for society" ("1" = "strongly disagree" to "5" = "strongly agree"). The two measures are included separately rather than combined as a ratio of risk to benefit, which has been common in previous comparisons of public and expert views, because although they are significantly negatively correlated, they can also relate in different ways to certain relevant attitudes and views.

We use hierarchical ordinary least squares (OLS) analyses to capture the effects of each variable. This analysis measures the relationships between independent variables and the dependent variables by analyzing "blocks" of the independent variables that are grouped by type and their assumed causal order. Demographics come before values, for example, because gender and age are more likely to affect political ideology than vice versa. The OLS regression model adds the blocks of similar variables (demographic variables, value variables, etc.) one at a time to see how much each type explains the total variance in individuals' responses to the dependent variable. In the public survey, we also controlled for experimental conditions present in the survey. The conditions are unrelated to this analysis but occurred before respondents answered the risk and benefit items used here and could have affected responses, so they were included as the first block in the regressions modeling public views.

For each dependent variable, we ran one analysis predicting just expert views and one predicting just public views, for a total of four regression models.

Demographic, Religiosity, and Ideology Differences Translate to Differences in Risk and Benefit Perceptions

The regression models indicate that the experts and public samples are similar in how some characteristics relate to their risk and benefit perceptions but also differ in some potentially important ways. Age and race are significantly related to perceiving less risk for both groups, with older and nonwhite respondents perceiving lower levels of risk than their peers (Table 2). In these tables listing the results of the regression models, the standardized beta indicates the direction of the relationship between each independent and dependent variable. Age's beta of −0.100 in the *expert* model column of Table 2, for example, indicates that for each one standard deviation increase in age, perceived riskiness decreases by 0.1 standard deviation. The asterisks next to this number indicate that the relationship between age and perceived riskiness is statistically significant.

Benefit perception also significantly relates to risk perception for both groups, with higher benefit perception predicting lower risk perception. The public sample's risk perception, however, is significantly correlated to religiosity and economic political ideology, while expert risk perception is not. Higher levels of religiosity and more conservative economic ideology predict higher levels of perceived risk for the public (Table 2).

Age is a significant predictor of benefit perception for only the expert group, with older experts significantly less optimistic about the benefits. Religion plays a significant role in the perceived benefits for both groups, with higher levels of

Table 2 OLS regression model predicting perceived riskiness of synthetic biology for society

Risk perception	Expert	Public
	Standardized betas (β)	
Block 1: Demographics		
Age	−0.100**	−0.078*
Gender (high = female)	−0.062	0.044
Race (high = nonwhite)	−0.148***	−0.125***
Incremental R^2 (percent)	1.6**	1.3**
Block 2: Value predispositions		
Religiosity	0.052	0.124**
Economic ideology (high = conservative)	−0.024	0.080*
Social ideology (high = conservative)	0.052	0.068
Incremental R^2 (percent)	0.6	1.7**
Block 3: Risk and benefit perceptions		
Beneficial	−0.232***	−0.135***
Incremental R^2 (percent)	5.1***	1.6***
Total adjusted R^2	7.3	4.6

In Tables 2 and 3, only final coefficients (standardized betas) of each independent variable are listed. For economic and political ideology, excluded variable coefficients and significance are reported to account for multicollinearity issues from having multiple similar variables in the same block

$^*p \leq 0.05$; $^{**}p \leq 0.01$; $^{***}p \leq 001$

religiosity predicting lower levels of perceived benefit from synthetic biology. Political ideology also predicts benefit perception but in significantly different ways for each group. As seen in Table 3, economic ideology significantly relates to public benefit perceptions (economically liberal respondents perceiving significantly greater benefits than moderates or conservatives do), but not experts' views. Social political ideology, however, significantly relates to both expert and public views, with liberals again perceiving greater benefit than do moderates and conservatives within each group (Table 3). The relationship of political ideology to the benefit perceptions of both groups is also illustrated in Figs. 1 and 2, which

Table 3 OLS regression model predicting perceived benefit of synthetic biology for society

Benefit perception	Expert	Public
Block 1: Demographics		
Age	−0.141***	0.057
Gender (high = female)	−0.062	−0.041
Race (high = nonwhite)	−0.034	−0.027
Incremental R^2 (percent)	1.3**	0.2
Block 2: Value predispositions		
Religiosity	−0.087*	−0.098*
Economic ideology (high = conservative)	0.007	−0.189***
Social ideology (high = conservative)	−0.102**	−0.271***
Incremental R^2 (percent)	1.9**	7.9***
Block 3: Risk and benefit perceptions		
Risky	−0.230***	−0.128***
Incremental R^2 (percent)	5.0***	1.4***
Total adjusted R^2	8.2	9.5

*$p \leq 0.05$; **$p \leq 0.01$; ***$p \leq 001$

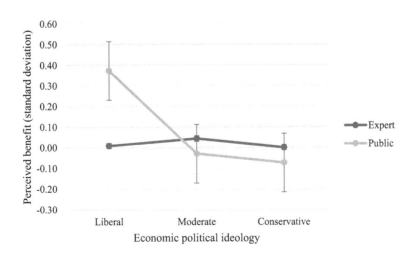

Fig. 1 Differences in US expert and public benefit perceptions of synthetic biology by economic political ideology

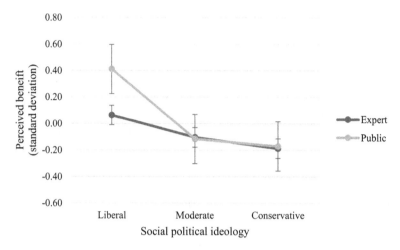

Fig. 2 Differences in US expert and public benefit perceptions of synthetic biology by social political ideology

show how much respondents who identify with each ideology differ in average benefit perception from the perceptions of others in their respective groups (expert or public).

The Implications for Public Engagement and Future Research: Building Across Divides

US publics and experts in synthetic biology do differ across several basic characteristics in what shapes their general risk and benefit perceptions of the field. Our analyses found substantial religious and ideological differences between experts and members of the public in the US. We also found that levels of religious guidance and political ideology relate to different perceptions of synthetic biology and in different ways depending on one's experience as either a researcher in the field or as a member of the public. Political ideology and religiosity do appear to have some similar relationships to both expert and public views of synthetic biology's risks and benefits. The fact that the public is more likely to be religious and politically conservative, however, suggests expert and public views could diverge over time based on these values and also that greater differences could materialize among different groups within the public. We found that public perceptions of synthetic biology more strongly relate to religion and political ideology than do experts' perceptions. This could result in these value differences having an even larger effect for public-expert differences in perceiving the risk and benefit of synthetic biology, particularly as variations in the perspectives of different parts of the public surface.

Public engagement in decision-making is valuable for a variety of reasons, ranging from democratic practices and inclusion to the quality of resulting decisions and the acceptability of those decisions. In this case, the significant relationships of both political ideology and religious guidance to perceptions of synthetic biology highlight how important incorporating public views with expert perspectives will be for representative decision-making that will receive broad public buy-in (see, e.g., Holdren et al. 2011, Trump et al. 2018).

Next Steps for Social Science Research and Public Engagement

These models, of course, do not provide the full level of detail we will need in public discourse and decision-making for a field as complex as synthetic biology. Stakeholders in these decisions include groups as diverse as the US public and pool of researchers across the private and public sectors, not to mention the international communities. The limitations in the model stem from the amount of available data on expert views of science, as well as limitations in the detail and understanding that can be extracted from close-ended questions on a survey. This study had a limited number of measures that appeared in both the expert and public surveys, the wording and scales on the risk and benefit perception measures differed slightly, and the public survey came out in 2014, 3 years before the expert survey and 3 years in which views of synthetic biology could change. The items that did overlap tended to be broader measures. If religion is going to be a factor that shapes views of synthetic biology in the US, for example, research will need greater specificity in measures of religious guidance and related values to look at the differences in the effects of different religions and their particular worldviews shaping perceptions of synthetic biology.

Overall, empirical social science research will have to closely examine the variety of views held by publics, experts, and other stakeholders, what shapes these views, and how these views and their associated values differ across and within groups. Further research using a range of methods can processes can examining differences in views across the many areas of synthetic biology and the specific risks, benefits, and options for progress in each of those areas can build on this analysis to facilitate discussion. This will be particularly necessary for facilitating discussion that makes the relevant values and assumptions shaping these perceptions explicit and addresses how different courses of action, or inaction, can create or address these risks and benefits. Given the speed of development in synthetic biology and other biotechnology fields, such as human genome editing, research will need to more consistently capture this data to have an accurate picture of both the field and the perceptions.

It is the work of further research and projects of public and science engagement and deliberation, using a range of methods and processes, to parse out more of factors that shape different views, as well as the views themselves. Social science can provide assessments of both shared and potentially divisive values or divided considerations and views. Public engagement and other forms of anticipatory governance can test and further explicate these divides, explore adjustments informed by engagement practices as well as additional social science research, and discern strategies that could foster improved alignment between science and societal goals. This research and on-the-ground work will be necessary for navigating the implications of synthetic biology and the different views shaping the research and its reception.

References

Akin, H., Rose, K. M., Scheufele, D. A., Simis-Wilkinson, M. J., Brossard, D., Xenos, M., & Corley, E. A. (2017). Mapping the landscape of public attitudes on synthetic biology. *Bioscience, 67*(3), 290–300.

Ansolabehere, S., & Konisky, D. M. (2009). Public attitudes toward construction of new power plants. *Public Opinion Quarterly, 73*(3), 566–577. https://doi.org/10.1093/poq/nfp041.

Barke, R. P., Jenkins-Smith, H., & Slovic, P. (1997). Risk perceptions of men and women scientists. *Social Science Quarterly, 78*(1), 167–176.

Bhattachary, D., Calitz, J. P., & Hunter, A. (2010). *Synthetic biology dialogue*. London: TNS-BMRB. https://bbsrc.ukri.org/documents/1006-synthetic-biology-dialogue-pdf.

Boldt, J. (2013). Creating life: Synthetic biology and ethics. In G. E. Kaebnick & T. H. Murray (Eds.), *Synthetic biology and morality* (pp. 35–49). Cambridge, MA: MIT Press.

Bord, R. J., & O'Connor, R. E. (1987). The gender gap in environmental attitudes: The case of perceived vulnerability to risk. *Social Science Quarterly, 78*, 830–840.

Brossard, D., & Lewenstein, B. V. (2010). A critical appraisal of models of public understanding of science: Using practice to inform theory. In L. Kahlor & P. Stout (Eds.), *Communicating science: New agendas in communication* (pp. 11–39). New York: Routledge.

Cummings, C. L., & Kuzma, J. (2017). Societal risk evaluation scheme (SRES): Scenario-based multi-criteria evaluation of synthetic biology applications. *PLoS One, 12*(1), e0168564. https://doi.org/10.1371/journal.pone.0168564.

Dohmen, T., Falk, A., Huffman, D., Sunde, U., Schupp, J., & Wagner, G. G. (2011). Individual risk attitudes: Measurement, determinants, and behavioral consequences. *Journal of the European Economic Association, 9*(3), 522–550. https://doi.org/10.1111/1524-4774.2011.01015.x.

Dowie, M. (1994). Pinto madness. In D. Birsch & J. Fielder (Eds.), *The Ford Pinto case* (pp. 15–36).

Dragojlovic, N., & Einsiedel, E. (2013). Playing god or just unnatural? Religious beliefs and approval of synthetic biology. *Public Understanding of Science, 22*(7), 869–885. https://doi.org/10.1177/0963662512445011.

Eisler, A. D., Eisler, H., & Yoshida, M. (2003). Perception of human ecology: Cross-cultural and gender comparisons. *Journal of Environmental Psychology, 23*(1), 89–101.

European Commission Directorate General for Health & Consumers. (2010). *Synthetic biology from science to governance*. Retrieved from http://ec.europa.eu/health//sites/health/files/dialogue_collaboration/docs/synbio_workshop_report_en.pdf

Evans, J. H. (2013). "Teaching humanness" claims in synthetic biology and public policy bioethics. In G. E. Kaebnick & T. H. Murray (Eds.), *Synthetic biology and morality* (pp. 177–203). Cambridge, MA: The MIT Press.

Fauci, A. S. (2010). *Advances in synthetic biology: Significance and implications*. Paper presented at the committee on energy and commerce in the United States house of representatives, Washington, D.C.

Festinger, L. (1962). *A theory of cognitive dissonance*. Stanford, CA: Stanford University Press.

Finucane, M. L., Slovic, P., Mertz, C. K., Flynn, J., & Satterfield, T. A. (2000). Gender, race, and perceived risk: The "white male" effect. *Health, Risk & Society, 2*(2), 159–172.

Fischhoff, B., Slovic, P., & Lichtenstein, S. (1979). Weighing the risks. *Environment: Science and Policy for Sustainable Development, 21*(4), 17–38. https://doi.org/10.1080/00139157.1979.99 29722.

Fleming, P., Townsend, E., van Hilten, J. A., Spence, A., & Ferguson, E. (2012). Expert relevance and the use of context-driven heuristic processes in risk perception. *Journal of Risk Research, 15*(7), 857–873. https://doi.org/10.1080/13669877.2012.666759.

Flynn, J., Slovic, P., & Mertz, C. K. (1993). Decidely different: Expert and public views of risks from a radioactive waste repository. *Risk Analysis, 13*(6), 643–648. https://doi.org/10.1111/j.1539-6924.1993.tb01326.x.

Flynn, J., Slovic, P., & Mertz, C. K. (1994). Gender, race, and perception of environmental health risks. *Risk Analysis, 14*(6), 1101–1108. https://doi.org/10.1111/j.1539-6924.1994.tb00082.x.

Goodman, M., & Hessel, A. (2013, May 28). The bio-crime prophecy: DNA hacking the biggest opportunity since cyber attacks. *Wired.* Retrieved from https://www.wired.co.uk/article/the-bio-crime-prophecy

Hart Research Associates. (2010). *Awareness & impressions of synthetic biology: A report of findings based on a national survey among adults.* Retrieved from http://www.synbioproject.org/publications/6456/

Hart Research Associates. (2013). *Awareness & impressions of synthetic biology: A report of findings based on a national survey among Adults.* Retrieved from SynBio Project: http://www.synbioproject.org/publications/6655/

Ho, S. S., Brossard, D., & Scheufele, D. A. (2008). Effects of value predispositions, mass media use, and knowledge on public attitudes toward embryonic stem cell research. *International Journal of Public Opinion Research, 20*(2), 171–192. https://doi.org/10.1093/ijpor/edn017.

Ho, S. S., Scheufele, D. A., & Corley, E. A. (2010). Making sense of policy choices: Understanding the roles of value predispositions, mass media, and cognitive processing in public attitudes toward nanotechnology. *Journal of Nanoparticle Research, 12*(8), 2703–2715. https://doi.org/10.1007/s11051-010-0038-8.

Ho, S. S., Scheufele, D. A., & Corley, E. A. (2011). Value predispositions, mass media, and attitudes toward nanotechnology: The interplay of public and experts. *Science Communication, 33*(2), 167–200. https://doi.org/10.1177/1075547010380386.

Hoffman, E., Hanson, J., & Thomas, J. (2017). *The principles for the oversight of synthetic biology.* Friends of the Earth U.S., International Center for Technology Assessment, and the ETC Group. Retrieved from http://www.etcgroup.org/content/principles-oversight-synthetic-biology

Holdren, J. P., Sunstein, C. R., & Siddiqui, I. A. (2011). *Memorandum: Principles for regulation and oversight of emerging technologies.* United States Office of Science and Technology Policy. Washington, D.C.

International Risk Governance Council. (2010). Policy brief – guidelines for the appropriate risk governance of synthetic biology. Retrieved from Geneva.

Jasanoff, S. (1990). *The fifth branch – scientific advisors as policymakers.* Cambridge, MA: Harvard University Press.

Jasanoff, S., Hurlbut, J. B., & Saha, K. (2015). CRISPR democracy: Gene editing and the need for inclusive deliberation. *Issues in Science & Technology, 32*(1), 25–32.

Jennings, B. (2013). Biotechnology as cultural meaning: Reflections on the moral reception of synthetic biology. In G. E. Kaebnick & T. H. Murray (Eds.), *Synthetic biology and morality* (pp. 149–175). Cambridge, MA: The MIT Press.

Kaebnick, G. E. (2013). Engineered microbes in industry and science: A new human relationship to nature? In G. E. Kaebnick & T. H. Murray (Eds.), *Synthetic biology and morality* (pp. 51–65). Cambridge, MA: MIT Press.

Kaebnick, G. E., & Murray, T. H. (2013). Introduction. In G. E. Kaebnick & T. H. Murray (Eds.), *Synthetic biology and morality.* Cambridge, MA: The MIT Press.

Kasperson, R. E., Renn, O., Slovic, P., Brown, H. S., Emel, J., Goble, R., et al. (1988). The social amplification of risk: A conceptual framework. *Risk Analysis, 8*(2), 177–187. https://doi.org/10.1111/j.1539-6924.1988.tb01168.x.

Kunda, Z. (1990). The case for motivated reasoning. *Psychological Bulletin, 108*(3), 480–498.

Leeper, T. J., & Slothuus, R. (2014). Political parties, motivated reasoning, and public opinion formation. *Political Psychology, 35*, 129–156. https://doi.org/10.1111/pops.12164.

Lord, C. G., Ross, L., & Lepper, M. R. (1979). Biased assimilation and attitude polarization: The effects of prior theories on subsequently considered evidence. *Journal of Personality and Social Psychology, 37*(11), 2090–2109. https://doi.org/10.1037/0022-3514.37.11.2098.

Lustig, A. (2013). Appeals to nature and the natural in debates about synthetic biology. In G. E. Kaebnick & T. H. Murray (Eds.), *Synthetic biology and morality* (pp. 15–33). Cambridge, MA: MIT Press.

Malka, A., Krosnick, J. A., & Langer, G. (2009). The association of knowledge with concern about global warming: Trusted information sources shape public thinking. *Risk Analysis, 29*(5), 633–647. https://doi.org/10.1111/j.1539-6924.2009.01220.x.

Marris, C. (2015). The construction of imaginaries of the public as a threat to synthetic biology. *Science as Culture, 24*(1), 83–98. https://doi.org/10.1080/09505431.2014.986320.

Pielke, R., Jr. (2007). *The honest broker – making sense of science in policy and politics.* Cambridge, UK: Cambridge University Press.

Presidential Commission for the Study of Bioethical Issues. (2010). *New directions: The ethics of synthetic biology and emerging technologies.* Retrieved from https://bioethicsarchive.george-town.edu/pcsbi/sites/default/files/PCSBI-Synthetic-Biology-Report-12.16.10_0.pdf

Renn, O. (1992). Concepts of risk: A classification. In S. Krimsky & D. Golding (Eds.), *Social theories of risk* (pp. 53–79). Westport: Praeger.

Rowe, G., & Wright, G. (2001). Differences in expert and lay judgments of risk: Myth or reality. *Risk Analysis, 21*(2), 341–356. https://doi.org/10.1111/0272-4332.212116.

Sarewitz, D. (2015). CRISPR: Science can't solve it. *Nature, 522*(7557), 413–414.

Savadori, L., Savio, S., Nicotra, E., Rumiati, R., Finucane, M., & Slovic, P. (2004). Expert and public perception of risk from biotechnology. *Risk Analysis, 24*(5), 1289–1299. https://doi.org/10.1111/j.0272-4332.2004.00526.x.

Scheufele, D. A., & Beier, D. (2017, May 18). Human genome editing: Who gets to decide? *Scientific American.* Retrieved from https://blogs.scientificamerican.com/observations/human-genome-editing-who-gets-to-decide/

Scheufele, D. A., Brossard, D., Dunwoody, S., Corley, E. A., Guston, D., & Peters, H. P. (2009, August 4). Are scientists really out of touch? *The Scientist.* Retrieved from https://www.the-scientist.com/daily-news/are-scientists-really-out-of-touch-43968

Schmidt, M., Ganguli-Mitra, A., Torgersen, H., Kelle, A., Deplazes, A., & Biller-Andorno, N. (2009). A priority paper for the societal and ethical aspects of synthetic biology. *Systems and Synthetic Biology, 3*, 3–7. https://doi.org/10.1007/s11693-009-9034-7.

Shapira, P., Kwon, S., & Youtie, J. (2017). Tracking the emergence of synthetic biology. *Scientometrics, 112*(3), 1439–1469. https://doi.org/10.1007/s11192-017-2452-5.

Shih, T. J., Scheufele, D. A., & Brossard, D. (2012). Disagreement and value predispositions: Understanding public opinion about stem cell research. *International Journal of Public Opinion Research, 25*(3), 357–367. https://doi.org/10.1093/ijpor/eds029.

Siegrist, M., Keller, C., Kastenholz, H., Frey, S., & Wiek, A. (2007). Laypeople's and experts' perception of nanotechnology hazards. *Risk Analysis, 27*(1), 59–69. https://doi.org/10.1111/j.1539-6924.2006.00859.x.

Slovic, P., Fischhoff, B., & Lichtenstein, S. (1979). Rating the risks. *Environment: Science and Policy for Sustainable Development, 21*(3), 14–39.

Slovic, P., Flynn, J., & Layman, M. (1991). Perceived risk, trust, and the politics of nuclear waste. *Science, 254*(5038), 1603–1607. https://doi.org/10.1126/science.254.5038.1603.

Slovic, P., Malmfors, T., Krewski, D., Mertz, C. K., Neil, N., & Bartlett, S. (1995). Intuitive toxicology. II. Expert and lay judgments of chemical risks in Canada. *Risk Analysis, 15*(6), 661–675. https://doi.org/10.1111/j.1539-6924.1995.tb01338.x.

Strobel, L. P. (1994). The Pinto documents. In D. Birsch & J. Fielder (Eds.), *The Ford Pinto case* (pp. 37–53). Albany, NY: State University of New York Press.

Su, L. Y., Cacciatore, M. A., Brossard, D., Corley, E. A., Scheufele, D. A., & Xenos, M. A. (2016). Attitudinal gaps: How experts and lay audiences form policy attitudes toward controversial science. *Science and Public Policy, 43*(2), 196–206. https://doi.org/10.1093/scipol/scv031.

Tesh, S. N. (1988). Vietnam veterans and agent orange. In *Hidden arguments: Political ideology and disease prevention policy*. New Brunswick: Rutgers University Press.

Trump, B. D., Cegan, J. C., Wells, E., Keisler, J., & Linkov, I. (2018). A critical juncture for synthetic biology: Lessons from nanotechnology could inform public discourse and further development of synthetic biology. *EMBO Reports, 19*(7), e46153.

Vincent, B. B. (2013). Ethical perspectives on synthetic biology. *Biological Theory (Thematic Issue: Synthesis), 8*, 368–375.

Whitfield, S. C., Rosa, E. A., Dan, A., & Dietz, T. (2009). The future of nuclear power: Value orientations and risk perception. *Risk Analysis, 29*(3), 425–437. https://doi. org/10.1111/j.1539-6924.2008.01155.x.

Wintle, B. C., Boehm, C. R., Rhodes, C., Molloy, J. C., Millett, P., Adam, L., et al. (2017). A transatlantic perspective on 20 emerging issues in biological engineering. *eLife, 6*, e30247. https:// doi.org/10.7554/eLife.30247.

Wright, G., Pearman, A., & Yardley, K. (2000). Risk perception in the U.K. oil and gas production industry: Are expert loss-prevention managers' perceptions different from those of the members of the public? *Risk Analysis, 20*(5), 681–690. https://doi.org/10.1111/0272-4332.205061.

Yeo, S. K., Cacciatore, M. A., Brossard, D., Scheufele, D. A., Runge, K., Su, L. Y., et al. (2014). Partisan amplification of risk: American perceptions of nuclear energy risk in the wake of the Fukushima Daiichi disaster. *Energy Policy, 67*, 727–736. https://doi.org/10.1016/j. enpol.2013.11.061.

Yeo, S. K., Cacciatore, M. A., & Scheufele, D. A. (2015). News selectivity and beyond: Motivated reasoning in a changing media environment. In *Publizistik und gesellschaftliche Verantwortung* (pp. 83–104).

Dignity as a Faith-Based Consideration in the Ethics of Human Genome Editing

Rev. Nicanor Pier Giorgio Austriaco

On July 26, 2018, the Pew Research Center released the results of a survey conducted among 2,537 adults in the United States to assess their views on the appropriateness of genome editing for babies (Funk and Hefferon 2018). A majority of the individuals surveyed (72%) favored gene editing that would treat a serious disease or condition, but a majority (80%) also thought that using these techniques to enhance a child's intelligence would take this technology "too far." One striking finding of the survey was that there is a large difference in acceptance of genome editing between those respondents who self-identify as highly religious and those who are less so, where religious Americans are more likely to view gene editing negatively. Where a significant number of respondents with high religious commitment (87%) thought that testing gene editing on human embryos was taking the technology "too far," for example, this number was significantly smaller (44%) among those with a low religious commitment. How do we explain these divergent views?

As a Catholic priest who is also a molecular biologist and a bioethicist, I am often asked to comment on the similarities and differences between a faith-based approach to bioethics and its secular counterpart. Though one could compare and contrast these two ethical traditions in many ways, I have come to see that the most fundamental difference involves their rival conceptions of human dignity. Where religious ethicists from the Judeo-Christian tradition – and I will focus here on the Catholic moral tradition – see human dignity, for the most part, as having both intrinsic and extrinsic dimensions, their secular counterparts only acknowledge dignity as an extrinsic value of the human agent. Though this difference may initially appear small and insignificant, it has far-ranging moral consequences. I will propose that this one disagreement can explain not only the divergent responses given by

N. P. G. Austriaco (✉)
Providence College, Providence, RI, USA
e-mail: naustria@providence.edu

© Springer Nature Switzerland AG 2020
B. D. Trump et al. (eds.), *Synthetic Biology 2020: Frontiers in Risk Analysis and Governance*, Risk, Systems and Decisions, https://doi.org/10.1007/978-3-030-27264-7_17

religious and secular Americans to the Pew survey on designer babies but also the other often-conflicting ethical claims made by these two rival groups of citizens about how we are to pursue every scientific and technological research program in our liberal and pluralistic society. My hope is that this chapter will help each of us, whether or not we are individuals of faith, to better understand the complex bioethical debates that accompany modern biological engineering.

Defining Human Dignity in the Catholic Moral Tradition

What is human dignity? Like every other philosophical claim, the principle of human dignity has a long, complex, and controversial history.[1] It should not be surprising therefore that Deryck Beyleveld and Roger Brownsword concluded a survey of how human dignity is used in international human rights documents in the following way:

> In sum, human dignity appears in various guises, sometimes as the source of human rights, at other times as itself a species of human right (particularly concerned with the conditions of self-respect); sometimes defining the subjects of human rights, at other times defining the objects to be protected; and, sometimes reinforcing, at other times limiting, rights of individual autonomy and self-determination. (Beyleveld and Brownsword 1998)

In my view, however, the debates over the meaning and extent of human dignity are inevitably disputes over the value of the human person. They are disagreements over how we are to answer the question: How much is each one of us worth?

For bioethicists working in the Catholic moral tradition and for many who embrace the Judeo-Christian worldview, human dignity has a twofold character. First, it is an intrinsic dignity that affirms that the human being has a worth that cannot be monetized. As Pope Francis has said, "Things have a price and can be sold, but people have a dignity; they are worth more than things and are above price" (Francis 2013). Each one of us is priceless. Each one of us is exceptional.

To say that human dignity is intrinsic is to say four things about human dignity and the human person. First, it is a claim that human dignity is inherent, essential, and proper to the human being. It is a dignity that is constitutive of human identity itself. It is a dignity that affirms that human beings are worthwhile because of the kind of things that we are and not because of what we can or cannot do. As such, it is a dignity that can only be possessed in an absolute sense – one either has it completely or does not have it at all – since one is either a human being or not. Understood as an intrinsic quality, there is no such thing as partial human dignity since there is no such thing as a partial human being.

Next, to say that human dignity is intrinsic is to say that human life is worthy of respect and has to be protected from all unjust attacks. As Pope St. John Paul II

---

[1] On this point, see the following: Rosen (2013), Christopher McCrudden (2013), Mieth and Braarvig (2014), and Debes (2017).

explained, "The inviolability of the person, which is a reflection of the absolute inviolability of God, finds its primary and fundamental expression in the inviolability of human life."[2] From a theological perspective, human life is inviolable because it is a gift from God. He alone is the Lord of life from its beginning until its end. Thus, no one can, in any circumstance, claim for himself the right directly to destroy an innocent human being.[3] The Bible expresses this truth in the divine commandment: "You shall not kill" (Ex. 20:13; Deut. 5:17). The Catholic Church's prohibitions against the destruction of human embryos, physician-assisted suicide, and euthanasia are grounded in her conviction that human beings have an intrinsic dignity that can never be violated.

Third, to say that human dignity is intrinsic is to say that the human being can never be treated as an object. In other words, as a person, the human being can never be treated purely as a means to an end or be used merely as tools to attain a goal. Instead, he has to be respected as a free moral agent capable of self-knowledge and self-determination in all the actions involving himself. Again, as Pope St. John Paul II forcefully declared, "The human individual cannot be subordinated as a pure means or a pure instrument either to the species or to society; he has value *per se*. He is a person. With his intellect and his will, he is capable of forming a relationship of communion, solidarity and self-giving with his peers."[4] We know this truth from our own experience. Individuals who discover that they have been manipulated often feel violated, humiliated, and diminished because they intuit that they are persons who have an intrinsic dignity that is attacked when they are used merely as objects of another's fancy.

Fourth, to say that human dignity is intrinsic is to say that all human beings are equal. All human beings as persons have an inestimable and thus equal worth. Our intrinsic dignity is the only reason for the fundamental equality among all human beings regardless of the biological, psychological, and spiritual differences that exist in every human population. Thus, as the Second Vatican Council of the Catholic Church taught in 1965, "Every form of social or cultural discrimination in fundamental personal rights on the grounds of sex, race, color, social conditions, language, or religion must be curbed and eradicated as incompatible with God's design."[5] Social discrimination is unjust precisely because it attacks the intrinsic and equal dignity of human persons.

In addition to intrinsic dignity, the Catholic tradition also affirms that every human being has an extrinsic dignity that is a measure of his worth in the eyes of his peers. It is contingent on how others value or do not value the individual. This extrinsic dignity is conferred and can be taken away. It can increase, decrease, and can even be lost through neglect, disease, or sin. This is the dignity to which we

---

[2] John Paul II, *Christifideles laici*, §38

[3] See the Congregation for the Doctrine of the Faith's Instruction on Respect for Human Life in its Origin and on the Dignity of Procreation, Replies to Certain Questions of the Day *Donum vitae* (22 February 1987), Introduction, §5.

[4] John Paul II (2003).

[5] Vatican II (1965).

refer when we say that someone is "dignified." In this sense, the judge can be considered to have greater worth than a buffoon, that is, he can have more extrinsic dignity than the buffoon, even though both also have an equal worth because of their intrinsic dignity. Today, in our consumerist society, the human being's extrinsic dignity is often benchmarked to his salary and accumulated wealth and the social status both usually bring with them.

From the perspective of the Catholic moral tradition and other traditions that emerge from the Judeo-Christian worldview, bioethics is grounded upon the fundamental principle that all human actions need to protect, preserve, and advance the dignity of the human person, especially his intrinsic dignity that is inviolable. Actions that fail to acknowledge the true inestimable worth of the person would be deemed out of bounds for a virtuous and just society.

Considering Human Dignity in the Ethics of Human Genome Editing

Given its foundational commitment to the advancement and protection of human dignity, it should not be surprising that the Catholic moral tradition approaches the ethical question of human genome editing by raising, what I call, dignity concerns. There are at least four dignity concerns applicable here that lead to four ethical guidelines for human genome editing.

First, there is the concern that we protect the human person from harm. To respect the dignity of the human person entails that we act to preserve his or her life and well-being from unjust attack. Therefore, genome editing should be permitted for therapeutic interventions that cure, delay, or prevent disability and disease, as long as there is reasonable assurance that the technology is safe.

Second, there is the concern that we protect the human person from being objectified or commodified. As we noted above, to respect the dignity of the human person entails that we never seek to treat him as mere means to an end. Therefore, genome editing should not be permitted that would allow anyone, parents included, to genetically engineer children according to their own subjective desires. This would reduce children to products designed to fulfill the dreams of their makers rather than treat them as persons who should have the freedom to pursue their own aspirations. Parents who design their son so that he will become a tall basketball player are not taking into account the possibility that he may want to pursue a career where height is a disability and not an advantage! A child should be welcomed and loved. He or she should not be designed and manufactured.

Third, there is the concern that we protect the human person from being marginalized. To respect the dignity of the human person entails that we never seek to treat him as less valuable or less worthwhile than his neighbors. Therefore, genome editing should not be permitted if it will exacerbate the divisions and inequalities already present in our societies, reinforce social stigmas, or encourage the eugenic temptations that our societies face to eradicate undesirable traits and tendencies. As such, genetic interventions for nontherapeutic reasons, reasons usually associated with enhancing the individual's personal or social opportunities, would be ruled out.

Finally, there is the concern that we protect the poor and vulnerable. To respect the dignity of the human person entails that we seek to respect the dignity of all persons regardless of their wealth or social status. Therefore, genome editing should not be permitted unless a genuine effort is made to ensure that there will be just access to this technology for everyone.

Debating Human Dignity in a Secular Society

For Catholic bioethicists, the claim for the intrinsic dignity of the human being can be justified in two ways. Philosophically, it is grounded in the nonreligious claim that the human person's capacity for thinking is determinate in a way no physical process can be (Feser 2013). As such, she must be a spiritual being whose ontological worth radically transcends the limited value of purely material things. Theologically, it is grounded in the faith-based claim that the human person is made in the image and likeness of God. In the words of Pope St. John Paul II,

> The dignity of the person is manifested in all its radiance when the person's origin and destiny are considered: created by God in his image and likeness as well as redeemed by the most precious blood of Christ, the person is called to be a "child in the Son" and a living temple of the Spirit, destined for eternal life of blessed communion with God. (John Paul II 1988)

This transcendent and eternal destiny, justified by both faith and reason, is the fundamental reason for the human being's intrinsic dignity, a personal dignity that is not dependent either upon his own or upon human society's recognition (Catechism of the Catholic Church 1997).

Given how the intrinsic dignity of the human person is justified by Catholic and other faith-based bioethicists, it should not be surprising that its existence has been rejected by many secular bioethicists. In a much-discussed essay, Ruth Macklin dismisses dignity as a useless concept that means nothing more than respect for persons or their autonomy (Macklin 2003). Macklin writes that "dignity seems to have no meaning beyond what is implied by the principle of medical ethics, respect for persons: the need to obtain voluntary, informed consent; the requirement to protect confidentiality; and the need to avoid discrimination and abusive practices." In addition, Macklin proposes that dignity "is nothing more than a capacity for rational thought and action, the central features conveyed in the principle of respect for autonomy" (Macklin 2003). She is not alone among secular thinkers who hold this view. In an essay where Steven Pinker condemns the "theocon" bioethicists who advocate a thick sense of human dignity, he writes, "The problem is that 'dignity' is a squishy, subjective notion, hardly up to the heavyweight moral demands assigned to it" (Pinker 2008). However, what is clear from his analysis is that Pinker, as do many other secular bioethicists, does not understand appeals to dignity among faith-based bioethicists because he fails to acknowledge the distinction between intrinsic and extrinsic dignity. For Pinker, dignity is only an extrinsic dimension of the human person that can be easily diminished or lost. It is a "phenomenon of human perception," he writes, "just as the smell of baking bread triggers a desire to eat it, and the sight of a baby's face triggers a desire to protect it, the appearance of dignity triggers a desire to esteem and respect the dignified person." In my view, no one who grasped the intrinsic nature of dignity of the human person would ever compare it with the smell of baking bread.

In fact, every one of us voluntarily and repeatedly relinquishes extrinsic dignity for other goods in life. Getting out of a small car is undignified. Having sex is undignified. Doffing your belt and spread-eagling to allow a security guard to slide a wand up your crotch is undignified. Most pointedly, modern medicine is a gantlet of indignities. Most readers of this article have undergone a pelvic or rectal examination, and many have had the pleasure of a colonoscopy as well (Pinker 2008).

Therefore, in Pinker's view, dignity is almost a useless concept. Once again, on his account, bioethics in a post-Christian and liberal society should be grounded not on respect for dignity but on respect for autonomy. It is autonomy that is inviolable, and not dignity.

From the perspective of the secular tradition of bioethics, therefore, it should not be surprising that for many ethicists, the guiding principle governing the ethics of human gene editing has inevitably emphasized protecting and preserving not the dignity of the human person but his autonomy, whether this is the parent's reproductive autonomy or the child's personal autonomy.[6] According to many who hold this view, human gene editing should be pursued to maximize the autonomy of the persons involved, again as long as it is safe and does not harm the health and well-being of another.

But in response, I have to challenge my secular colleagues: Why is autonomy inviolable? Why should it be respected and maximized? Why is it intrinsically valuable such that it trumps all other concerns, including many of the dignity concerns that Catholic bioethicists have proposed?[7] Some have argued that autonomy is intrinsically valuable because every reasonable person will always choose autonomy over heteronomy.[8] A person would always choose to make his or her own decisions rather than delegate those choices to others.

---

[6] For a sense of the reproductive autonomy and designer baby debate, see the following essays: Robertson (1996), Huber (2006), Mameli (2007), Parker (2007), Massmann (2018).

[7] For a critique of the intrinsic value of autonomy from the secular perspective, see Varelius (2006). Instead, Varelius argues that autonomy only has instrumental value in promoting the well-being of the individual. But if this is true, then autonomous individuals should not be allowed to act in ways that undermine their health and well-being or the health and well-being of other human beings. This certainly is not the view of the vast number of contemporary ethicists who believe that autonomy should play a primary role in bioethics and public policy, again because they presuppose that it has some intrinsic, inviolable worth.

[8] On this point, see Glover (1977).

However, from the perspective of evolutionary theory, the capacities to think and to choose are mere evolutionary adaptations that are no more valuable than any other evolutionary adaptations in nature. In fact, expanding John Rawl's proposal of a veil of ignorance to encompass all the species of the planet, it is not clear that it would always be reasonable to choose the human capacities of thought and choice over one of the other evolved adaptations, if one did not know the ecological niche one would find oneself once the veil is lifted (Rawls 1971). For instance, if I found myself in the Siberian tundra, I think that it would be more reasonable to want to be a polar bear with the capacity of hunting and fishing among the ice floes rather than to be a human being with the capacity of thinking and choosing yet naked and utterly helpless. And yet if I found myself in Boston, it would be more reasonable to want to be a human being rather than a bear. This thought experiment suggests that from the perspective of a post-Christian and materialist worldview, the capacity to be an autonomous agent – indeed, the capacity to be a person too – is not intrinsically more valuable than any other evolved capacity in nature. Its worth is relative because its value is dependent upon the environmental niche of the organism that may or may not have it. But if this is the case, why then should autonomy be respected and preserved? Why should it be inviolable?

I have not yet found a secular response to these questions that adequately explains why autonomy should be defended as an intrinsically valuable good. I do not think that one is forthcoming. In my view, Immanuel Kant and the other Enlightenment philosophers who invented autonomy (Schneewind 1997) simply presupposed a Christian worldview that proclaimed the exceptionalism of the human person made in the image and likeness of God.[9] Autonomy is inviolable because the human person who, in his very nature, thinks and chooses is inviolable. Autonomy is worthwhile only because it is an emergent capacity of the human person who is inherently worthwhile. Thus, *pace* Macklin and Pinker, respecting the intrinsic dignity of the human person is not the same as respecting their autonomy. The former justifies the latter, and the latter cannot stand without the former. In at least this one way, the tradition of secular bioethics is reliant for its own internal coherence and intelligibility, upon the rival tradition of Christian bioethics that it dismisses and rejects.

In conclusion, this chapter began with a discussion of the recent Pew Research Center study that revealed the striking differences between religious and secular attitudes toward the genetic engineering of the genomes of babies where religious

---

[9] It is striking that I have Buddhist friends who think that claims to the inviolable nature of human autonomy are incoherent within their worldview, especially when this view is permeated by a Confucian philosophy that is inherently communitarian in nature. On this, see Florida (1998). Florida writes, "Confucian philosophy that is inherently communitarian in nature. Florida writes, "In traditional Buddhist ethics, autonomy is not featured as a major category. The Buddhist emphasis on the responsibility of each person for his or her own karma or moral character implies something like this notion; however, there is something in the modern Western insistence on autonomy that goes against the Buddhist grain. … [I]ndividualistic autonomy is contrary to the central Buddhist insight of co-conditioned causality, which insists on the interdependency of all beings. It is particularly at odds with the bodhisattva ideal of sacrificing self for others that is at the heart of the *Lotus Sūtra*".

Americans are more likely to view gene editing of babies as taking medical technology too far. Some of these disagreements, in my view, can be traced to the divergent accounts of dignity presupposed by the respondents. Accordingly, religious Americans, most of whom belong to the Judeo-Christian tradition, would be wary of any technology that could undermine the dignity of the human person, especially his or her intrinsic dignity that is inviolable. In contrast, the secular American tradition may be more permissive of technological advancement in the name of preserving individual liberty and reproductive autonomy. Though there are those who think that these two perspectives are incommensurable and therefore irreconcilable, I have proposed that the secular tradition actually needs an account of intrinsic dignity to justify its autonomy claims. As such, to remain coherent and intelligible, both sides should be able to acknowledge that in cases where dignity concerns apparently come into conflict with autonomy concerns, dignity should trump autonomy because the former in the end grounds the latter. Dignity makes autonomy valuable.

References

Beyleveld, D., & Brownsword, R. (1998). Human dignity, human rights, and human genetics. *The Modern Law Review, 61*(5), 661–680, 665.

Catechism of the Catholic Church (1997) Libreria Editrice Vaticana, Vatican City, §356.

Congregation for the Doctrine of the Faith. Instruction on Respect for Human Life in Its Origin and on the Dignity of Procreation. Replies to Certain Questions of the Day. Donum vitae (February 22, 1987) Introduction, §5.

Debes, R. (Ed.). (2017). *Dignity: A history*. Oxford: Oxford University Press.

Feser, E. (2013). Kripke, ross, and the immaterial aspects of thought. *American Catholic Philosophical Quarterly, 87*(1), 1–32.

Florida, R. E. (1998). The *Lotus Sūtra* and health care ethics. *Journal of Buddhist Ethics, 5*, 170–189.

Francis. (2013). Address of Holy Father Francis to participants in the meeting organized by the International Federation of Catholic Medical Associations, September 20, 2013. https://w2.vatican.va/content/francesco/en/speeches/2013/september/documents/papa-francesco_20130920_associazioni-medici-cattolici.html

Funk, C., & Hefferon, M. (2018, July 26). Public views of gene editing for babies depend on how it would be used. Pew Research Center. http://www.pewinternet.org/2018/07/26/public-views-of-gene-editing-for-babies-depend-on-how-it-would-be-used/

Glover, J. (1977). *Causing death and saving lives* (p. 81). Harmondsworth: Penguin Books.

Huber, J. (2006). Do germline interventions justify the restriction of reproductive autonomy? In J. Niewonhner & C. Tannert (Eds.), *Gene therapy: Prospective technology assessment in its societal context* (pp. 109–121). New York: Elsevier Science.

John Paul II (1988). *Christifideles laici, Post-synodal Apostolic Exhortation of his Holiness John Paul II on the Vocation and the Mission of the Lay Faithful in the Church and in the World*. Vatican City: Libreria Editrice Vaticana, §38.

John Paul II (2003). Address to the Plenary Session on the Subject 'The Origins and Early Evolution of Life', October 22, 1996. *In Papal addresses to the Pontifical Academy of Sciences 1917-2002 and to the Pontifical Academy of Social Sciences 1994–2002*, Scripta Varia 100 (pp. 370–374). Vatican City: Pontifical Academy of Sciences.

Macklin, R. (2003). Dignity is a useless concept. *British Medical Journal, 327*(7429), 1419–1420.

Mameli, M. (2007). Reproductive Cloning, genetic engineering and the autonomy of the child: The moral agent and the open future. *Journal of Medical Ethics, 33*(2), 87–93.

Massmann, A. (2018, May 19). Genetic enhancements and relational autonomy: Christian ethics and the child's autonomy in vulnerability. *Studies in Christian Ethics*. Published online. https://doi.org/10.1177/0953946818775558.

McCrudden, C. (Ed.). (2013). *Understanding human dignity*. Oxford: Oxford University Press.

Mieth, D., & Braarvig, J. (2014). *The Cambridge handbook of human dignity: Interdisciplinary perspectives*. Cambridge: Cambridge University Press.

Parker, M. (2007). The best possible child. *Journal of Medical Ethics, 33*(5), 279–283.

Pinker, S. (2008, May 28) The stupidity of dignity. *The New Republic*. Available at: https://newrepublic.com/article/64674/the-stupidity-dignity

Rawls, J. (1971). *A theory of justice*. Cambridge: Belknap Press of Harvard University Press.

Robertson, J. A. (1996). *Children of choice: Freedom and the new reproductive technologies*. Princeton: Princeton University Press.

Rosen, M. (2013). *Dignity: Its history and meaning*. Cambridge: Harvard University Press.

Schneewind, J. B. (1997). *The invention of autonomy*. Cambridge: Cambridge University Press.

Varelius, J. (2006). The value of autonomy in medical ethics. *Medicine, Health Care and Philosophy, 9*, 377–388.

Vatican II. (1965, December 7). Pastoral constitution on the church in the modern world, *Gaudium et spes*, §29.

Highlights on the Risk Governance for Key Enabling Technologies: From Risk Denial to Ethics

Myriam Merad

Introduction

Discovering new processes, investing in new products, expanding into new markets, innovating, and differentiating are all about taking risks. Although risk-taking and scientific advancement are at the heart of technology development and commercialization, other social and humanitarian factors are also critical to the success or failure of a technology's widespread adoption and use. Notably, this includes consideration of ethical, legal, and social implications (ELSI) that frame a technology in terms of its normative societal value – not just by its technical capabilities (Calvert & Martin 2009).

A lack of consideration of ELSI-related issues for emerging technology development and governance can contribute to critical failures in its ability to acquire regulatory approval or societal acceptance – making it essential to understand what such implications are as well as what can be done to address them (Douglas & Stemerding 2014; Merad & Trump 2020; Palma-Oliveira et al. 2018). As synthetic biology is a technology that is already being shoehorned into historical debates of genetically modified organisms, relevant developers and practitioners in this space should take proactive steps to address ELSI-related issues *before* they threaten to derail broader discussion of how scientific advancements in this biological space contribute to vastly improved quality and standards of living worldwide (Torgersen 2009).

M. Merad (✉)
CNRS, Paris, France

UMR ESPACE (Université de Nice), Nice, France

UMR LAMSADE (Université Paris-Dauphine) –PSL, Paris, France
e-mail: myriam.merad@unice.fr

© Springer Nature Switzerland AG 2020   399
B. D. Trump et al. (eds.), *Synthetic Biology 2020: Frontiers in Risk Analysis
and Governance*, Risk, Systems and Decisions, https://doi.org/10.1007/978-3-030-27264-7_18

The crux of the matter is to understand how the integration of technical scientific progress alongside implications-related discussion influences the political and institutional mechanisms behind the assessment and approval of emerging technologies at the regulatory and governance levels (Finkel et al. 2018). Ultimately, the legitimacy and validity of such regulatory assessment processes are dependent upon how well technical risks are governed alongside social concerns and ELSI-related implications. Synthetic biology is no exception to this rule but does carry heightened complexity relative to ongoing global concerns related to genetic engineering as well as the inconsistent global norms and standards related to the governance of synthetic biology technologies and products (Trump et al. 2018a).

In this chapter, we turn to the lessons of history, literature, and decision analytical sciences to draw on experiences of responsible and non-responsible risk-taking. Lessons drawn from such examples are likely relevant to the process of synthetic biology's governance and decision-making (Trump 2017) and in many instances have already begun to shape the debate regarding what forms of synthetic biology research and commodification are socially approved and validated and what others are unacceptable due to concerns of risk, ethics, or morals.

The Challenge of Crafting and Executing Just and Valid Decisions

The polymath François Rabelais lived in an era that in many ways resembles ours. It was a time of transition when new technologies, in this case printing, were revolutionizing cultural practices by allowing a wide dissemination of the new ideas of the Renaissance. In chapter 8 of Rabelais' work *Pantagruel* (1532), he writes:

> The whole world is full of learned people, very doctrinal preceptors, very vast libraries... I see the bandits, executioners, adventurers, grooms of today more learned than the doctors and preachers of my time.[1]

This is the case in our time, when information on various disciplinary fields is available to large numbers and it is difficult to account for scientific validity and consistency. Science and counter-science are then intertwined in the public sphere, leaving it open to the denial of scientific facts. For synthetic biology, this includes passionate yet often hyperbolic debate regarding the technology's benefits and biosafety risks (Schmidt 2008) – leaving little room for more technical discussion related to hazard characterization or exposure assessment among technical experts and key policymakers.

---

[1] Le monde entier est plein de gens savants, de précepteurs très doctes, de bibliothèques très vastes [...] Je vois les brigands, les bourreaux, les aventuriers, les palefreniers d'aujourd'hui plus savants que les docteurs et les prêcheurs de mon temps.

This situation has repeated itself around the world in different time periods. The most contemporary are the rise of creationism in the 1980s, the debate on genetic engineering regulation and regulatory protection in the 1990s, and the work of Swedish researchers funded by a tobacco company to deconstruct the scientific facts on the risks associated with tobacco.

In these circumstances, the denial of scientific facts and the denial of risks have different motivations. These may be risk perception biases and factors of cognitive or even sociocultural nature affecting all segments of civil society. Denying risk may stem from uncertainty and ignorance where the removal of access restrictions for "scientists" and "expertise" makes the relationship to scientific and technological validity increasingly unclear.

Simultaneously, denial may be a rhetorical tactic to provide an appearance of scientific and technical arguments, facts, or debates.

In his analysis of controversies around climate change issues, Mark Hoofnagle (2009) identified more than five mechanisms: conspiracy theory, cherry-picking (choosing questionable scientific papers that support their theories and question research on the subject), false experts, changing the rules of the game and the problem, and other logical errors (e.g., consequences as argument, false leads and diversions, etc.).

Risk denial can have positive effects during creative or entrepreneurial action, namely, by encouraging actors to take risks. This positive risk-taking has been at the heart of some management and leadership trends. However, this risk-taking can also lead to sinking promising companies.

In his books *From Good to Great* published in 2001 and *How the Mighty Fall* published in 2009, Jim Collins, a renowned management professor, has observed more than 60 companies in different sectors of activity. From this observation, Collins proposed a five-step model that moves companies from success to decline (the sin of arrogance, "always-plus," risk denial, lethal weapon, and surrender). Risk denial is the third stage where, locked in its arrogance and its logic of blind growth, the company no longer perceives internal and external signals.

Internally, no one dares to contradict the hierarchical superior, who himself does not dare to discuss his choices. The atmosphere becomes deleterious, collaboration is reduced, and individualistic behaviors become the panacea where one ends up looking for the culprits.

Externally, the information received is partial and the methods of participation in decisions promise more than they offer (e.g., an information process that is renamed the consultation process). These analyses of governance modes, decision-making processes, and organizational configurations complete the explanatory arsenal of risk denial mechanisms (Linkov et al. 2018).

Indeed, risk denial can be at a certain degree positive when there is a balance between the perceived expected "negatives" and "positives" and when negatives are fairly distributed among shareholders and stakeholders. When this equation is weakened and evidences and facts lean toward the existence of "major negatives" (risks that will have nonreversible impacts), risk denial is a selfish "willful blindness" and became negative.

Atonic and Monotonic Shifts

Rabelais invites us to consider that knowledge is guided by conscience and ethics (see Fig. 1). In Pantagruel (1532), a letter to a religious authority, "Solomon," ends with an aphorism:

> Science without conscience is but the ruin of the soul.[2]

One may wonder about the modalities of "conscience" to be implemented to prevent risk denial when it may be likely to endanger society, the environment, and companies.

This topic brings its own scholarship and experience. Learning from experiences and developing, like the work proposed by Nicolas Dechy (*in* Dechy et al. 2010), "risk denial cases" are essential. Similarly, gathering knowledge from different scientific disciplines such as human and social sciences, life sciences, and engineering can inform bias mechanisms, and the contextual and organizational configurations of denial, as well as the governance modes acting on risk denial at individual, collective, organizational, institutional, and cultural levels.

There are many prescriptive approaches to overcome risk denial once its factors have been identified. Societal responsibility approaches (see Merad 2013), nudges (*see* Thaler and Sunstein 2008), participatory and concerted governance frameworks, or regulation and control frameworks are all possible solutions.

Figure 1 presents three issues related to ethical considerations of scientific innovation:

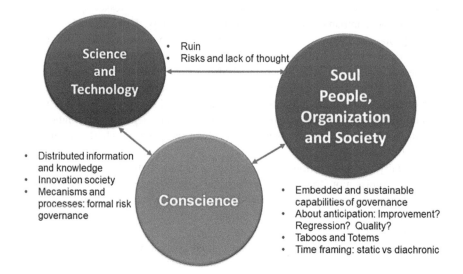

Fig. 1 Ethical consideration around issues of science and innovation

---

[2] Science sans conscience n'est que ruine de l'âme.

- *Science and technology* specifies the scientific disciplines that are mobilized (or not), the nature of uncertainties that are tackled, the nature of problems that are considered, and their dynamic of change over the time. These also comprise the controversies and polemics that emerge and special cultural, economic, and societal characteristics that diachronically give knowledge about technological ambitions or fears.
- *People, organizations, and society* and their micro, meso, and macro levels provide anthropocentric insights about the culture, the values, the structure, and the regulatory mechanisms that exist and change across time and territories.
- *Infra- and supra-mechanisms that contribute to micro, meso, and macro level of consciousness.* By "consciousness," we mean considering, beside patents, science and society, and expected positive outcomes, some latent issues such as unexpected negatives that can threaten the anthropo-system and more largely the ecosystem. In other terms, consciousness examines how expected positives within the short and medium terms (e.g., improvement of health and living conditions) can turn out to negatives at short, medium, and/or long terms.

The interactions between these three issues reveal a set of questions and considerations that can be summarized as follows:

- What are the potential known and unknown negatives that can be expected within the short, medium, and long terms? Among them, which are irreversible and can cause the ruin of the system and which are reversible?
- Which domains have encouraged willful blindness (*see* Heffernan 2011) and which lack considerations from the scientific community, regulatory bodies, and civil society?
- How are information and knowledge distributed and shared among people, organizations, and the society? Are there potential reversible and nonreversible inequities that can emerge? How is societal justice considered at diverse time and space scales?
- How prone and averse is the society to innovation? What can we learn from past positive and negative societal experiences (Greer and Trump 2019) ?
- What are the formal/explicit and informal/implicit mechanisms and processes of risk and uncertainty governance? What are their modes of genesis and dynamic of evolution and reform?
- Are there embedded and sustainable capabilities of risk and uncertainty in governance? What are they? What are the existing resilience mechanisms?
- What are the existing mechanisms of anticipation and foresight? Is there regression or progression of the capabilities and the quality of the societal system to encourage foresight?
- What are the taboos and totems that operate at the collective, organizational, and societal levels around the issues of science and innovation?
- How does the time framing (static vs. diachronic) influence those last considerations?

Emerging risks are often denied. These questions are at the core of ethical consideration when it comes to emerging risks.

Innovation and New Technologies: New and Emergent Risks

As stated earlier, each technology presents a promise of positive outcomes and with it a share of potential negatives. One common misuse of language is to link emerging technologies to emerging risks. This wandering language suggests that risks can be emerging by nature, de facto inducing a singularity of attention: as they are new, they raise new societal and ethical issues that are unique in history (on a specific context and territory). However, it may be that these risks are newly manifesting or detected without being truly new.

A more detailed analysis of the subbases of innovations would reveal that they are not only technical but also societal. Indeed, what is qualified as emergent or not is the result of a social dynamic and construct that will create a "shift in attention," or a turning point (Goldthorpe 1997; Abbott 2001), where evidence and low signals will receive a new kind of attention. This is undoubtedly true for synthetic biology, where diverging signals and discussion priorities have independently arisen from the natural sciences (i.e., technical capacity) as well as the social science (i.e., risk assessment, governance, communication) (Trump et al. 2018b).

Chateauraynaud (2011) suggests a pragmatic theory on how argument dynamics induce a variety of social coping capacities that he, respectively, named emergence, controversy, polemics, political mobilization, and normalization (Fig. 2).

This theory suggests that what are considered as emergent risks are the result of processes that repeat themselves for each new technology. Furthermore, what are considered to be new ethical considerations are just new expressions of past ones.

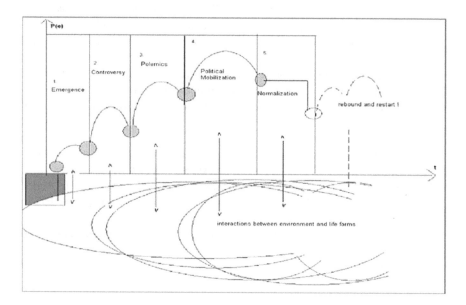

Fig. 2 Sociological ballistic. (Translated from Chateauraynaud 2011)

To illustrate the induced pitfalls, Jean-Pierre Dupuy (2007) has drawn a parallel to jokes in a comedy club where there jokes are so well known that members save themselves the trouble of telling them because they get the same effect by listing the number designating them. Dupuy specifies that the same is true for certain issues related to the ethics of science and technology: there are recurrent problems with recurrent rhetoric and induced suggested solutions.

Some Perspectives and Open Questions About Participatory and Deliberative Governance for Key Enabling Technologies

More than 13 years ago, the debates around nanotechnologies culminated on issues related to emergent risks (*see* Allhoff et al. 2007). Participatory democracy and deliberative democracy were promoted as the remedy for key ethical questions (*see* Farrelly in Allhoff et al. 2007). This includes Key Enabling Technology communications with stakeholders, understanding of how stakeholders perceive new technologies, and thinking about risk culture. Since then, it is difficult to assess what the improvements are. Accordingly, it became obvious that the object and the objectives of the improvement in the nature of the links and interactions between science and society are neither clear nor similar for the scientific community in the field and for the practitioners.

The review of the literature in the field and the interview of policymakers, regulators, and risk analysis and risk communication practitioners (*see* Merad and Trump 2018; Merad et al. 2016; Merad et Carriot 2015; Merad and Carriot 2013) show that issues related to the "assessment of the improvements induced by the stakeholder's engagement processes" and to the "unintended manipulative effects of the stakeholder's involvement structures and processes" have not been sufficiently investigated. With that respect, we identify three major challenges and a list of related open questions to structure future research on these issues.

1. Where is the field of Key Enabling Technology (KET) Communication and Stakeholder Engagement going?

   • What are the specific concepts, methods, and guidelines that have been developed?
   • What has been done in practice? How can we bridge the gap between researchers, practitioners, and regulators?
   • Are there differences from one country to another (e.g., depending on cultures, on mistrust on science and innovation, etc.)?
   • Are there differences from one field to another (micro and nanoelectronics, nanotechnology, industrial biotechnology, advanced materials, photonics, and advanced manufacturing technologies, etc.)?

2. What practical suggestions and recommendations can address the "framing effect" of risk communication and stakeholders' engagement?

- Shall we consider sustainability and risk issues separately within the communication and engagement processes?
- Are participatory problems thinking through and structuring promising leads?
- How can stakeholders engage in the innovation process and how they codesign the unknown (e.g., C-K theory)?
- What are practical recommendations about nudging (see nudge theory) practices to influence group decision-making?

3. Is there a need to reinforce or rethink the issue of risk culture and its impact on innovation and sustainability cultures?

- What are the new challenges in teaching science and technology?
- How could we develop enabling conditions to a responsible social innovation?

Conclusion

While at odds with past technologies, NETs are part of a social and historical context where a trace of controversy can persist and taint the novelty (e.g., genetically modified organisms and synthetic biology).

New technologies such as synthetic biology are often driven by a desire to improve the quality of life of society –by improving health, wealth, industry, entertainment, or similar capabilities. This improvement also comes with its accompanying risks, of which the carriers and regulators of these technologies are not necessarily aware. It is clear that innovation is about taking risks and getting out of the box. However, when these risks may have irreversible consequences or are unfairly distributed, it is necessary that they be carefully considered.

It is for this reason that regulatory mechanisms must operate (Kuzma 2015). These mechanisms can be of the technocratic type (e.g., setting up agencies and institutions in charge of risk assessment) or of the "public scrutiny" type (e.g., setting up a participatory and deliberative approach to risk management and governance). They can also take the form of "self-regulation" (e.g., raising awareness of decision biases such as the example of the procedural bias provided by Collins' analysis) (Tucker & Zilinskas 2006; Malloy et al. 2016).

Whatever the type or combination of mechanisms chosen, they act as a "consciousness" (or even a meta-consciousness according to Husserl) to reduce the propensity of innovators or risk takers to deny irreversible or major risks. This is in a way an awareness of a responsibility that goes beyond the legal one and is the basis for an ethical reflection.

In recent years, and under the impetus of environmental democracy movements (e.g., the Aarhus Convention in 1998), the "public scrutiny" regulatory mechanism has been widely acclaimed. We agree with this, but we differ from current approaches by arguing for a need for reflection on the consistency and contribution of these approaches to reducing and preventing risks in general and emerging risks in particular.

Ultimately, the contribution and consistency of these approaches should focus on characterizing compliance with two conditions. The first is related to the validity of the participatory and deliberative approaches and methods put in place. The second is related to their legitimacy and legitimization by the actors involved and/or impacted by these risks. There is still a long way to go to achieve this. As such, synthetic biology research from various biological platforms or product areas should include a research agenda based on three key questions, in other words addressing these issues of validity and legitimacy through a mixture of disciplinary and vocational perspectives (Vincent 2013; Trump et al. 2019).

References

Abbott, A. (2001). On the concept of turning point. Chapter 8. In *Time matters - on theory and method* (pp. 240–260). Chicago: The University of Chicago Press Books.

Allhoff, F., Lin, P., Moor, J. A. H., Weckert, J., & Roco, M. C. (2007). *Nanoethics: The ethical and social implications of nanotechnology* (p. 416). Hoboken: Wiley-Interscience. ISBN: 978-0-470-08417-5.

Calvert, J., & Martin, P. (2009). The role of social scientists in synthetic biology. *EMBO Reports, 10*(3), 201–204.

Chateauraynaud, F. (2011). *Argumenter dans un champ de forces. essai de balistique sociologique* (p. 477). Paris: Editions Petra, coll. « Pragmatismes », ISBN: 9782847430394.

Collins J. (2001). From good to great: Why some companies make the leap... And others Don't. ISBN: 978-0-06-662099-2. William Collins editor. pp. 320.

Collins, J. (2009). *How the mighty fall: And why some companies never give in.* ISBN: 9780977326419 (p. 240). London: Random House Audio Books.

Dechy N., Dien Y., Llory M. (2010). Pour une culture des accidents au service de la sécurité industrielle. Maîtrise des Risques et de Sécurité de Fonctionnement, Lambda-Mu 17 conference, Oct 2010: https://hal-ineris.archives-ouvertes.fr/ineris-00973593/document

Douglas, C. M., & Stemerding, D. (2014). Challenges for the European governance of synthetic biology for human health. *Life Sciences, Society and Policy, 10*(1), 6.

Dupuy, J.-P. (2007). Some pitfalls in the philosophical foundations of nanoethics. *Journal of Medicine and Philosophy, 32*, 237–261, 2007.

Farrelly C. (2007). Deliberative democracy and nanotechnology. IN Allhoff F., Lin P., Moor JA. H., Weckert J., Roco M. C. Nanoethics: The ethical and social implications of nanotechnology. Hoboken: Wiley-Interscience 416. ISBN: 978-0-470-08417-5.

Finkel, A. M., Trump, B. D., Bowman, D., & Maynard, A. (2018). A "solution-focused" comparative risk assessment of conventional and synthetic biology approaches to control mosquitoes carrying the dengue fever virus. *Environment systems and decisions, 38*(2), 177–197.

Goldthorpe, J. (1997). Current issues in comparative macrosociology. *Social Comparative Research, 16*, 1–26.

Greer, S. L., & Trump, B. (2019). Regulation and regime: the comparative politics of adaptive regulation in synthetic biology. Policy Sciences, 1–20.

Heffernan, M. (2011). *Willful blindness: Why we ignore the obvious* (p. 304). New York: Walker & Company.

Hoofnagle M. (2009).*Climate change deniers: Failsafe tips on how to post them*. The Guardian. https://www.theguardian.com/environment/blog/2009/mar/10/climate-change-denier

Kuzma, J. (2015). Translational governance research for synthetic biology. *Journal of Responsible Innovation, 2*(1), 109–112.

Linkov, I., Trump, B. D., Anklam, E., Berube, D., Boisseasu, P., Cummings, C., et al. (2018). Comparative, collaborative, and integrative risk governance for emerging technologies. *Environment Systems and Decisions, 38*(2), 170–176.

Malloy, T., Trump, B. D., & Linkov, I. (2016). Risk-based and prevention-based governance for emerging materials. *Environmental Science & Technology., 50*, 6822.

Merad, M., & Trump, B. D. (2020). *Expertise under scrutiny: 21st century decision making for environmental health and safety*. Cham: Springer International Publishing. https://doi.org/10.1007/978-3-030-20532-4.

Merad, M., & Trump, B. (2018). The legitimacy principle within French risk public policy: A reflective contribution to policy analytics. *Science of the Total Environment, 645*, 1309–1322. https://doi.org/10.1016/j.scitotenv.2018.07.144.

Merad M., Dechy N., Dehouck L., Lassagne M. (2016). Risques majeurs, incertitudes et decisions – Approche pluridisciplinaire et multisectorielle. Ma edition. ISBN: 9782822404303.

Merad, M., & Carriot, P. (2015). *Evaluer la concertation dans le domaine des risqué et de l'environnement- Eléments méthodologiques- Livre Blanc*. Paris: Afite, le réseau d'experts pour l'environnement, DL. ISBN: 978-2-9545398-2-9.

Merad M., Carriot P. (2013). Médiation et concertation environnementales - Un accompagnement à la pratique. Collection « Références »- Editions AFITE. ISBN: 978-2-9545398-0-5.

Merad M. (2013). *Organisations hautement durables: Gouvernance, risques et critères d'apprentissage Editions Lavoisier*. ISBN: 978-2-7430-1535-0.

Palma-Oliveira, J. M., Trump, B. D., Wood, M. D., & Linkov, I. (2018). Community-driven hypothesis testing: A solution for the tragedy of the anticommons. *Risk Analysis, 38*(3), 620–634.

Rabelais. (1532). Pantagruel.

Schmidt, M. (2008). Diffusion of synthetic biology: A challenge to biosafety. *Systems and Synthetic Biology, 2*(1–2), 1–6.

Thaler, R., & Sunstein, C. (2008). *Nudge: Improving decisions about health, wealth, and happiness* (p. 312). New Haven: Yale Press.

Torgersen, H. (2009). Synthetic biology in society: Learning from past experience? *Systems and Synthetic Biology, 3*(1–4), 9.

Trump, B. D., Cegan, J., Wells, E., Poinsatte-Jones, K., Rycroft, T., Warner, C., et al. (2019). Co-evolution of physical and social sciences in synthetic biology. *Critical Reviews in Biotechnology, 39*(3), 351–365.

Trump, B. D., Foran, C., Rycroft, T., Wood, M. D., Bandolin, N., Cains, M., et al. (2018a). Development of community of practice to support quantitative risk assessment for synthetic biology products: Contaminant bioremediation and invasive carp control as cases. *Environment Systems and Decisions, 38*(4), 517–527.

Trump, B. D., Cegan, J. C., Wells, E., Keisler, J., & Linkov, I. (2018b). A critical juncture for synthetic biology: Lessons from nanotechnology could inform public discourse and further development of synthetic biology. *EMBO Reports, 19*(7), e46153.

Trump, B. D. (2017). Synthetic biology regulation and governance: Lessons from TAPIC for the United States, European Union, and Singapore. *Health Policy, 121*(11), 1139–1146.

Tucker, J. B., & Zilinskas, R. A. (2006). The promise and perils of synthetic biology. *The New Atlantis, 12*, 25–45.

Vincent, B. B. (2013). Ethical perspectives on synthetic biology. *Biological Theory, 8*(4), 368–375.

Index

© Springer Nature Switzerland AG 2020
B. D. Trump et al. (eds.), *Synthetic Biology 2020: Frontiers in Risk Analysis and Governance*, Risk, Systems and Decisions, https://doi.org/10.1007/978-3-030-27264-7